TORRANCE PUBLIC LIBRARY
3 2111 00549 2795

P9-CLO-160

WITHDRAWN

DEC
DATE DUE
26 97

SUN AND EARTH

Herbert Friedman

SCIENTIFIC
AMERICAN
LIBRARY

An imprint of Scientific American Books, Inc.
New York

This book is number 15 in a series.
Earlier volumes in the series are unnumbered.

Library of Congress Cataloging in Publication Data

Friedman, Herbert, 1916–
Sun and Earth.

Bibliography: p.
Includes index.
1. Sun—Popular works. 2. Earth—Popular works.
I. Title.

QB521.4.F75 1985 523.7 85-14295
ISBN 0-7167-5012-0

Printed in the United States of America

Book design by Malcolm Grear Designers

Scientific American Library
An imprint of Scientific American Books, Inc.
New York

Distributed by W. H. Freeman and Company,
41 Madison Avenue, New York, New York 10010,
and 20 Beaumont Street, Oxford OX1 2NQ, England.

3 4 5 6 7 8 9 KP 5 4 3 2 1 0 8 9 8 7

To my wife, Gertrude, with love

Contents

Preface ix

1 From Stonehenge to Skylab 1

2 Inside the Sun: Core to Surface 23

3 The Solar Atmosphere 55

4 An Inconstant Sun: Electromagnetic Radiation 87

5 An Inconstant Sun: Solar Particle Radiation 147

6 In Search of a Climate Connection 189

7 Origins and Endings 225

For Further Reading 237

Sources of the Illustrations 241

Index 244

Preface

Generations of natural scientists from the time of Newton have been enthralled by the subtle beauties of color and light in the sky. The rainbow and even the cerulean blue of the daytime sky remained tantalizing problems in physical optics until recent years. But the relation of sun and earth has fascinated humankind since long before Newton. Small wonder. No spectacle in nature can match the transient glory of a solar eclipse. And what could be more obvious than that the light and warmth of the sun sustain all life on earth?

Not until late in the nineteenth century did scientists fully appreciate the awesome power of the sun, and they were at a loss to account for it. To warm us, feed us, and drive the circulation of the air and sea, the earth intercepts only a billionth part of the sun's total radiation, but this tiny fraction amounts to 5 million horsepower per square mile. This prodigious outpouring of energy was finally understood through the insights of Einstein, Eddington, Bethe, and other pioneers of nuclear physics.

The flow of solar power is steady. Its variations are normally so small that only the most advanced sensors now flown on satellites can detect them. Changes in the total annual insolation of the earth are very small, yet variations that reflect the earth's orbital geometry are large enough to drive substantial seasonal climatic change. Cyclic variations in the direction of the earth's spin axis, with the variation of its orbit from circular to slightly elliptical, have been sufficient to drive the advance and retreat of mile-high glaciers in cycles lasting a hundred thousand years.

The prevailing thought until the middle of the twentieth century was that the earth and sun faced each other across a near-perfect vacuum traversed by a calm and invariable flow of heat and light. In the last several decades we have learned that the earth is afloat in a sea of particle radiation and meteoritic debris, is constantly buffeted by a wind of gas streaming from the sun, and often is bombarded by storms of energetic protons, ultraviolet light, and X rays from the same source. Yet we on the surface of the earth are well protected by its magnetic umbrella and atmospheric blanket. The variability of the sun is most dramatic at times of explosive flares, when a flash of solar X rays reaches deep into the earth's atmosphere to convert the ionosphere, normally a mirror for short radio waves, into a black absorber. For the next day or more, the full blast of magnetized plasma expelled by the flare sets compass needles oscillating wildly and turns on the spectacle of green and red auroral lights. As the high-latitude disturbance grows into a worldwide ionospheric storm, it garbles radio communications and generates high winds in the upper atmosphere. When the storm subsides and all the elements of the sun-earth system have run through their responses to a great flare, the last sign is a faint red airglow that suffuses the night sky. This final message from sun to earth is the death rattle that signals the exhaustion of the great solar upheaval, yet its faint whisper carries the energy of dozens of hydrogen bombs.

I have written this book not only to describe what is known of the workings of the sun-earth system but to convey the excitement and high adventure of discovery as the pieces of the scientific puzzle were fit together. Almost all of this story belongs to the twentieth century, and the pace of scientific progress has accelerated with each decade. The classical methods of ground-based science have been greatly augmented by space technology: instruments on balloons, rockets, and satellites serve as proxy eyes to behold secrets of the sun and earth that nature never intended us to view, outside the narrow wavelength band of visible light. From models of the evolution of stars, astrophysicists have derived the history of the sun and earth from birth to middle age and can project its future until the sun will die. Sun and earth have taken on new significance in our new view from space. Looking earthward, the first astronauts photographed the delicate beauty of our emerald planet set against black emptiness. That image stirs a deep sense of loneliness and fragility. The balance between sun and earth is delicate, and that balance is now threatened by human tampering with its controls. To understand the sun-earth system is not only a major scientific challenge, it may help us to avoid human follies that could threaten the benign environment of life on earth. The sun-earth system is the home of humanity, and we must explore its every facet.

We shall not cease from exploration.
And the end of all our exploring
Will be to arrive where we started
And know the place for the first time.

T. S. Eliot, *Four Quartets*

In writing this book, I have made many references to individual scientists and their experiences, although these are only a few of the many contributors to the scientific literature of the field. The scope of the subject matter is so broad that detailed scientific treatment could not be attempted within the pages allotted to the book. Readers are referred to the list of further readings to satisfy their interest in a more complete development of the scientific story.

I am happy to acknowledge the assistance that I received from the staff of Scientific American Books: Cheryl Kupper's editing of the text, Travis Amos's diligent pursuit of illustrations, and Karen McDermott's management of the publication were enormously helpful.

Herbert Friedman
May, 1985

1

From Stonehenge to Skylab

I am the eye with which the Universe
Beholds itself and knows itself divine;

. . .

I feed the clouds, the rainbows and the flowers
With their ethereal colours; the moon's globe
And the pure stars in their eternal bowers
Are cinctured with my powers as with a robe;
Whatever lamps on Earth or Heaven may shine
Are portions of one power which is mine.

"Hymn to Apollo"
Percy Bysshe Shelley

The megaliths of Stonehenge rise above England's Salisbury Plain. Erected around 2600 B.C. by Neolithic people, the pillars are positioned so that the courses of the sun and moon could be tracked with the seasons of the year. The setting may also have served as a site of sun worship. Watercolor by John Constable.

Bright Eye of Day, Giver of Life

The vast congregation of stars that populates our galaxy fills a disk so large that light, traveling at 6 trillion miles per year, takes a hundred thousand years to cross it. On a clear, dark night, we see the vaulting arch of the Milky Way, a veil of starlight that defines the plane of the galaxy in which our sun holds an undistinguished place, twice as far from the center as from the edge. If we could travel by spaceship a few hundred light-years from the earth, the sun would fade into the sparkling background of stars that punctuate the blackness of space. Among the hundred billion stars of our galaxy, it is unexceptional in size and power. It ranks midway between the hot, blue stars and the cool, red stars. Although it appears to us a hundred billion times brighter than any other star, it would almost disappear alongside a blue giant such as Rigel, which has a luminosity 15,000 times that of the sun. Thirty-six million suns would fit inside Antares, a red supergiant. Yet for us, it is by far the most important object in the heavens.

The sun sustains all life on the earth. It lights our day and gives warmth to the soil, the oceans, and the atmosphere. It controls climate, bringing droughts and ice ages, and it drives the winds that blow across the earth and create our weather. Its storms disturb radio communications, cause electrical power surges, and even mark the rings of trees with radioactivity.

Only in this century have astronomers begun to understand the complex relationship of sun, earth, and life in any detail. But the dependence of life upon the sun's rays has been apparent since our ancestors first thought about the world around them. Small wonder, then, that ancient myth is full of accounts of sun worship and that astronomy was the mother science. The realization that the sun is a star took a long time to develop. No confirmed records of speculation that the sun and the stars are similar bodies exist before the seventeenth century. The sun's brilliant disk set it apart from nighttime stars, which were thought to be mere points of light fixed on a celestial sphere that enclosed the world. Even Copernicus and Kepler were constrained by this medieval conception.

When an eclipse erased the sun from the sky, ancient peoples took it as an expression of the sun god's displeasure with human behavior, and the resurrection of the sun required urgent prayer and sacrifice.

Strong biblical admonitions against sun worship testify to the tenacity of this practice. Moses commanded the Israelites to "take ye therefore good heed unto yourselves lest thou lift up thine eyes unto

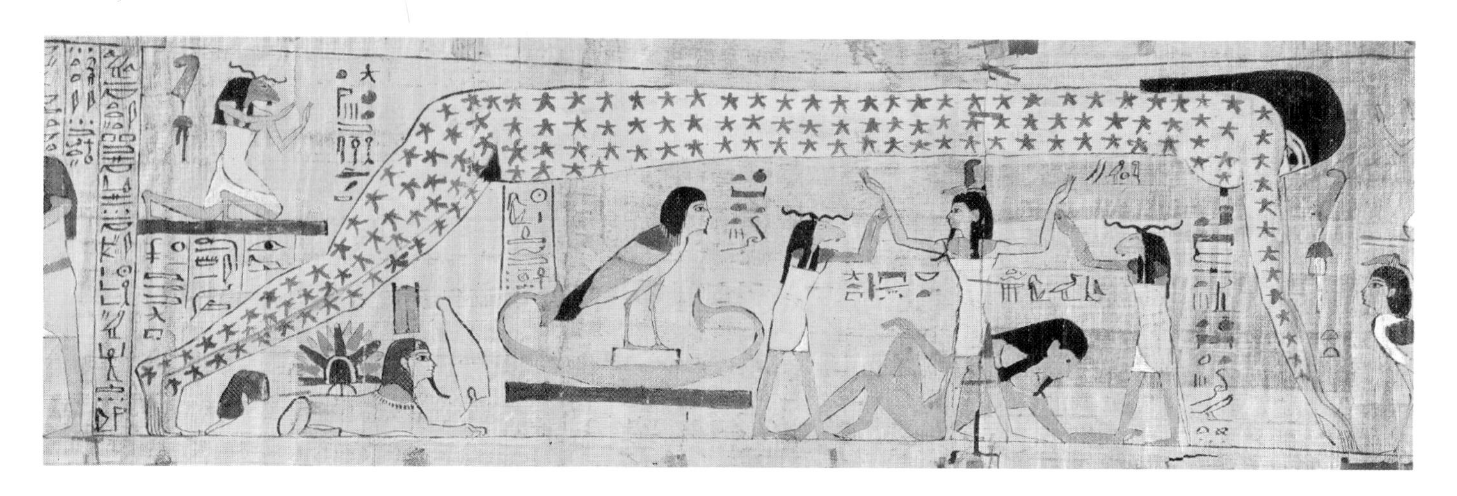

Three thousand years ago, the Egyptians believed that when the sun disappeared below the horizon in the west, it was swallowed by the sky goddess Nut. During the night, it traveled through her star-studded body that formed the vault of the heavens, to be born again at dawn.

heaven and when thou seest the sun, and the moon, and the stars, even all the host of heaven, shouldst be driven to worship them and serve them" (Deuteronomy 4:15–19). The papal astronomer Father Angelo Secchi was somewhat more forgiving than Moses. In 1875, he called sun worship "an error perhaps less degrading than many another since this star is the most perfect image of the Divine, the instrument whereby the Creator communicates all his blessings in the physical sphere."

Out of this fascination with the sun, astronomy was born. Five thousand years ago, the Tower of Babel, a ziggurat built as a watchtower, very likely served the Babylonians as an astronomical observatory as well. Priests kept detailed astronomical diaries and predicted future juxtapositions of the sun, moon, and stars. They recognized that eclipses come in cycles and that their general conditions repeat every nineteen years.

Orientation to the sun was a basic element in the design of Egyptian temples, which were constructed to celebrate the order of the world—the regularity of day and night, the seasons, and the calendar. The great pyramids of Egypt appear to have been aligned to reveal the changing geometry of sun and earth over the course of the year. In summer, the point of sunrise crept slowly south each successive morning; in winter, it worked its way back to the north again. Under the gaze of the Sphinx, the sides of the pyramids of Giza, built about 2800 B.C., line up with the rising sun at the vernal equinox.

The majestic temples erected at Karnak were not only places to worship Ra, the sun god, but also astronomical observatories. According to British astronomer Sir J. Norman Lockyer,

The sides of the great Egyptian pyramids appear to have been built to align with the rising sun at the vernal equinox. Their position reveals the changing relationship of sun and earth over the seasons of the year.

Every part of the temple was built . . . to limit the light which fell on its front into a narrow beam . . . so that once a year when the Sun set at the solstice the light is passed without interruption along the whole length of the temple, finally illuminating the sanctuary in most resplendent fashion. . . . The ray of light was narrowed as it progressed inward by a series of doors, ingeniously arranged, that acted as the diaphragms of the telescope tube in concentrating the light rays. . . . The temple was virtually an astronomical observatory and the idea was to obtain exactly the precise time of the solstice.

Far to the north, on bleak Salisbury Plain in England, Neolithic people began to construct Stonehenge in about 2600 B.C. They apparently arranged the stones so that they could track the course of the sun and moon, probably in order to follow the seasons, plan agricultural activities, and predict future eclipses. Nearly four thousand years later, across the Atlantic, the Mayans, obsessed by the need to record the passage of time, developed a calendar that fixed dates hundreds of millions of years in the mythic past. The Dresden Codex, a hieroglyphic inscription that survived the ravages of Spanish invaders, lists 1034 eclipse predictions within a 450-year period.

These monumental efforts to follow the sun must have required expenditures of labor and treasure at least as taxing to early societies as the construction of great astronomical observatories and the assembling of space missions are to ours. Unlike modern astronomical efforts, early astronomers directed their investigations almost solely toward charting the locations of the sun in relation to the earth and learning to predict changes in daily and seasonal cycles from them.

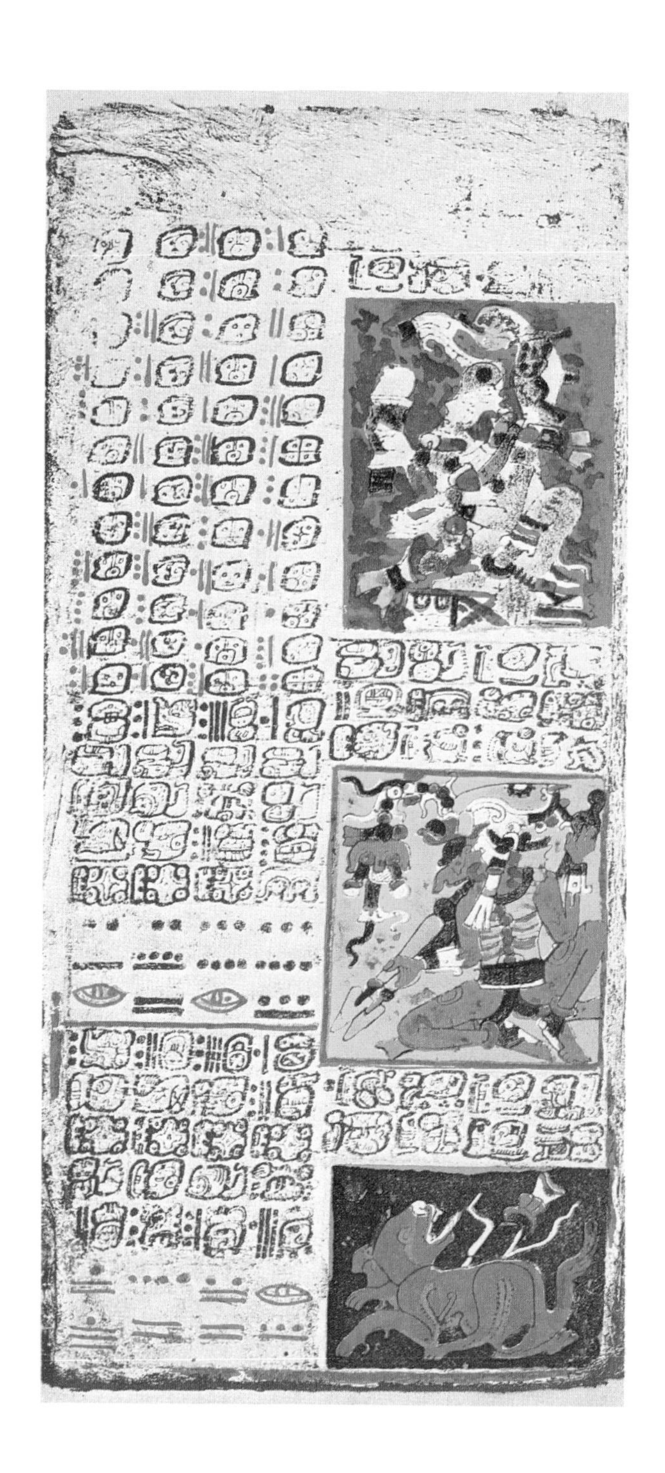

The Dresden Codex, dating from about A.D. 1300, is a manuscript containing records of astronomical observations of the Mayan civilization. Astronomer-priests of this pre-Columbian culture accumulated a remarkable listing of solar eclipses and the orbital cycles of the sun, the moon, and the planet Venus.

Chaco Canyon, New Mexico, was the principal site of the Anasazi culture, predating the present Pueblo Indians. The great "tower" kiva at Casa Rinconada is a roofless ceremonial temple, dating from the eleventh century. On June 21, the summer solstice, a shaft of light enters a sole window at dawn and moves slowly across a special niche in the wall.

Even Copernicus, whose calculations enabled him to show that the earth may best be viewed as another planet revolving about the sun and not as the center of the universe, concerned himself primarily with position and time.

Since the days of the ancients advances in astronomy have largely followed technological innovations. The foremost of these was the telescope, which Galileo Galilei used in 1611 to detect spots on the sun and to chart the sun's rotation. Distortion of the earth's atmosphere, however, severely limits all telescopic observation from the ground. Only four decades ago, the astronomer Henry Norris Russell captured the problem in the observation that "all good astronomers go to the moon when they die so that they may observe the universe without the interference of a dirty atmosphere."

Rocket and satellite technology developed after World War II has enabled all astronomers, both good and bad, to gain a clear view of the heavens before death. Telescopes installed in rockets and satellites traveling in the near-vacuum of space a hundred miles overhead allow astronomers to see the sun with its wavelengths unfiltered by the atmosphere and without any shimmering caused by turbulence. This second giant technological breakthrough has given us much of what we now know about the sun and its effects on the earth.

The McMath telescope atop Kitt Peak in Arizona is the world's largest solar telescope. Its vertical tower rises 110 feet to support a heliostat, or tracking mirror, that directs sunlight 500 feet down the diagonal tunnel to the telescope mirror. The image is reflected back to a third mirror that sends the light to the observing room.

The space missions are justly heralded for the spectacular pictures of the sun and other celestial bodies that they beam to earth. The sheer beauty of these images, however, is secondary to the meaning that astronomers can read in them. Advances in astrophysics now allow astronomers, with spectroscopic analysis, to make sense of what they discover. If all we had were telescopes, though, even extraterrestrial ones, astronomy would still suffer the constraints described by August Comte in 1844:

> The stars are only accessible to us by a distant visual exploration. This inevitable restriction not only prevents us from speculating about life on all these great bodies, but also forbids the superior inorganic speculations relative to their chemical or even their physical natures.

Comte concluded that we were forever doomed to ignorance about the chemical constitution of the stars. But he was wrong. We can discover the chemical and physical natures of the stars by analyzing the wavelengths of starlight, and we can now view stars by formerly unimaginable means, such as radio telescopes and neutrino detectors.

Visible and Invisible Starry Messages

Galileo could determine neither what the sun is made of nor how hot it is, not because he lacked a strong telescope but because he had no way of analyzing the "starry messenger"—the light seen by the naked eye—or of knowing about invisible radiation. Photographic film and electronic sensors, along with modern analytical techniques, now help astronomers identify the presence of every element in the sun from a distance of 93 million miles just as precisely as if a sample of solar material were brought down to earth. In order to comprehend modern astronomical observations of the sun, however, one must first know something about the nature and meaning of the electromagnetic spectrum, of which light forms a part.

In modern understanding, light has both a wave and a particle character. It is a form of electromagnetic radiation, waves of alternating electric and magnetic force traveling through space at 186,000 miles per second (300,000 kilometers per second). Like ripples in a pond, light waves have a characteristic wavelength—the distance from

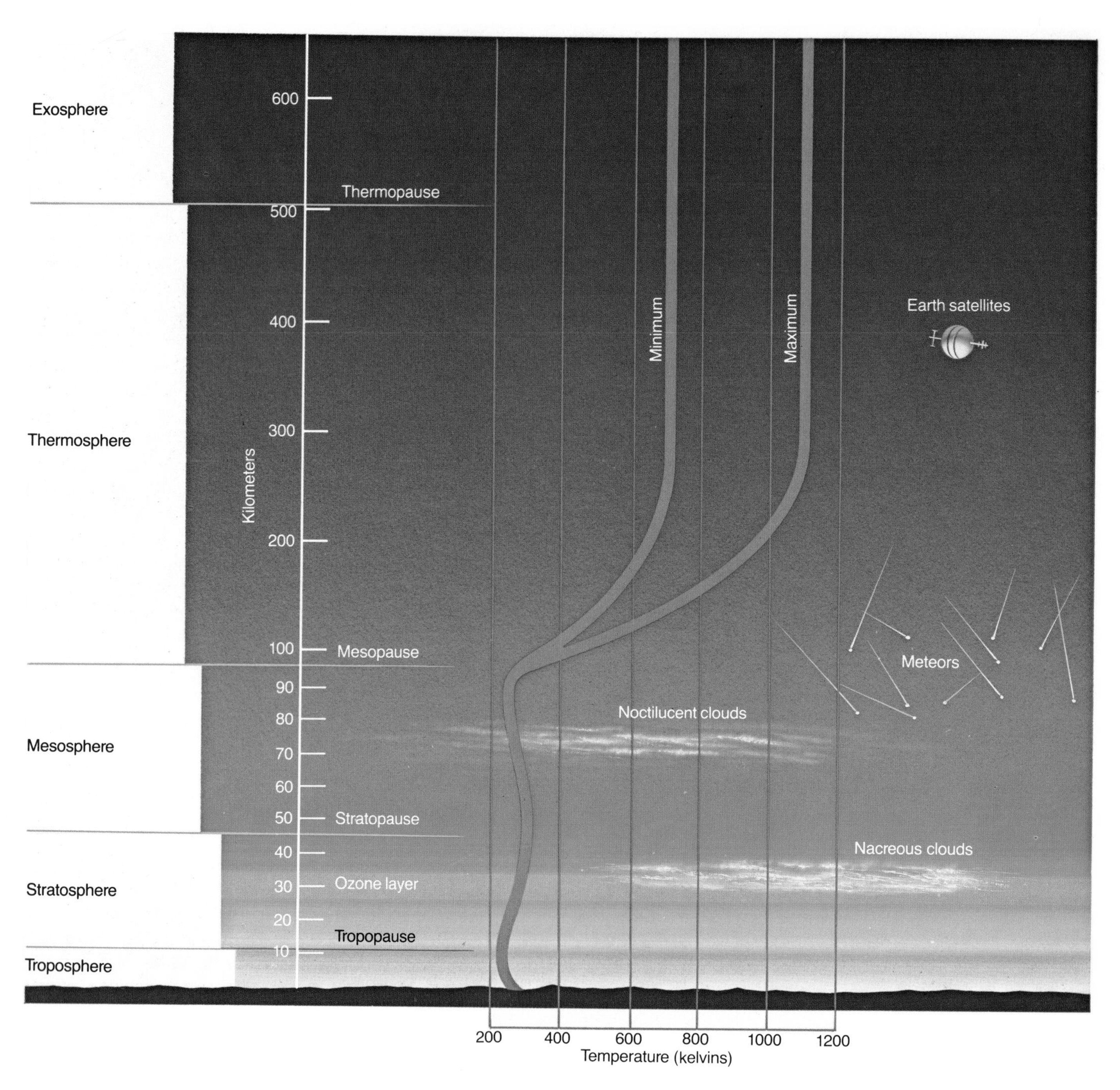

Ionized component of the atmosphere is produced by cosmic rays, gamma rays, X rays, and ultraviolet rays, which have maximum ionization rates in the D, E, F1, and F2 regions. The maximum concentration of electrons occurs in the F2 region, where the ionized component is about 1/1000 of the neutral-particle density. Shortwave radio signals are reflected from the ionosphere. X rays from solar flares produce radio blackout at an altitude of 60 to 75 kilometers.

Microwaves penetrate the ionosphere for satellite communications. Temperature inversions define the tropopause, stratopause, and mesopause. The atmosphere is mixed, and the composition of major constituents is essentially constant up to the mesopause. Ozone concentrates in a thin layer in the stratosphere. At higher levels, molecules dissociate and lighter elements separate out by diffusion. Temperatures in the thermosphere maximize at sunspot maximum.

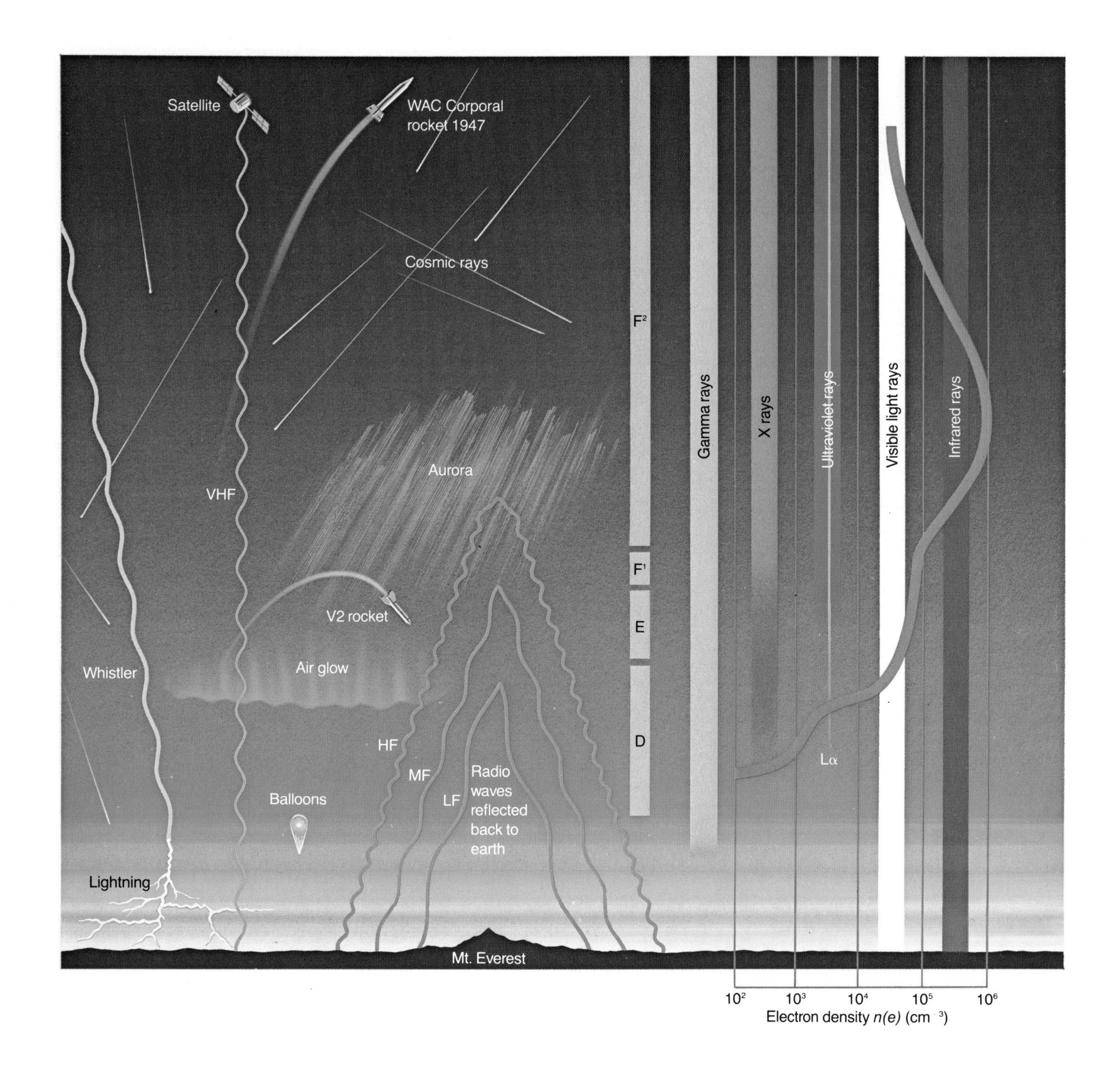

Satellite
WAC Corporal rocket 1947
Cosmic rays
Aurora
VHF
V2 rocket
Whistler
Air glow
HF
MF
LF
Radio waves reflected back to earth
Balloons
Lightning
Mt. Everest
F2
F1
E
D
Gamma rays
X rays
Ultraviolet rays
Visible light rays
Infrared rays
Lα
10²
10³
10⁴
10⁵
10⁶
Electron density n(e) (cm⁻³)

crest to crest—and a characteristic frequency—the number of crests that pass a given point each second. The wavelength and frequency of light are simply related by the formula:

$$\text{wavelength} = \frac{\text{velocity of light}}{\text{frequency}}$$

If an electron vibrates a million times a second (one megacycle per second), it radiates a train of a million electromagnetic waves each second. At the speed of light, the wave train extends 186,000 miles each second, and the wavelength is 1230 feet. This particular frequency and wavelength correspond to radio waves in the broadcast band. Higher frequencies progress through the television band, the infrared band, and the visible band. Look up at the blue sky for one second, and the electrons in the retina of your eye will dance back and forth 500,000,000,000,000 (5×10^{14}) times. At still higher frequencies come the invisible rays—ultraviolet, X rays, and finally gamma rays at frequencies as high as 10^{30} cycles per second.

Light also consists of quantum packets of energy, or photons, that are characterized by frequencies directly proportional to the quanta of energy:

$$\text{quantum energy} = h \times \text{frequency}$$

where h is a very small number, the Planck constant, and each quantum is a tiny bit of energy. For example, a 60-watt light bulb emits 10^{20} photons per second. Each elemental atom can absorb and radiate a very specific set of photon energies. This property of atomic radiation is the key to spectrochemical analysis of stars.

Over ages of evolution, the human eye has adapted most harmoniously to only a small fraction of the full electromagnetic spectrum of solar radiation, that of the sun's light as it reaches the earth's surface, a range of color from violet to deep red. We see precisely this range because the rainbow band of light passes freely through the atmosphere, whereas deeper violet and farther red are absorbed high above the earth. An eye with color sensitivity peaked in the yellow is the best receiver of sunlight at ground level. (Actually, the sun is most intense in the blue-green range, but preferential scattering by dust and aerosols subtracts blue light, leaving a yellow sun.) If life had evolved on a planet with a different atmosphere, eyes, if such organs existed at all, might have been sensitive to infrared or ultraviolet rays.

The electromagnetic spectrum, now observed from the ground and from space, spans all frequencies from very long radio waves to ultrashort-wavelength gamma rays. Gamma-ray frequencies as high as 10^{30} hertz have been detected via secondary light (Cerenkov radiation) produced by interactions with the molecules of the high atmosphere. (1 electron volt [eV] = 1.6×10^{-12} erg; 1 angstrom = 10^{-8} centimeter.)

The Electromagnetic Spectrum

Name of region	Opacity of atmosphere	Wavelength (cm)	Frequency (hertz)	Energy per quantum (electron volts)
Gamma rays		10^{-20} … 10^{-9}	10^{30} … 3×10^{19}	10^{16}
X rays		10^{-8} (1 angstrom) … 10^{-6}	3×10^{16}	10^{4}
Ultraviolet		3×10^{-5}	10^{15}	
Visible (Violet, Blue, Green, Yellow, Orange, Red)				2
Infrared		10^{-4} … 10^{-1}	3×10^{11}	
Microwaves		1	3×10^{10}	10^{-4}
Spacecraft		1 inch 10^{2}	3×10^{8}	
Television		10^{3}	3×10^{7}	
FM				
Shortwave		10^{4}	3×10^{6}	
(AM)		10^{5}	3×10^{5}	10^{-9}
(VLF)		> 1 mile		

Opaque

Partially transparent

Transparent

By analogy with pitch on the musical scale, visible colors span not quite one octave, a frequency factor of 2. The full electromagnetic spectrum of detectable solar radiation from the deepest base frequencies to the ultrahigh squeal of gamma rays spreads over about 70 octaves. If the "music of the spheres" is written for the range of 10 piano keyboards, an astronomer's one-octave eyesight can hardly be considered a high-fidelity receiver for the cosmic symphony. As so often happens in science, discoveries of more and more octaves of solar radiation and of ways in which to analyze them came in bits and pieces that only later astronomers were able to stitch together.

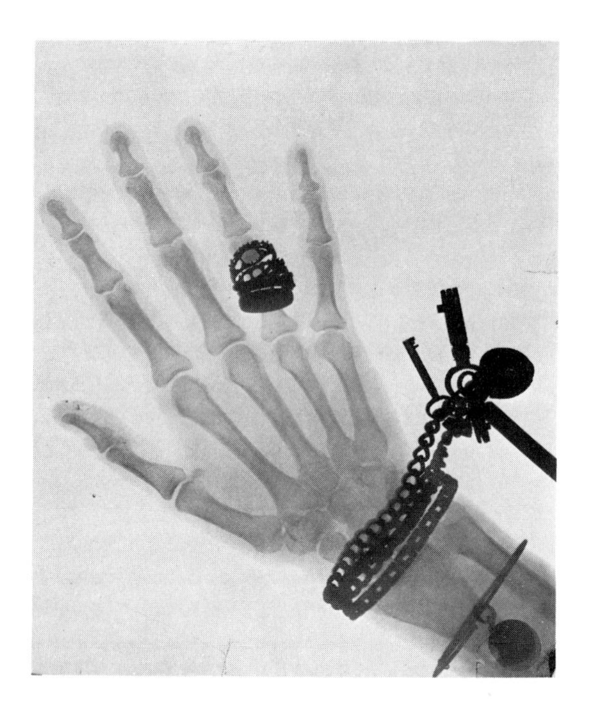

Roentgen's discovery of X rays and their ability to penetrate flesh but not bones led to photographs, such as this picture of a hand, that created a popular sensation.

The discovery of invisible radiation took place in the nineteenth century. Newton believed that the sun's light contained its heat, but in about 1800, Sir William Herschel discovered that the sun also emits radiation that is invisible to the naked eye. Experimenting with colored-glass filters, Herschel sensed warmth on his eyeball even when the sun's direct light was blocked out. By positioning a thermometer in the various colors dispersed by a prism, he found very little temperature increase in the blue, green, and yellow. But as he moved the thermometer beyond the visible red, where nothing could be seen, the temperature rose rapidly. He had discovered invisible infrared rays. A year later, invisible ultraviolet rays were discovered by J. W. Ritter, who observed that silver chloride, placed beyond the violet end of the spectrum, turned black.

Almost a century after Herschel, Wilhelm Roentgen, working in Würzburg, Germany, experimented with the conduction of electricity through a highly exhausted vacuum tube. In 1895, he noticed that a fluorescent screen of barium platinocyanide that happened to be near his discharge tube was glowing. Further investigation revealed that whatever radiation it was that caused the fluorescence could pass through opaque materials and affect photographic plates. This invisible radiation also penetrated flesh and cast shadows only of bones. He doubted that these mysterious rays were related in any way to light rays, and named them X rays. Roentgen's discovery immediately attracted great attention, and within two weeks he was invited to make a personal demonstration before Kaiser Wilhelm II.

When Roentgen reported his research to the French Academy of Sciences in 1896, his audience included a physicist, Henri Becquerel, who suspected that the fluorescence of the glass wall of Roentgen's tube itself produced the X rays. He therefore proposed that crystals of uranium sulfate, which were known to phosphoresce strongly when exposed to sunlight, might also be a source of X rays. While testing this hypothesis with uranium salts, he observed that heavily wrapped photographic plates stored nearby had become fogged. The fogging occurred during a prolonged dark and dismal rainy stretch in Paris that ruled out sunlight leaks, and Becquerel realized that some spontaneous and continuous emission of radiation must come from the uranium sulfate crystals themselves. In many respects, this new radiation that blackened photographic plates after penetrating sheets of metal resembled Roentgen's X rays. Becquerel had discovered the radioactivity of uranium and the accompanying emission of extremely short-wavelength gamma rays.

In the mid-nineteenth century, James Clerk Maxwell developed the modern theory of light as an electromagnetic wave, thereby unifying new knowledge of invisible light with existing knowledge of visible light. Maxwell's theory predicted that it should be possible to generate man-made electromagnetic waves of very long wavelength, and in 1886, seven years after Maxwell's death, Heinrich Hertz succeeded in the first experimental verification of Maxwell's theory. Connecting a Leyden jar condenser to a coil, he caused sparks to jump between a pair of brass balls. Each spark produced radiofrequency oscillations at about 50 megacycles per second. Using a similar spark gap as a receiver, Hertz could detect radio waves through a wooden door, a glass window, and other nonconducting materials, but not through a metal wall. And to confirm the similarity to light waves, Hertz focused his radio waves with metal mirrors and refracted them with prisms of "coal-tar pitch." With these simple tests, Hertz established the fundamental electromagnetic properties of radio waves.

By the end of the nineteenth century, the full range of electromagnetic radiation had been discovered and a theory invented to account for it. Only optical radiation, however, had found application in astronomy.

Unlocking the Chemistry of the Universe

Optical spectroscopy began with Sir Isaac Newton. In 1766, he wrote:

> I produced me a triangular glass prism, to try therewith the celebrated phenomena of colours. And in order thereto having darkened my chamber and made a hole in my window-shuts, to let in a quantity of the sun's light, I placed my prism at its entrance, that it might thereby be refracted to the opposite wall. . . . It was at first a very pleasing divertisement to view the vivid and intense colours produced thereby.

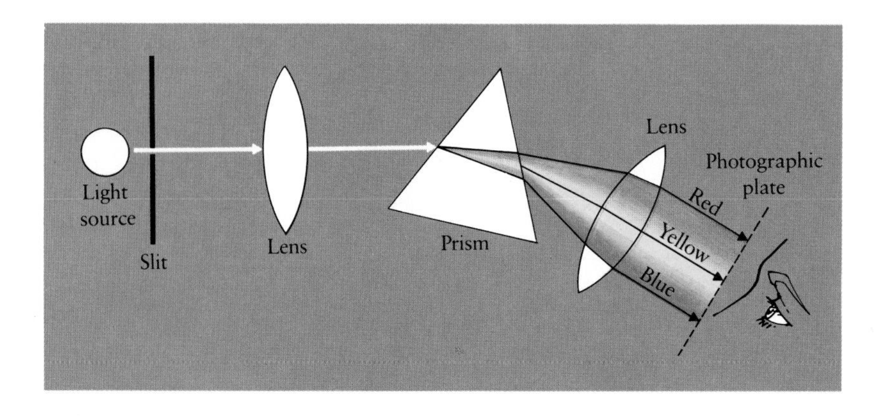

Components of a slit spectroscope. Parallel rays of light are refracted according to color in passing through the prism, once on entering and again on leaving. The separated colors are focused by the lens onto the eye, a viewing screen, a photographic plate, or some other sensor.

Johann von Fraunhofer used a prism spectroscope to disperse the colors of sunlight in order to test the refraction of optical glasses. He discovered that in each part of the rainbow continuum, the spectrum was channeled by black lines at hundreds of specific wavelengths.

As early as 1814, Johann von Fraunhofer, a young lens designer, stumbled upon a surprising discovery. While trying to isolate pure colors from sunlight to test the refraction of his optical glasses, he used a prism to disperse the colors, just as Newton had. He improved on Newton's arrangement, however, by making the light enter the prism through a fine slit. When the light was of one color, he could focus a sharp image of the slit in that color after the light had been refracted by the prism. Such a combination of slit-source, prism, and lens forms a slit spectroscope and produces slit images, or spectrum "lines," in the component colors of the refracted light.

When Fraunhofer looked at the spectrum of sunlight with his instrument, he discovered that each of the rainbow colors contained thousands of fine black lines, like a ladder of closely spaced rungs, interrupting the smooth progression of color from red to violet. At first, he attributed these "imperfections" to flaws in the color transparency of his glass, but soon it became clear that the dark lines were a real feature of the solar spectrum. Fraunhofer found exactly the same lines in the spectra of the moon and planets, which shine by reflected sunlight, while an incandescent light source in the laboratory produced a pure rainbow spectrum completely free of dark lines. These observations proved beyond doubt that the lines were of solar origin. Fraunhofer counted 574 dark lines and labeled the clearest with the letters A to H, but he had no intuition about the true significance of the lines, intending to use them merely as identifications of standard colors for his optical tests.

The mystery of Fraunhofer lines lay unsolved for more than three decades. An observation of Fraunhofer's, however, contained the clue that led to the solution and, eventually, to the chemical analysis of the

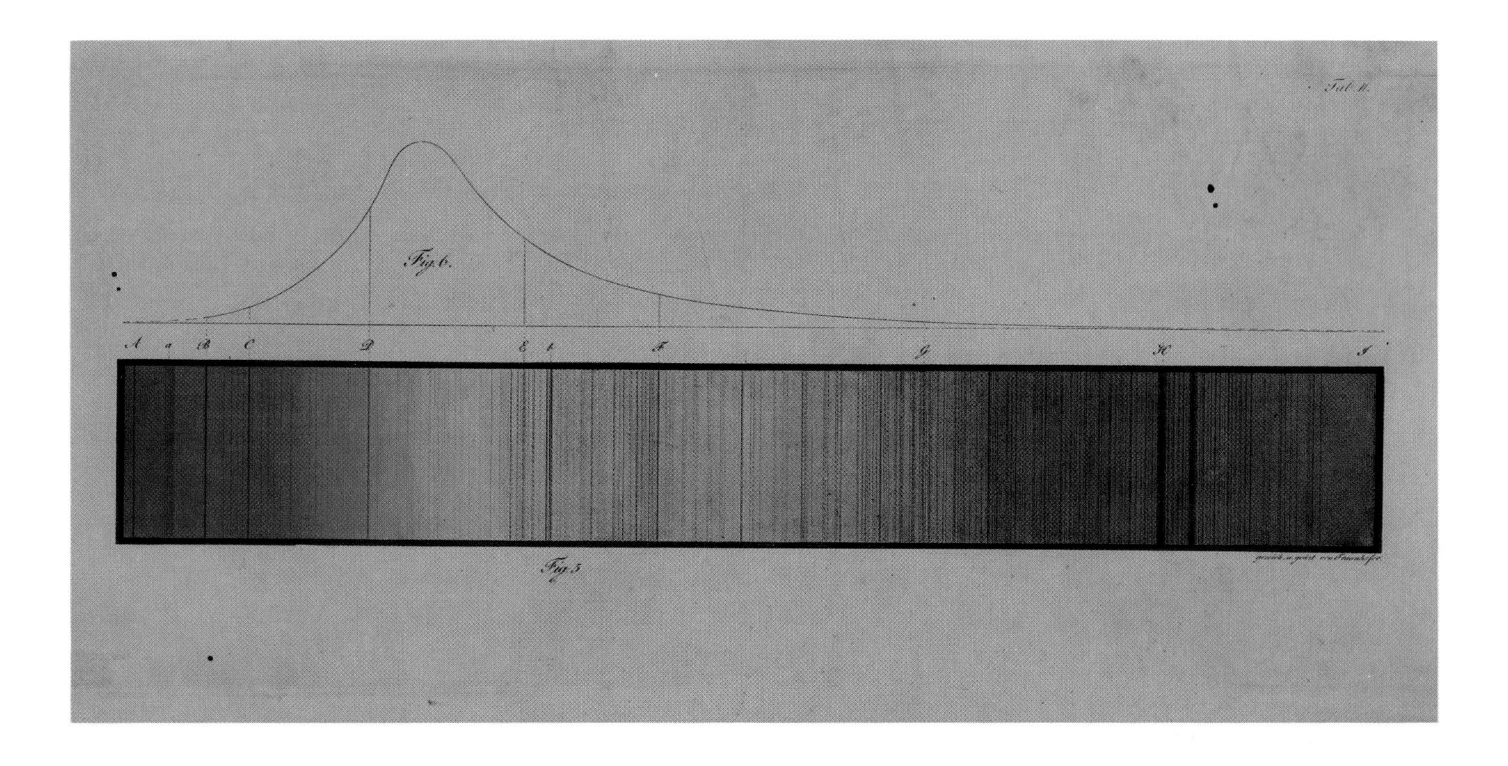

Fraunhofer counted 574 dark lines in his early spectrum. Today, more than 25,000 lines in the sun have been identified. Nearly every element known on earth produces its own set of lines that "fingerprint" its presence in the sun.

solar system. He had noticed the coincidence of a particularly prominent dark line in the yellow region of the solar spectrum, which he labeled with the letter D, with the bright yellow line almost always present in the spectra of flames. By the 1850s, scientists realized that the yellow line in flames was attributable to sodium, but they were slow to associate sodium with the Fraunhofer D line in the sun. Gustav Kirchhoff and Robert Bunsen, the inventor of the well-known Bunsen burner, finally made the great leap from laboratory studies of absorption spectra to the interpretation of Fraunhofer lines in the sun.

A note in the 1902 issue of *Nature* suggests how Kirchhoff and Bunsen may have stumbled on this insight. As the two men pointed a spectroscope from their laboratory window in Heidelberg at the flames of a fire raging in Mannheim, they detected spectral lines of barium and strontium. Later, while they strolled along the "Philosopher's Walk" through the wooded hills above Heidelberg, Bunsen mused that if they could analyze the fires at Mannheim, they should be able to do the same for the sun. "But," he added, "people would think we were mad to dream of such a thing."

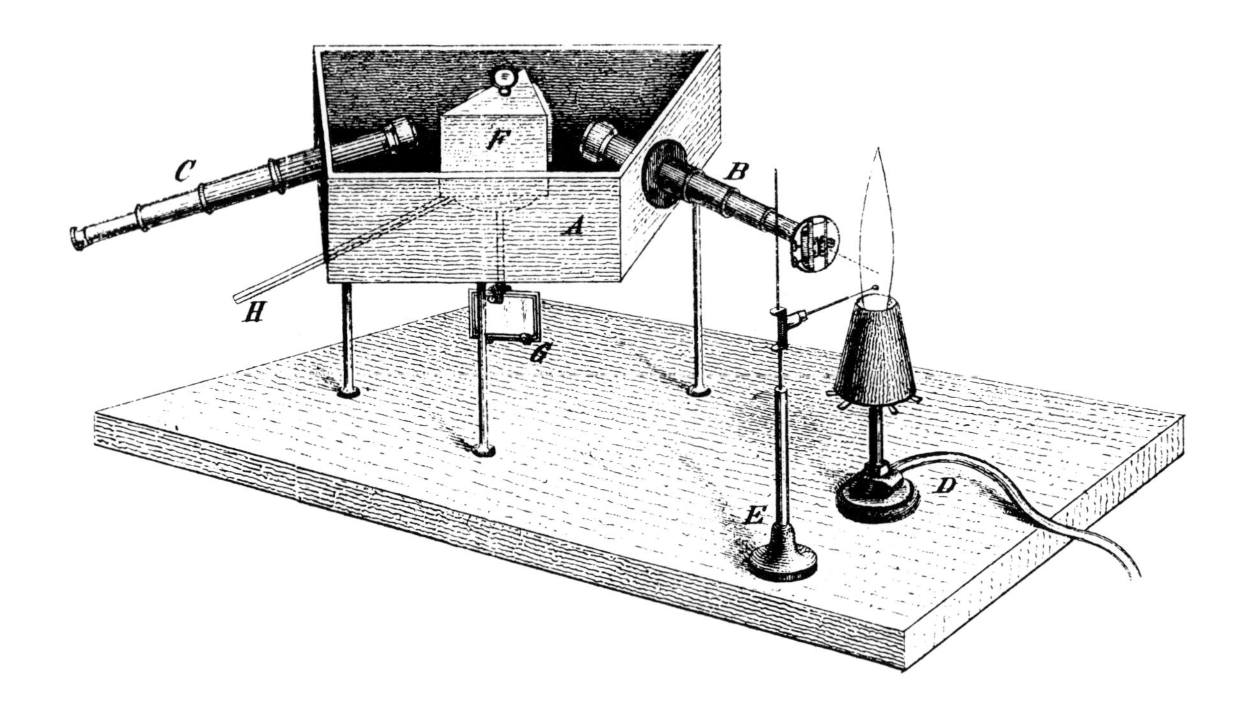

Robert Bunsen invented his gas burner (D) in 1850. Its colorless flame was ideal for identifying chemical elements (E) from the colors that they imparted to the flame. When he and Gustov Kirchhoff assembled a collimator (B) a prism (F) and eyepiece (I) to make a new spectroscope (C), the size of a cigar box for chemical analysis, they discovered the unique connection between chemical elements and their spectral lines.

Today, the two men are honored for their madness. On the main street in Heidelberg, a plaque reads: "In this building, Kirchhoff, together with Bunsen, founded spectral analysis, turned it to the sun and stars, and with it unlocked the chemistry of the universe."

When Kirchhoff put a sodium flame before the slit of his spectroscope, a bright yellow spectrum line appeared. Such a bright line is called an emission line. Light from an incandescent source showed a continuous rainbow band of color. But, he discovered, if the light passed through a sodium flame before entering the spectroscope, the yellow D line was absorbed and appeared as a black line in the rainbow. Quickly producing emission-line spectra of many common elements, Kirchhoff found that he could match lines of hydrogen, magnesium, calcium, iron, and others with Fraunhofer's black lines. He concluded that each element, when heated to vaporization, absorbs those wavelengths, and only those wavelengths, that it can itself emit.

The solar Fraunhofer-line spectrum is thus a fingerprint of the elements that occur in the sun. The dark lines originate in the upper part of the solar atmosphere where the gas is still very hot but much cooler than the underlying body of gas that radiates "white light."

Today, more than 25,000 Fraunhofer lines, including the ultraviolet, have been identified, and dark lines for nearly every element now known on earth have been detected in the sun. As Ralph Waldo Emerson once wrote, "There is nothing on Earth which is not in the heavens in a heavenly form, and nothing in the heavens which is not on the Earth in an earthly form."

Furthermore, the character of the spectral lines—whether they appear sharp or fuzzy, dark or only half shaded, slightly shifted toward the red or toward the blue—permits the astrophysicist to deduce the temperature, pressure, density, composition, strength of gravity, electric force, magnetic force, degree of turbulence, and convective movements in the region of the sun where the spectrum line is produced.

The New Vision of Sun and Earth

Many scientists at the turn of the century still thought, much as the ancients had, that the sun burns like fire. But if a pile of coal as big as the sun had been ignited at the time of the pharaohs, it would by now have been totally consumed. Only with Einstein's revolutionary theory of the equivalence of mass and energy have we come to understand how sunlight is produced by nuclear energy. The sun is a steady, reliable thermonuclear reactor, and it will deliver power to the earth for billions of years to come.

Until recently, the astronomer's view of the heavens had been marred by a murky, shimmering atmosphere that distorts light rays and blankets the sun's X rays and much of its ultraviolet and infrared radiation. With the advent of extraterrestrial observatories, we can now describe a new sun as human beings in previous generations had never seen it. Since 1946, when rockets first went aloft carrying small telescopes and spectrographs above the atmosphere, more and more sophisticated technology culminated in the Skylab mission of 1973. Since Skylab, solar studies have continued with improved instruments aboard the free-flying Solar Maximum Mission and on the Space Shuttle, as well as on smaller satellites and rockets. In the next decade, astronomers hope to fly a Solar Optical Telescope (SOT), 1.25 meters in diameter, that will resolve surface features as small as 70 kilometers across, almost 10 times better than the resolution achieved with telescopes on the ground.

Skylab

Skylab floats against a black daytime sky. The Apollo Telescope Mount and its four solar panels face the sun. During manned operation the solar telescopes were controlled from within the multiple docking adapter, a cylindrical structure seen here with its round entrance door face on. To replace film magazines, the astronauts performed EVAs (extravehicular activities), climbing to the circular platform at the top to gain access to the telescopes.

The most ambitious project to study the sun in modern times was the Skylab mission, the first American space station and manned scientific workshop. Bigger than a boxcar, Skylab measured 118 feet long and 22 feet in diameter and weighed 100 tons. The mission used a Saturn V rocket left over from the Apollo moon program to place the laboratory in a 270-mile-high circular orbit, where it arrived on May 14, 1973, in badly crippled condition. One of its two large solar panels, intended to provide electrical power, had been ripped off, and the other was jammed in its folded launch position. On the verge of failure, the billion-dollar mission was rescued by astronauts Pete Conrad and Joe Kerwin, who climbed over the derelict spacecraft and freed its jammed solar panel.

Aboard Skylab, high-resolution X-ray and ultraviolet cameras and spectrographs, the most advanced hardware ever assembled for a campaign of solar observations, were carried in the Apollo Telescope Mount (ATM). The instrumentation captured photographs of finer detail than ever before and, being above the atmosphere, covered all wavelengths from visible light to X rays. The film and film vaults on Skylab weighed 4760 pounds, more than the total weight of all previous spacecraft devoted to astronomical observations.

The team-up of scientist-astronauts in space with colleagues on the ground was an outstanding success. At the end of each day's work, scientists at the Johnson Space Center in Houston evaluated the information from Skylab, planned operations for the following day, and set longer-term strategies. Guided by TV instruments that monitored the sun continuously for incipient

As modern science has advanced our knowledge of the sun, so has it transformed our understanding of space between sun and earth, all the way down to the space in which we live. Before the launching of deep space probes, the interplanetary space was thought to be almost a complete void. But the earliest Explorer spacecraft missions discovered an invisible wind of superheated solar gas that flows variably but continuously out of the sun. So thin is the wind that particles traverse

After removal from one of the Skylab solar telescopes the film magazine is attached to a long manipulator boom that transports it to a waiting astronaut at the airlock door. To protect it from radiation, the film was placed in a lead box container. At the time of this EVA, the Skylab was orbiting at a height of 435 kilometers and a speed of 29,000 kilometers per hour.

flare and prominence activity, the scientist-astronauts trained their telescopes on active regions, using on-the-spot judgments to initiate rapid sequences of photographic exposures. Astronaut Edward G. Gibson discovered that the entire disk is speckled with "bright points" of X-ray emission, each about the size of the earth. Some 1500 emerge per day and last about eight hours. As they break out, they bring internal magnetic fields to the surface. Gibson learned to read these bright points as clues to the outbreak of flares, and he succeeded in pointing the spectroheliograph to a major flare just as it began to flash toward its peak brilliance. Numerous images were subsequently captured of explosive flaring and other short-lived events.

Skylab was manned by three crews. During the 171 total days of their visits, they bagged 160,000 solar images, enough for many years of analytical study. Their ATM observations ran the gamut of transient solar activity phenomena: the structures and contortions of magnetic fields, huge expanding prominence loops and surges, billion-ton bubbles of escaping high-velocity gas, brilliant flares, coronal holes, microflaring bright points, jetting spicules, and more. About 150 scientists in 17 countries tuned in daily to the Skylab operations and coordinated their ground-based observations to complement the space studies most effectively.

The original plan, based on a predicted level of solar activity, called for leaving Skylab in orbit and returning to it as soon as the Space Shuttle was ready. Unfortunately, work on the shuttle took longer then anticipated, and higher-than-expected air drag on Skylab brought the satellite down several years earlier than anticipated. Skylab reentered the atmosphere and burned up in 1978. The first flight of space shuttle Columbia took place in 1981. Had the shuttle flights started sooner, Skylab could have been boosted to a higher orbit where it would have been subject to much less drag and where it might still be performing as a unique solar observatory.

the 93 million miles from sun to earth with rarely a collision. (By contrast, molecules in the air around us can move barely millionths of an inch without bumping one another.) Ephemeral as it may be, the solar wind exerts a profound influence on the terrestrial environment. Added to the wind, at times, are colossal bubbles of gas weighing billions of tons that blow off the sun and impact the earth at speeds of millions of miles per hour.

The earth itself nestles in a protective magnetic cocoon that carves a cavity out of the solar wind. On the sunward side, the magnetic shield is blunted by the pressure of the solar wind. Downwind, away from the sun, it is swept into a long tail, past the orbit of the moon, that reaches perhaps as far as 1000 earth radii. The huge volume of electrified gas confined inside this magnetic bag is called the magnetosphere. It encompasses the Van Allen radiation belts, and it is filled with charged particles of all energies. When the electrified wind blows across the magnetized polar caps, it creates a gigantic natural dynamo that generates power equal to all the kilowatts consumed in the United States. Sudden gusts of wind cause the magnetosphere to quiver like a mass of jelly, sending to the earth magnetic shivers that adversely affect transpolar radio tranmissions and disturb the orbital stability of satellites.

The coupling between sun and climate is, in theory, so delicate that a variation in solar brightness of only a few tenths of a percent can create flourishing biological productivity or freeze the earth in the grip of an ice age. Glaciers have advanced and retreated roughly every hundred thousand years in synchrony with the stretching of the earth's orbit from circular to just slightly elliptical and back to circular. Why pronounced climate changes are so finely tuned to changes in exposure to the sun as minute as those produced by orbital variations is a challenging scientific question. Important clues may be written in the records of tree rings, coral reefs, ice cores, and the skeletal remains of plankton that accumulate layer upon layer in the ooze of the ocean bed. Study of past climate from these proxy records and probing the interaction of a host of other terrestrial factors, such as snow and ice cover and continental drift, is fundamental to understanding the sensitivity of climate to solar variability.

In a statistical sense, the earth is only an insignificant speck of rock in the vastness of the cosmos. Viewed from the moon, the earth is an emerald sphere of delicate beauty, a rare cosmic jewel magically suspended in the sun's gravitational grip. We live within a thin membrane of air and water surrounding the earth, protected from the harshness of space and nourished by a benign sun. As we learn more and more about the narrowness of life-supporting ecological zones in solar-planetary configurations, we must be profoundly impressed with the delicate balance between sun and earth that spells the difference between the development of life on earth and its apparent absence elsewhere in the solar system. Nature has brought us over eons of time to our present harmony with the environment. We must strive to understand the sun-earth relationship in all its complexity lest, in ignorance, we tamper with it to our eventual sorrow.

2

Inside the Sun: Core to Surface

That orbed continent, the fire that severs
day from night.

William Shakespeare, *Twelfth Night*

Sunrise III, by Arthur G. Dove.

The furnace in the heart of the sun is a 15-million-degree pressure cooker that makes helium out of hydrogen by thermonuclear fusion. Since it turned on 5 billion years ago, our nuclear power plant in the sky has burned 5 million tons of hydrogen each second to feed life-sustaining energy to the earth. It was only 200 years ago that scientists first learned how to take the measure of solar power, and the source of its energy remained unfathomable to this century.

The Power of Sunlight

In the late eighteenth century, the British astronomer Sir William Herschel performed a very ordinary experiment. From measurements of the temperature of water standing in the sun, he computed that the noonday sun would melt one inch of ice in two hours and 12 minutes. This simple result implied solar radiation of astounding power. As C. A. Young used to describe it in an introductory lecture to his Princeton students:

> Since there is every reason to believe that the sun's radiation is equal in all directions, it follows that, if the sun were surrounded by a great shell of ice, one inch thick and 186 million miles in diameter, its rays would melt the whole in the same time. If, now, we suppose this shell to shrink in diameter, retaining, however, the same quantity of ice by increasing its thickness, it would still be melted in the same time. Let the shrinkage continue until the inner surface touches the photosphere . . . the ice-envelope would become more than a mile thick, through which the solar fire would still thaw out its way in the same time.

The law of the conservation of energy—energy cannot be created or destroyed, but only transformed—was enunciated in 1842 by Julius Robert Mayer, the City Physician of Heilbron. Heat can create motion, as steam from a boiler drives a piston that causes a steam engine to move. Conversely, mechanical motion can create heat, as pumping air into a tire makes it warm. It occurred to Mayer that the sun might absorb enough energy to maintain its temperature by steadily swallowing up meteorites. But calculation showed that the mass of meteorites captured in this fashion every 100 years would need to equal the mass of the earth. If the sun grew by that mass each century, its increasing gravity would speed up the orbital velocity of the earth by nearly two seconds a year. Kepler, in his time, would easily have detected such a change. Finally, Mayer's theory had to be abandoned.

At one time in his career, Sir William Thomson, later Lord Kelvin, conceived of a far more dramatic variation on Mayer's meteorite theory. If meteorites were insufficient to keep the sun burning, perhaps it consumed whole planets, one by one. As interplanetary gas dragged on the planets and shrank their orbits, the first to spiral into the sun and disappear would be Mercury. This would deliver enough energy for nearly seven years. Venus would contribute about 84 years, and so on, until the digestion of Neptune gave the sun perhaps 2000 final years. To his disappointment, Thompson found that all the planets together would contribute only 46,000 years, a conclusion that dampened any remaining enthusiasm for the capture of matter as a source of solar energy. (In searching for possible sources of variation in the sun's luminosity, however, some astronomers have proposed that the sun accrues mass from collisions with interstellar gas clouds as it journeys around the center of the galaxy.)

William Thomson, later Lord Kelvin, discoverer of the second law of thermodynamics, originator of the absolute temperature scale, and inventor of numerous refinements in electrical instruments and submarine telegraphy.

The German physicist Hermann von Helmholtz recognized the deficiency in Mayer's theory almost immediately. In 1854, he proposed that, rather than consuming in-falling matter, the sun extracted energy from its own slow collapse. Later, along with Kelvin, he estimated that a decrease in diameter of only 150 feet per year would suffice to produce the energy radiated. Such a slow rate of contraction would be just perceptible in a few thousand years, and the sun could have been shining at its present rate for about 22 million years. From fossil history, however, geologists figured that 22 million years was hardly enough. Contemporary paleontologists wanted 250 million years, and today we know the sun has radiated at near its present intensity for more than 4 billion years. A mechanism of solar energy production 10 million times more efficient than that offered by the Kelvin-Helmholtz shrinkage theory was needed.

Modern Concepts

After Henri Becquerel discovered radioactivity in 1896, speculations about subatomic particles began to appear in the scientific literature. In 1905, Einstein put forth his fundamental thesis that matter and energy are interchangeable:

$$E = mc^2$$

or energy equals mass multiplied by the square of the velocity of light. With c a very large number, 3×10^{10} centimeters per second, even as

Sir Arthur Eddington, pioneer theorist of the internal structure of stars.

little as one gram (1⁄27 ounce) of mass, multiplied by c^2, becomes 21,500,000,000,000 calories. It would take 670,000 gallons of burning gasoline to produce that much energy.

Astrophysicists had no alternative but to accept that nuclear conversion of mass into energy was the mysterious source of stellar power. Nobody knew, though, how the process worked inside a star. The details came step by step, beginning with the remarkable discoveries of nuclear transmutations by Lord Ernest Rutherford at the Cavendish Laboratory.

In 1919, Rutherford demonstrated the artificial transmutation of nuclei by energetic particle collisions. When he bombarded nitrogen nuclei with helium nuclei (alpha particles) released by radioactive bismuth, he found to his surprise that every once in a while, a nitrogen nucleus absorbed an alpha particle and only a single proton escaped, leaving behind a nucleus of oxygen. James Chadwick and others quickly joined him in exploring a host of nuclear transformations that occur when particles collide with very high energy.

By the time Sir Arthur Eddington published *Internal Constitution of Stars* in 1926, he was able to infer the temperature of the solar core from very elementary considerations. Gravitational force that pulls all matter toward the center holds the sun together. A counterpressure proportional to the temperature of gas pushes matter outward. For the sun to assume a stable size, its temperature must be high enough for gas pressure to balance gravitational force. Eddington estimated the central temperature of the sun to be 40 million degrees Celsius. Today we believe that its temperature is close to 15 million degrees, but many nuclear physicists in the 1920s thought that even 40 million degrees was inadequate to sustain nuclear energy conversion.

At Eddington's calculated temperature, atoms in the solar furnace would have boiled off their electrons. Hydrogen nuclei (protons) and free electrons would race through the interior at average speeds of 1000 kilometers per second. Even at that speed, physicists thought that fusion of hydrogen into helium could not occur. Rather, they estimated temperatures as high as tens of billions of degrees. Eddington stubbornly defended his model. Certain not only that nuclear energy was the only possible source of stellar power but also that his proposed temperature had to be correct, he wrote in defiance,

> We do not argue with the critic who urges that the stars are not hot enough for this process; we tell him to go and find a hotter place. . . . What is possible in the Cavendish Laboratory may not be too difficult in the Sun.

In 1932, Chadwick discovered that the atomic nucleus of any element heavier than hydrogen contains protons *and* neutrons. If an electron and a proton collide with enough force, they stick together and form a neutron. Two protons and two neutrons can combine to form a helium nucleus. In that process, a small amount of mass is converted into energy, and therein lies the secret of thermonuclear power.

Two chains of nuclear reactions operate in the solar furnace, the proton cycle and the carbon cycle. In addition to the familiar proton, neutron, and electron, these chains also include ghostlike particles called neutrinos, which have zero or near-zero mass—neutrinos are discussed later in more detail—and antiparticles, which deserve a quick look here. In 1920, the English physicist P. A. M. Dirac theorized that for every particle, positively or negatively charged, there ought to exist an antiparticle with the opposite charge. For example, a positron is an antielectron with the same mass but a positive charge exactly equal to the electron's negative charge. When a particle and an antiparticle meet, they annihilate each other, and the total mass of the pair appears as photon energy. If an electron and positron collide, for example, their combined mass is converted into gamma radiation. This is how most nuclear energy is converted into radiation in the core of the sun.

To return to nuclear reaction chains in the sun, however, the proton cycle, which may account for as much as 99 percent of the nuclear power generated in the sun, starts with the fusion of two protons to form heavy hydrogen, or deuterium. A deuteron, (hydrogen 2, or ^{2}H) consists of a proton plus a neutron. In the act of fusion, one proton must switch in essence instantaneously (10^{-21} second) into being a neutron, with the release of a positron and a neutrino. The positron very quickly encounters a free electron, and the pair "annihilates," with the release of two gamma rays.

Proton-proton fusion requires a very high temperature to raise collisional energy enough to overcome the electrical repulsion of the positive charge on each particle. Only when the temperature exceeds 13,000,000°C can the fusion process go forward. The higher the temperature, the faster the reaction. At its core, the sun supports its overlying mass against gravitational collapse at a temperature of 15,000,000°C.

It takes about 14 billion years, on average, to form a deuteron. By contrast, the next step follows with little delay. In about 0.6 seconds,

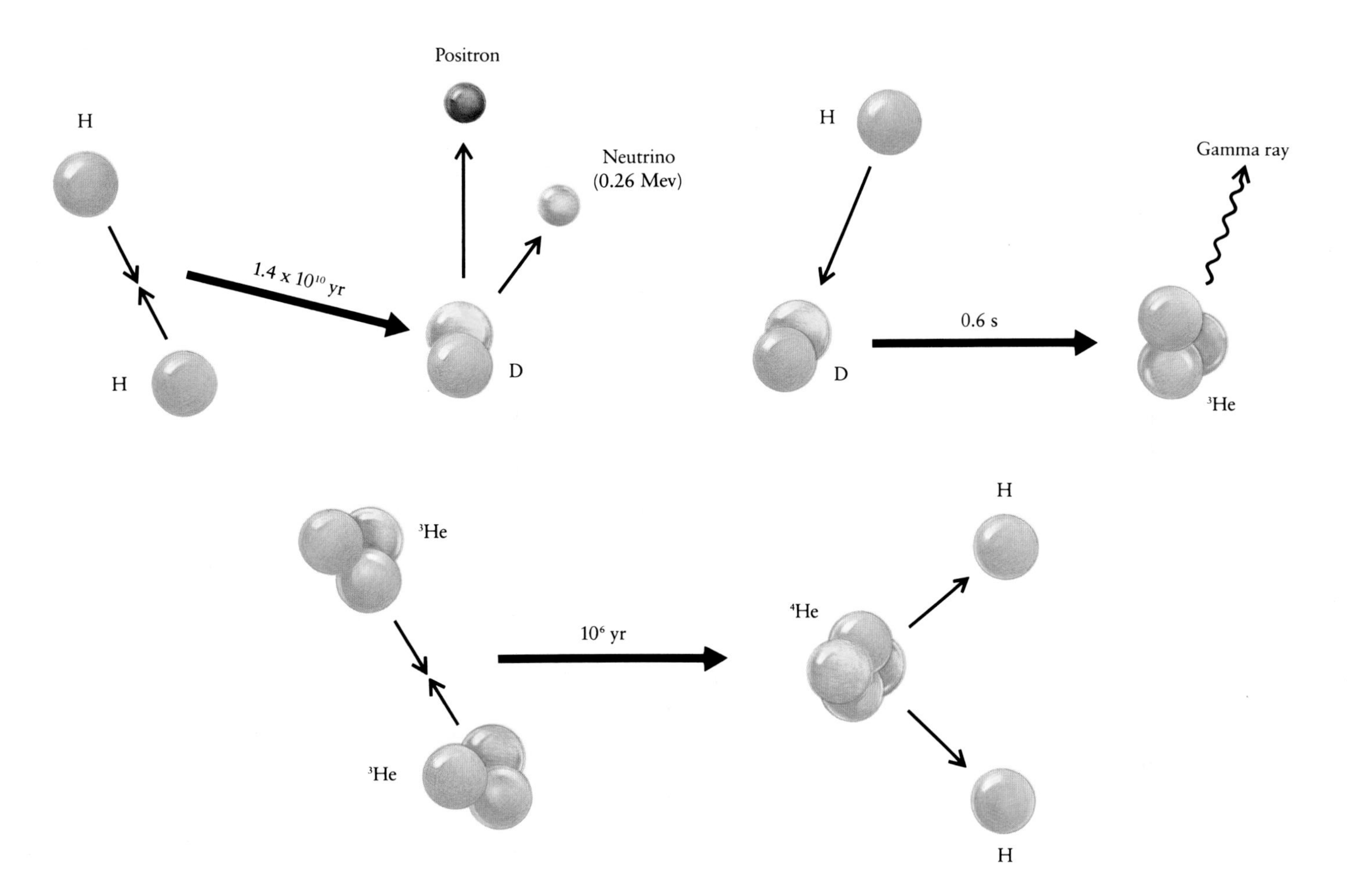

Fusion of hydrogen into helium in the proton-proton cycle. The reaction times are given at each step, as is the energy of the neutrino. The first two reactions must be counted twice to produce the two ^{3}He nuclei for the final fusion. Protons are shown in red, neutrons in green.

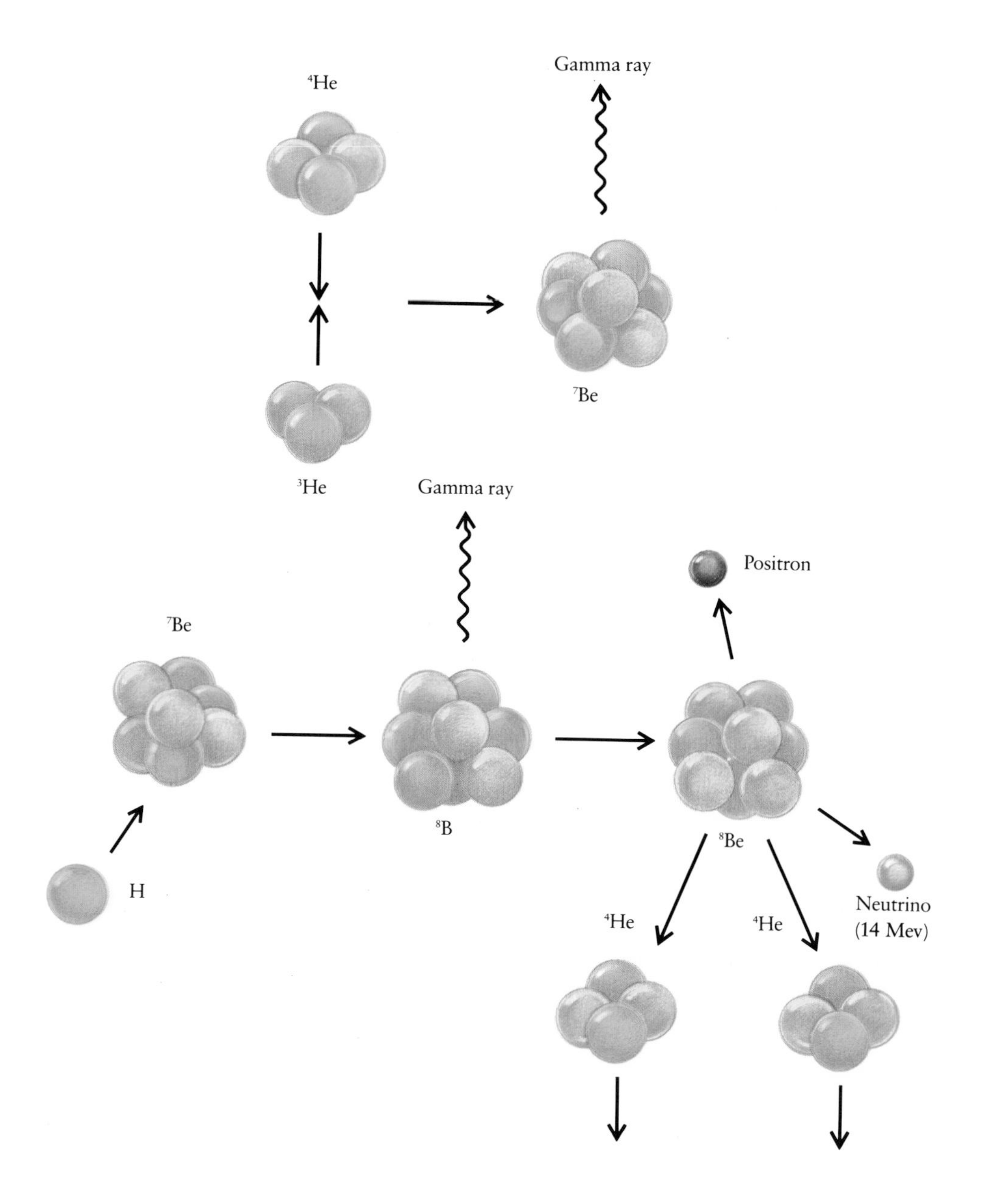

Gamma ray
⁴He
⁷Be
³He
Gamma ray
Positron
⁷Be
⁸B
⁸Be
H
⁴He
⁴He
Neutrino
(14 Mev)

A minor side reaction of the proton-proton cycle produces beryllium 7, which captures a proton to make boron 8, which undergoes radioactive decay to beryllium 8 with emission of a positron (e^+) and a high-energy (14-MeV) neutrino. This is the neutrino that Davis' experiment detects. ^{8}Be decays to two ^{4}He nuclei. This chain occurs only about 5 percent as often as the basic proton-proton reaction.

the deuteron collides with another proton to from a nucleus of "light helium" (helium 3, or ^{3}He) and releases a gamma ray. In the third step, which takes about a million years, two ^{3}He nuclei fuse to form the nucleus of normal helium, ^{4}He, also called an alpha particle. In this reaction, two protons are returned to the solar gas, and another gamma ray is emitted.

Additional steps involving the formation of boron and beryllium occur in the proton-proton cycle, which also produce helium. At each step in the chain, energy is derived from the conversion of mass, and the sun gets lighter and lighter. Altogether, 0.7 percent of the mass of four protons is lost and 4.2×10^{-5} ergs of energy is produced in making one alpha particle. (One erg is about as much energy as a fly delivers when it lands on your nose.) For the entire sun, 5 million tons of mass per second is converted into energy—around 600 times the weight of water flowing over Niagara Falls each second. Since the sun was born, about 5.5×10^{23} tons, or $1/4000$ of its mass, has disappeared in the form of radiation, and about 4 percent of its hydrogen has been processed into helium. If no other factors influenced the evolution of the sun, it could continue to fuse hydrogen into helium for more than a hundred billion years. (The sun's evolution will depart radically from its present track in another 4 or 5 billion years, however, and the sun will then balloon to giant proportions and rapidly burn itself out, as we shall see in Chapter 7.)

Hans Bethe proposed the carbon cycle of thermonuclear reactions to explain the energy of the sun.

A small percentage of solar energy may be produced by the carbon cycle, which requires a higher core temperature. Hans Bethe is said to have worked out the complicated details of the carbon cycle to relieve the boredom of the train ride to Ithaca, New York, following a conference in Washington in 1938. (That same year, he and his colleague Charles Critchfield discovered how helium could be formed from hydrogen in the sun at the comparatively low temperature of 15 million degrees.) Of those details, which Bethe published in the United States and Carl Friedrich von Weizsächer published in Germany in 1938, it is enough to say here that ordinary carbon (carbon 12) acts as a catalyst to bring about the buildup of nitrogen from nitrogen 13 to nitrogen 15 by successive fusions of protons. Finally, nitrogen 15 captures a proton and restores carbon 12 with the release of an alpha particle.

The proton-proton cycle and the carbon cycle are widely accepted by solar physicists, yet nobody knows with absolute certainty that they are correct because nobody, as yet, has been able to see into the core of the sun. Both cycles produce neutrinos, however, and therefore neutrino astronomy presents one possibility for observational confirmation. In fact, the findings of neutrino astronomy have raised serious questions about current models of thermonuclear fusion in the sun.

The nuclear transformations that convert hydrogen into helium via the carbon cycle. Carbon and nitrogen nuclei, created out of protons early in the history of the universe, act as catalysts. The reaction times at each step are shown, as are the energies of the neutrinos. Above 15 million kelvins, the carbon cycle begins to dominate over the proton-proton cycle.

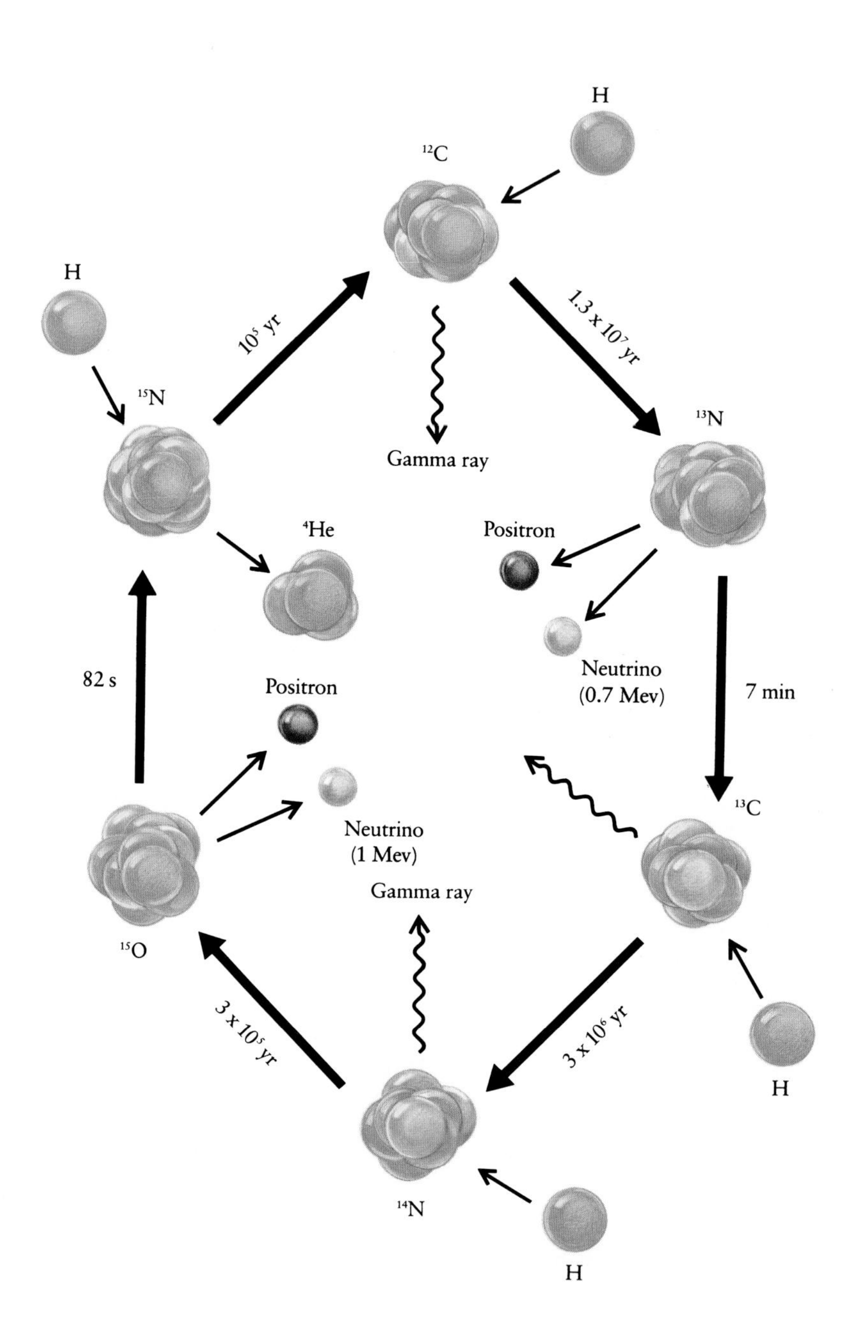

Solar Neutrinos

Neutrinos they are very small.
They have no charge and have no mass
And do not interact at all.
The earth is just a silly ball
To them, through which they simply pass
Like dustmaids down a drafty hall.

John Updike, from "Cosmic Gall"

Neutrinos released in thermonuclear reactions are uncharged, subatomic particles that have zero or very nearly zero mass. They move with the speed of light and interact almost negligibly with matter. Of all the particles in nature's subatomic zoo, neutrinos are by far the most elusive. A neutrino of average energy can penetrate 3500 light-years of lead with only a 50/50 chance of being stopped by a collision with a lead nucleus. Trillions of neutrinos shot through every person on earth in the time it took to read this sentence.

The sun is an enormous neutrino factory. Every fusion of a helium nucleus releases a neutrino, and the sun's nuclear furnace generates about 2×10^{38} neutrinos per second—about one-tenth as much neutrino power as visible-light power radiated from the solar surface. Neutrinos escape the sun with only one chance in a thousand billion of being absorbed: the earth itself is almost completely transparent to them. Thus, if a neutrino telescope could be constructed, we could look into the very heart of the sun, but the experimental problem of designing a telescope that could detect even a few neutrinos a day is mind-boggling.

Although the neutrino was first theoretically postulated in 1931 by the Austrian physicist Wolfgang Pauli, its actual existence was not established until 1956, when Frederick Reines and Clyde L. Cowan set up an enormous detection apparatus, filled with 1000 pounds of water as absorber, directly in the stream of neutrinos from the AEC's Savannah River Project nuclear reactor. Estimates put the neutrino flow from this source at about 30 times that expected from the sun. Reines and Cowan succeeded in detecting one or two collisions of neutrinos with hydrogen nuclei in the water tank each hour—a feeble response, but clearly positive.

Solar neutrino astronomy became a reality in 1964 when Raymond Davis, Jr., and his colleagues at the Brookhaven National Laboratory took up this extraordinary challenge. Carrying the experiment out 1500 meters underground to shield the apparatus from cosmic rays, Davis installed a detector tank containing 400,000 liters of C_2Cl_4 (perchlorethylene, a cleaning fluid) in the Homestake mine at Lead, South Dakota. Some 7000 tons of rock had to be removed, and the tank had to be constructed in sections in order to be handled on the main hoist. After the sections were welded together down in the chamber, it took 144 trips down the shaft to fill the great vessel.

This Olympic-size pool of cleaning fluid contains about 10^{30} atoms of the common chlorine-37 isotope (about 25 percent of natural chlorine). Chlorine 37 is made up of 17 protons and 20 neutrons. A very energetic solar neutrino has a certain chance of colliding with one of the neutrons in chlorine 37 and converting it into a proton. The resulting nucleus of argon 37 has 18 protons and 19 neutrons. Within 34 days on the average, one of the argon-37 protons combines with an electron orbiting the nucleus to form a neutron, with a release of energy. This radioactive argon must be reclaimed from the tank and detected before it decays. Davis' ingenious method of removing argon requires flushing helium through the liquid. The argon bubbles out with it, and the two are then relatively easily separated—test samples of argon injected into the tank are recovered with 90 percent efficiency. Finally, the amount of radioactive argon is measured with an electronic particle counter. Although the experimental community has scrutinized this work for years, no defect in the experiment has been uncovered.

For convenience in dealing with the small numbers involved, John Bahcall of the Institute for Advanced Research introduced the solar neutrino unit (SNU) to indicate 10^{-36} captures per second per target atom. In Davis' tank, a rate of one SNU corresponds to about one neutrino capture by chlorine 37 per week. In the best current theoretical model of the sun, the predicted rate is about 5.5 SNU, or almost one per day. But Davis observes less than two SNU, or about one every three days. Is something wrong with the sun or with the experiment? The case of the missing neutrinos has become something of a *cause célèbre* for astrophysicists.

The dilemma could be resolved if the sun's core temperature were a million degrees cooler than that predicted by the current astrophysical model. In effect, the neutrino telescope is an extremely sensitive thermometer for measuring the central temperature of the sun. Neutrinos of the energy detected in the chlorine reaction are generated at a rate proportional to temperature raised to the 13th power. With some

A mile underground in a South Dakota gold mine, Raymond Davis, Jr., constructed this neutrino detector. A tank as large as an Olympic swimming pool is filled with 100,000 gallons of cleaning fluid (perchlorethylene). The capture of a neutrino converts chlorine 37 into argon 37, which in turn is flushed from the tank and detected by its radioactivity.

small adjustments in theoretical reaction rates, in estimates of the efficiency of thermal convection in the sun's outer regions, and in other parameters of the solar model, can the calculated core temperature be made to fit the neutrino flux? Unfortunately, even small adjustments could upset all modern theories. At stake, therefore, is the entire astrophysical theory of the architecture of stellar interiors.

Dilhan Ezer and Alastair Cameron have speculated that somehow in the past something stirred up the sun's outer and inner regions so that more helium-3 fuel was brought to the core. As the reactor burned brighter, the core expanded and the temperature fell, with a dramatic decrease in neutrino flux. If such a stirring occurred every 100 million years or so, it would be followed each time by a drop in core temperature and a lowered neutrino rate for the ensuing 10 million years. The Ezer-Cameron model is only one of several hypotheses put forth to explain away the apparent discrepancy in the results of Davis' experiment. Perhaps solar chemical composition varies from the surface to the interior, for example. In its 4.5-billion-year life, the sun has traveled through many interstellar clouds, and it may have accreted substantial amounts of material that is richer in heavier elements than its interior material is. Fewer heavier elements in the center of the sun than we infer from its surface would imply a cooler interior because more radiation would escape faster.

The most advanced theories of elementary particle physics describe three types of electrons and their corresponding neutrinos. Pauli theorized about electron neutrinos, the kind produced when a neutron decays into a proton and an electron. But just as the electron has two heavy relatives, the muon particle and the tau particle, so there are two more kinds of neutrinos, the muon neutrino and the tau neutrino. Some physicists now suspect that the three types of neutrinos change back and forth between electron, muon, and tau forms millions of times per second as they travel. Fred Reines likens these transformations to a dog's molecules switching to form a cat as it trots along, "incredible but not inconceivable." Although the nuclear reactions in the sun should produce only electron neutrinos, neutrino oscillations might randomize the mixture even before they escape the sun. Davis' detector measures only electron neutrinos, and the existence of undetected muon and tau neutrinos may explain the factor-of-3 discrepancy between observation and theory. For these oscillations between types to occur, neutrinos must have a finite mass. Several experiments to measure their mass are in progress but the results so far have been controversial.

Solar physicists believe that the answer to the neutrino problem will require a broader-gauged experimental attack than Davis has mounted. The number of SNUs for the chlorine experiment is predicted on the basis of a side reaction of the proton-proton cycle that produces beryllium 8, a radioactive isotope that emits a positron and a high-energy neutrino. More sensitivity to lower-energy neutrinos in the proton-proton reaction can be obtained by using gallium. But a detector capable of sensing one event per day would require 50 tons of gallium, about five years' worth of the world's current supply, at a cost of $25 million. Perhaps physicists could negotiate an international trade agreement: the gallium could be used for several years and then returned to the world market. Lithium and indium would detect neutrinos of still other energies, and the combined results would almost amount to neutrino spectroscopy. The entire enterprise is so complex, however, that experts talk of a 20-year research program.

After this basic introduction to the nature of energy generation in the sun, we can now examine how this energy is produced in the solar interior.

The Architecture of the Sun

Solar physicists approach the architecture of the sun very much after the manner of a character described by Jonathan Swift: "There was a most ingenious architect who contrived a new method for building houses by beginning at the roof and working downwards to the foundation." Like Swift's architect, astrophysicists start with the roof of the sun, its photosphere. Without ever seeing deeper into the sun than a few hundred miles, they can construct a model of the sun's interior architecture—its pressure, temperature, and density—all the way down to its cellar, the thermonuclear core. Sir Arthur Eddington described this deductive process as "intellectual boring," and he took great pride in how far he could pursue the structure of a star from the comfort of his armchair.

The sun's interior is usually envisioned as a series of thin shells, each of which must satisfy certain conditions. First, each successive shell must support the overlying shells and must, in turn, be held up from below. In other words, at every point in the sun the gravitational attraction inward of the interior mass must be exactly counterbalanced by the pressure of hot gas and light photons that tend toward outward expansion. Second, temperature and density gradients must lead to the observed flow of sunlight. Once mass, radius, and energy outflow have been fixed, the solar model can have only one unique architecture.

Mass, Radius, and Luminosity

Early in the century, astronomers used various trigonometric parallax methods to measure and weigh the sun. One of the more precise methods required measuring the earth by well-known surveying methods and then, having determined the distance between two points on the earth, triangulating on celestial objects to find their distances.

The moon is close enough so that when observed simultaneously from opposite ends of the earth, the angular displacement is about two degrees. This "parallax" is easily measured, and even the ancient astronomers could calculate that the moon's distance is about a quarter of a million miles. But the sun is 400 times farther away, and the much smaller angular displacement is exceedingly difficult to measure. To complicate the problem, the bright sky near the sun hides the background stars that make useful reference points. These difficulties were circumvented by measuring the distance to Venus, or Mars, or the asteroid Eros when it is nearest the earth. Eros comes as close as 14 million miles, and, because it appears as a point in the sky, like a star, it can be positioned accurately against the starry background. The orbits of planets and asteroids are known, so that the establishment of the distance to any one of them fixes the scale of the solar system. (Today the distance to Venus is measured most accurately by radar. We now know that the earth-to-sun distance is 93 million miles, and the solar diameter is 865,000 miles.)

To weigh the sun, the astronomer first weighs the earth. Newton's law of gravitation states that two spherical bodies attract each other with a gravitational force proportional to the product of their masses and inversely as the square of the distance between their centers. When the pull of the earth on a small mass is measured using a sensitive balance, the mass of the earth is found to be 200 trillion trillion tons (2×10^{32} grams).

Next, the astronomer compares the earth's gravitational pull with that of the sun. Galileo found that when he dropped weights from the Tower of Pisa, they fell 16.1 feet in the first second. Under the sun's gravitational pull, the earth itself falls continuously around the sun, as though tied to it by an invisible string. If gravity were suddenly to disappear, the earth would fly away in straight line. In circling the sun once a year, its orbit deviates from that hypothetical straight line by ⅛ inch every second. The ratio of these two distances, 16.1 feet and ⅛ inch, is the ratio of the gravitational acceleration produced by the mass of the earth at its surface (4000 miles from its center) and the mass of the sun at a distance of 93 million miles. Applying Newton's law, we find that the sun is 329,390 times as massive as the earth.

The solar energy received at earth has been measured by instruments—thermopiles, bolometers, and radiometers of one sort or another—that detect the temperature rise produced by the warming effect of incoming radiation. As we saw, Sir William Herschel measured the sun's radiant energy by timing the melting of a layer of ice. Earlier devices were primitive but effective. In 1837, C. S. Poulet painted a copper pot black so that it would be highly absorbent and nonreflecting, filled it with water, and inserted a thermometer. From the rate at which the temperature rose, he computed the intensity of incident solar radiation to an accuracy of about 10 percent. A sophisticated cavity radiometer flown on a Nimbus satellite today can measure the solar energy flow to better than 0.01 percent.

Although evidence points to solar variability on the order of a fraction of a percent, the rate of energy input to the earth—1.36 kilowatts per square meter—is referred to as the solar constant. The total power delivered to earth is about 10^{14} kilowatts, tens of thousands of times greater than all the human-generated power consumed on earth.

The Shells of the Sun

The interior of the sun can be described as a set of nested shells. Reversing Swift's architect's approach, we shall start in the cellar, or deepest interior: the energy-generating core. A relatively quiescent radiation zone reaches from the core out to about 70 percent of the sun's radius. Relatively little gross mass movement of gas takes place in this zone. Core radiation slowly diffuses through this shell until it reaches the boundary of a convection zone, where the temperature has fallen sufficiently to make the gas highly opaque. Turbulent convection then takes over, carrying the flow of energy all the way to the sun's visible surface. The surface itself is a thin shell, the photosphere, only a few hundred miles thick. Although we have no unobstructed window to the interior, the newly developing science of helioseismology, the study of solar vibrations, promises effective new means for probing the interior shells.

A semblance of this shell structure extends beyond the photosphere in the outer solar atmosphere: a thin, spiky chromosphere is embraced by a corona that reaches tens of solar radii into interplanetary space. These regions are highly irregular, marked by dynamic weather patterns, violent streaming of solar gas, and erupting bursts of radiation.

The solar model requires a central pressure of about 300 billion atmospheres (one atmosphere equals 14.7 pounds per square inch, the weight of the entire column of air above one square inch of the earth's surface). To produce this fantastic pressure, the gaseous core must be heated to a temperature of about 15 million degrees Celsius. If the sun were as cool throughout its body as it is at its surface, the outer parts would crush the interior regions, leaving a dwarf star smaller than the earth. The concept of such a high central temperature is almost impossible to convey. In *The Universe Around Us*, Sir James Jeans explained that a pinhead of matter at the temperature of the solar core "would emit enough heat to kill anyone who ventured within a thousand miles of it." (Jeans based his calculation on a theoretical temperature of 55 million degrees, but the image still holds true.) At the sun's central temperature and density, conditions are right for the generation of nuclear fusion by the processing of helium nuclei out of hydrogen nuclei.

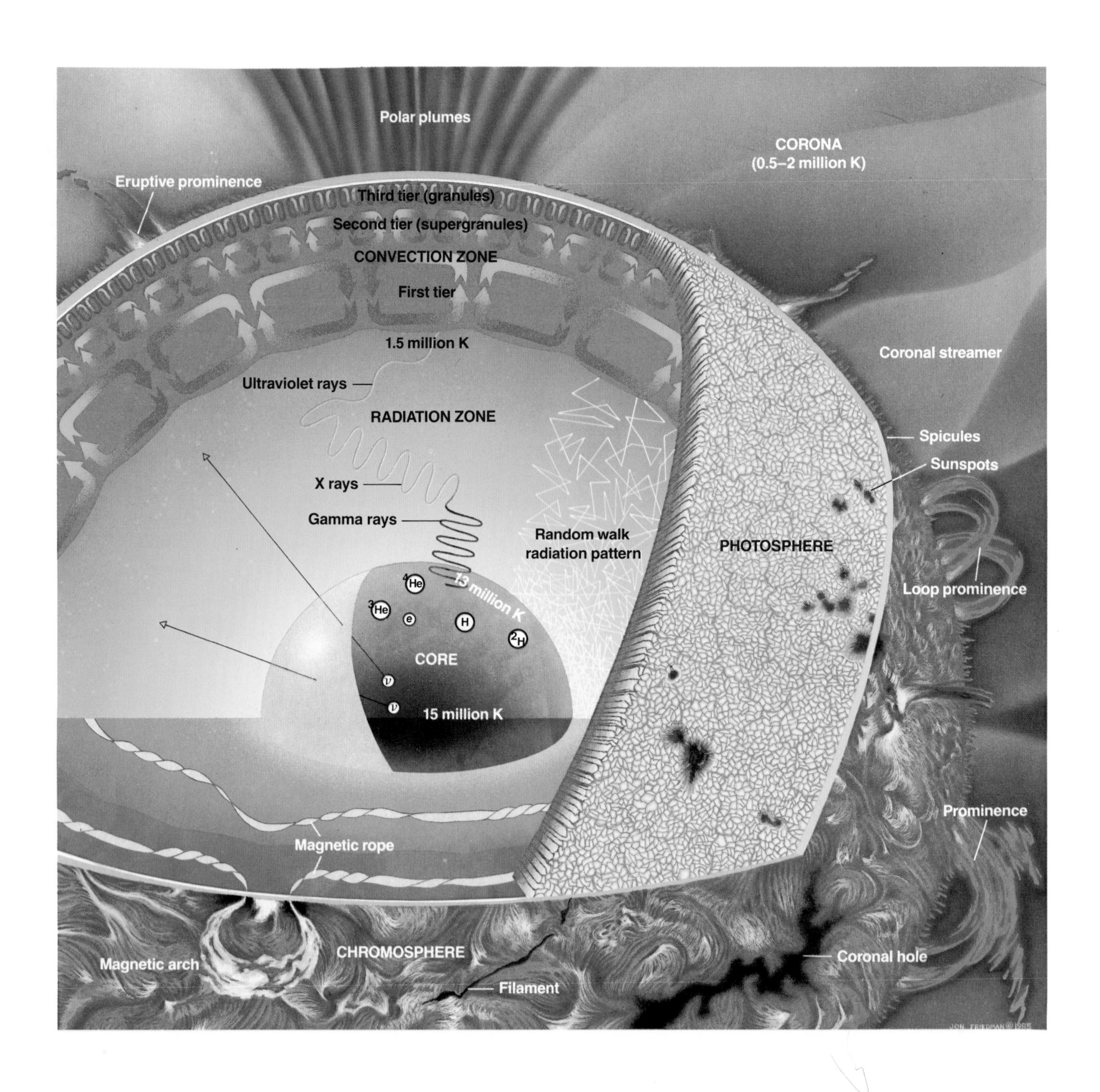
Polar plumes
CORONA
(0.5–2 million K)
Eruptive prominence
Third tier (granules)
Second tier (supergranules)
CONVECTION ZONE
First tier
1.5 million K
Coronal streamer
Ultraviolet rays
RADIATION ZONE
Spicules
Sunspots
X rays
Gamma rays
Random walk
radiation pattern
PHOTOSPHERE
4He
13 million K
Loop prominence
3He
e
H
2H
CORE
ν
ν
15 million K
Prominence
Magnetic rope
CHROMOSPHERE
Magnetic arch
Coronal hole
Filament

The architecture of the sun. Thermonuclear power is generated in the inner core, shown dark because its radiation is in the form of invisible X rays and gamma rays. Nucleons involved in the core reactions are neutrons (N), protons (^{1}H), deuterons (^{2}H), helium 3 (^{3}He), and alpha particles (^{4}He). Electrons, positrons, and neutrinos also play a role. Outside the core energy is carried by radiation that begins as gamma and X rays and by a series of collisions is steadily redshifted to ultraviolet light. In the convection zone energy is carried by turbulent motion in a tiered pattern of large-to-small convection cells. The visible surface has a bubbly appearance of small cells randomly organized in a chromospheric network. Magnetic arches, jetting spicules, looping prominences, flares, and streamers project above the surface. A dark coronal hole offers an escape route for solar wind.

Although the density of the sun's nuclear-power zone (158 grams per cubic centimeter) is about 12 times that of solid lead, the sun remains gaseous throughout its interior. In fact, the region of the core behaves like a perfect gas, but with abnormal atoms. Most of the atoms' outer electrons have been buffeted off by constant collisions and the bullet-like hits of X rays. All the hydrogen and helium nuclei are stripped bare, and a heavy atom such as iron may be left with only a few of its normal 26 electrons. Normal atoms cannot get closer than one-hundred-millionth of an inch before their outer electrons touch. But with their electrons stripped, nuclei can approach 100,000 times closer.

As Eddington vividly describes the scene:

> Crowded together within a cubic centimeter there are more than a trillion trillion atoms, about twice as many free electrons and 20 billion trillion X rays. The X rays are traveling with the speed of light and the electrons at 10,000 miles per second. Most of the atoms are . . . simply protons travelling at 300 miles a second. Here and there will be heavier atoms, such as iron, lumbering along at 40 miles a second. I have told you the speeds and the state of congestion of the road; and I will leave you to imagine the collisions.

Roughly 90 percent of the energy that eventually floods through the solar surface and into space is produced within a central core that reaches only one-quarter of the distance to the surface but contains about 40 percent of the total mass of the sun, at an average temperature of about 14 million degrees Celsius. How does this energy get from the core to the surface?

The three familiar modes of energy transport are radiation, conduction, and convection. Radiant energy is carried directly by waves moving with the speed of light across empty space, as one feels the heat of a fire at a distance. Conduction occurs when, for example, a metal rod is heated at one end and the heat travels rapidly toward the cooler end. An organized metallic lattice is required to transfer energy from atom to atom down the length of the rod. Convection takes place when chaotic masses of gas flow in turbulent fashion, each atom carrying its own parcel of energy all the way.

Obviously, conduction plays little part in the transfer of energy within the body of the sun. Inside the core, nuclear-fusion reactions create energetic particles and radiation. Nuclear fires burn fiercely from the center of the sun, at about 15 million kelvins (K), to the outer fringe of the core, at about 13 million kelvins, where nuclear fusion

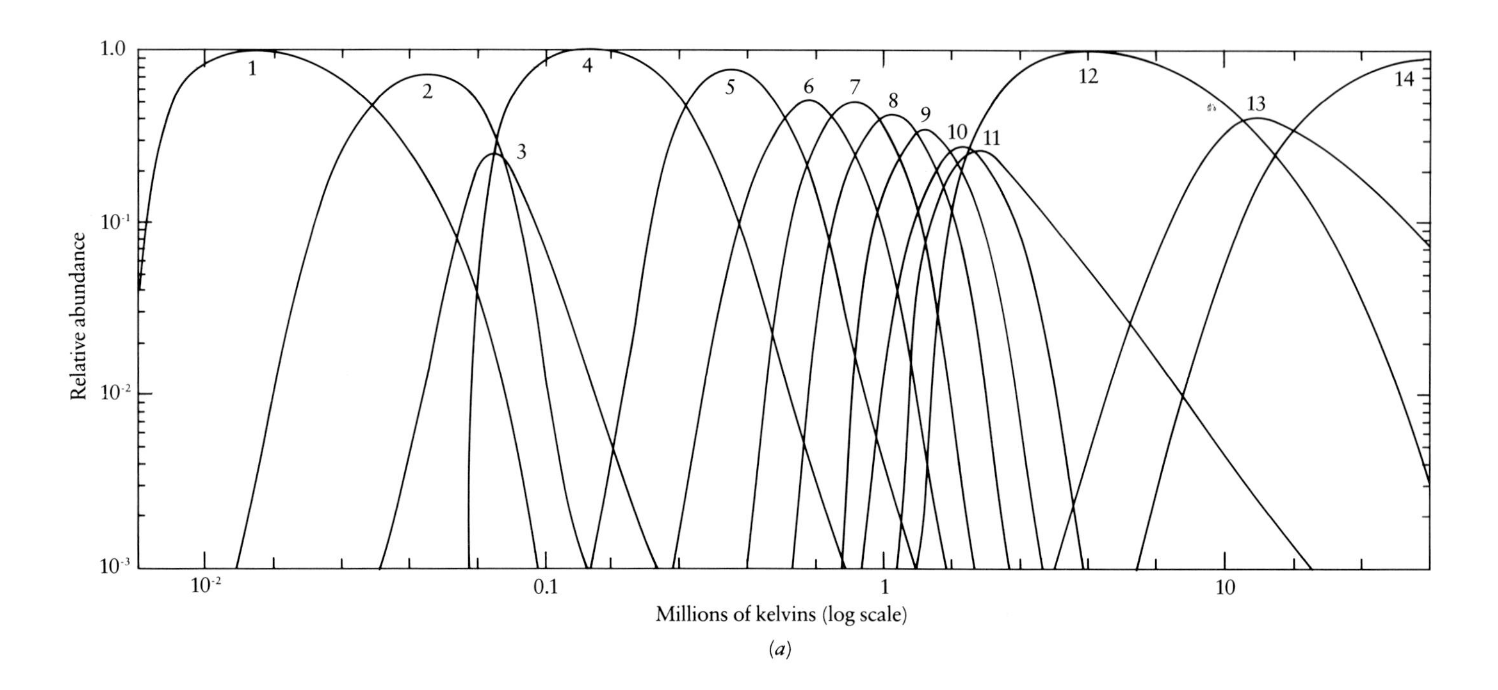
1.0
10⁻¹
10⁻²
10⁻³
Relative abundance
1
2
3
4
5
6
7
8
9
10
11
12
13
14
10⁻²
0.1
1
10
Millions of kelvins (log scale)

(*a*)

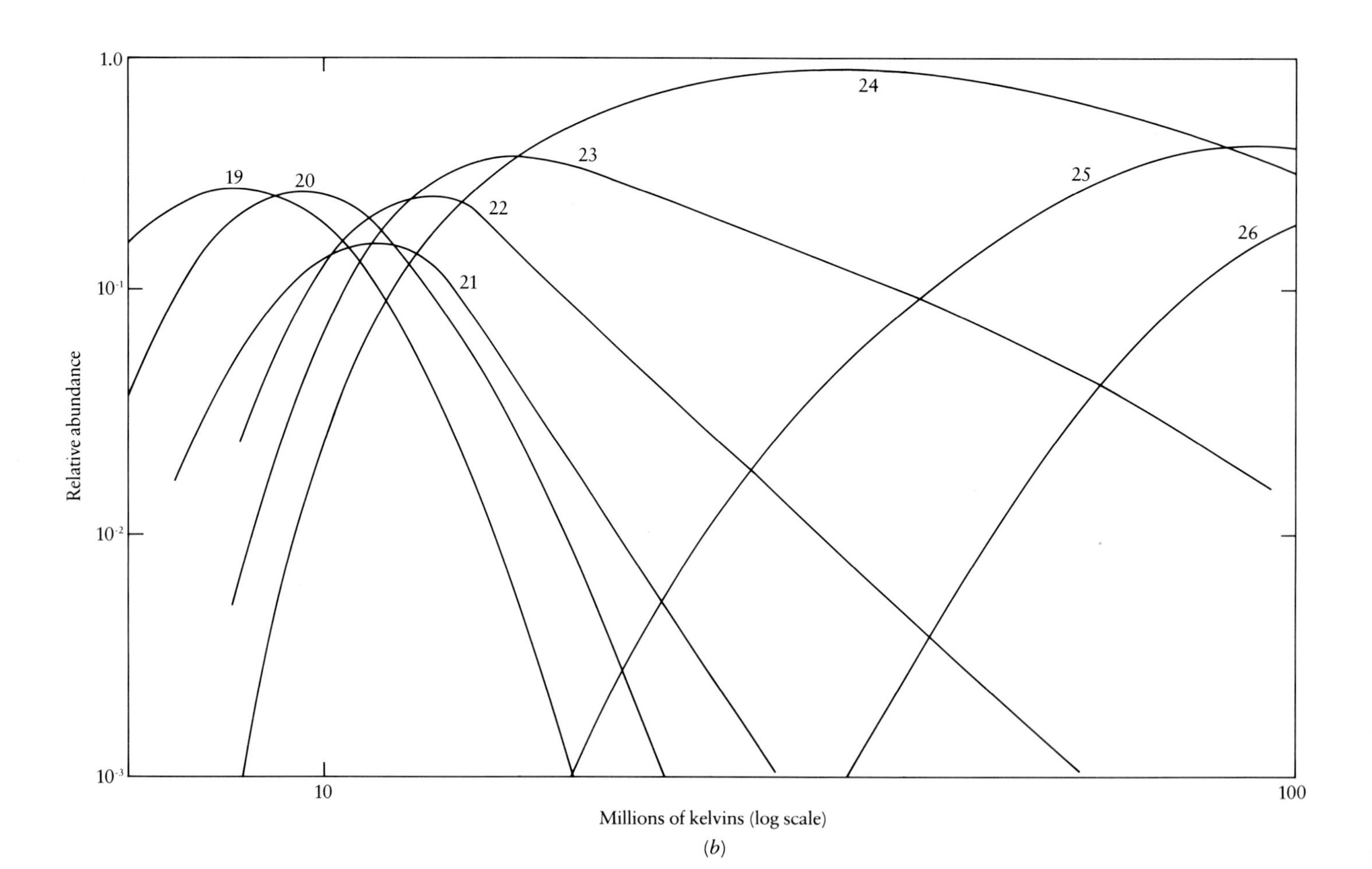
1.0
10⁻¹
10⁻²
10⁻³
Relative abundance
19
20
21
22
23
24
25
26
10
100
Millions of kelvins (log scale)

(*b*)

In the maelstrom of hot plasma in both the solar interior and the corona, the buffeting of silicon atoms strips them of many of their normal 14 electrons (*a*). The various curves are labeled with the number of electrons lost by the nucleus at increasing temperatures. At the 2 million kelvins typical in the corona, most of the iron ions are 12 to 16 times ionized (*b*). In the plasma of a great flare at 20 million kelvins, most of the iron atoms have been stripped of 24 electrons.

ceases. (Absolute zero on the Kelvin scale equals −273°C.) About one-quarter of the way out from the center to the surface, radiant energy transfer becomes dominant, primarily in the form of X rays. If, in fantasy, we could travel to this interior region, it would take X-ray vision to observe the optically pitch-black, superheated gas. Even though the X-rays travel with the speed of light, their outward diffusion is very slow because the average distance traveled between collisions is only a matter of centimeters. Their zigzag progress is like that of a man weaving his way through a dense crowd of people, bumping and being bumped from side to side while edging forward. At each passing brush with an atom, the X-ray wavelength lengthens, or redshifts, toward the visible.

At about nine-tenths of a solar radius toward the surface, the gas has cooled to about half a million kelvins, and its density has fallen to about 1 percent that of water. Here the gas begins to convect, like boiling water in a kettle or the hot air rising from a radiator, and energy seethes to the surface in turbulent flow. At the edge of the sun, heat energy is transferred into space as visible and infrared radiation. From the moment of generation in the core until its escape into space, the energy-transfer process may take millions of years.

The luminous shell that forms the sharp, visible edge of the sun is called the photosphere. Its thickness is only about 0.1 percent of the solar radius, and its 10^{17} tons of matter is only a 20-billionth part of the sun's total mass. Stretched over the enormous surface of the sun, this amounts to only a few grams of mass above each square centimeter, a small fraction of the density of a similar column of air above the earth's surface. The density of the sun's atmosphere falls off rapidly with height; for approximately each 55 miles of additional altitude, the total amount of overlying gas decreases by 50 percent. While the opacity of the photosphere hides the sun's interior, its blinding brilliance separates the body of the sun from its tenuous, far-reaching, outer fringes, which are almost invisible against the yellow disk.

Having come to the roof of the sun, its radiant surface, we may wonder why, if the sun is gaseous throughout, the edge of its disk is so sharp. Ordinary unpolluted air is highly transparent to visible light. If the solar gas were like air, we should be able to see deep into the interior. But when gas is heated to thousands of degrees Celsius, hotter than the flame of an acetylene torch, it becomes very opaque. Because of its high temperature, even the thin solar atmosphere becomes optically thick when its density begins to exceed a millionth that of ordinary water. An important physical mechanism that contributes to this opacity involves negative hydrogen ions. Ordinary hydrogen consists of a proton and an electron, but the neutral hydrogen atom in a nar-

The variation of energy generation, temperature, and pressure with fractional radial distance from the center of the sun to its surface is charted. Nearly all thermonuclear energy production takes place in a core region extending to about one-quarter of the solar radius. The base of the convection zone is reached at 0.7 solar radius, where the temperature is about 1 million kelvins. In the photosphere, pressure drops to about 0.1 atmosphere, off the scale of the graph.

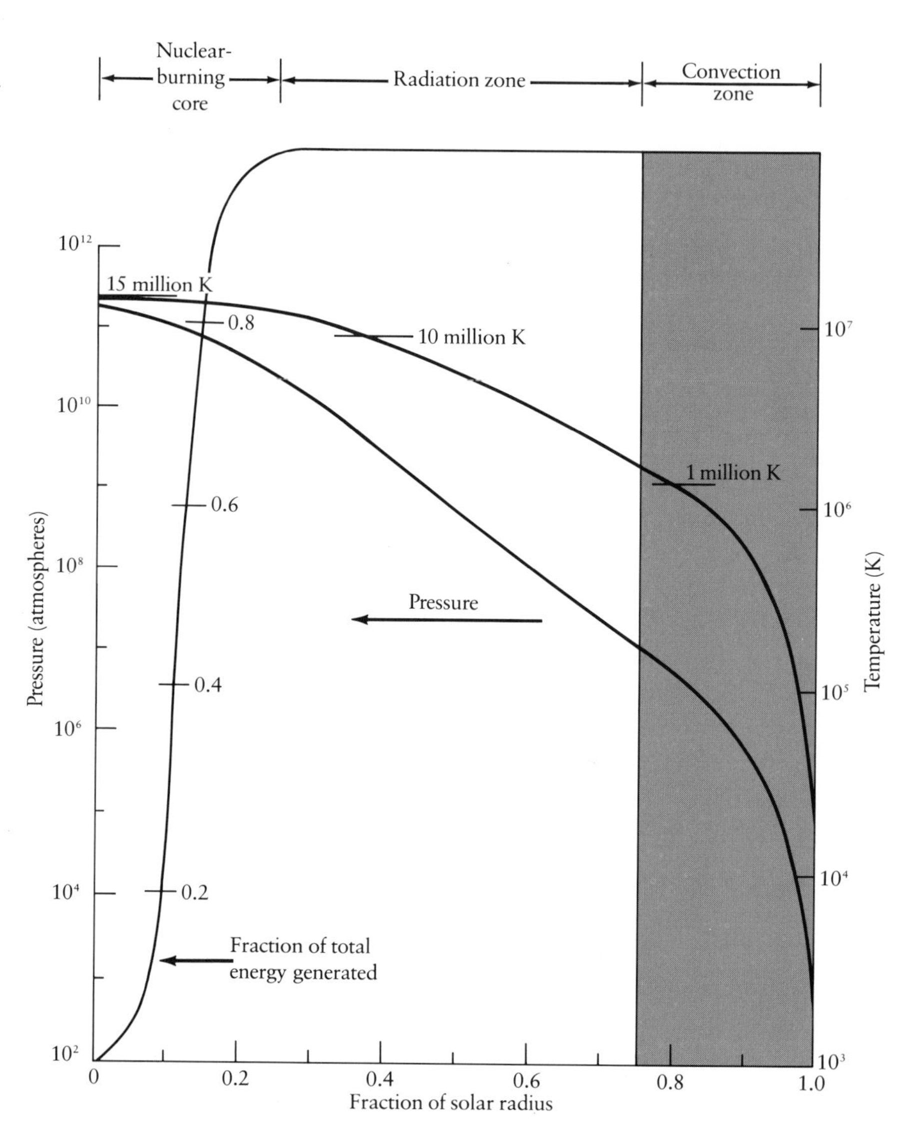

row temperature range can attach a free electron to itself and become a negative ion. As radiation filters through the sun's outer layer, most of the opacity is provided by this negative hydrogen ion, even though its concentration is only a millionth that of the normal hydrogen atom. Then, rather abruptly, the negative-ion region is passed, and radiation escapes. The suddenness of this change in the concentration of negative ions gives us a sharp-edged sun.

The shell structure of the sun beneath the photosphere is known only in theory. At the surface and above, in the chromosphere and corona, the sun becomes less inscrutable. Marked by constant, dynamic activity, its face is mottled by granules, blotched by inky sunspots, streaked by frothy, twisting, fiery prominences, and, frequently, burst by brilliant flares.

The Constantly Changing Face of the Sun

Since the time of Galileo, the characteristic markings on the bright disk of the sun have been the subject of study. Every wavelength reveals a different face and provides specific clues to the sun's interior metabolism.

In white light, we see mainly the lower levels of the photosphere,

Photograph of the sun in the hydrogen-alpha red line, taken on September 19, 1966, at the Sacramento Peak Observatory, Sunspot, New Mexico. Light of this wavelength originates high in the chromosphere. The large sunspot near the center of the disk is about the size of the earth. The long, dark filaments, also called dark flocculi, are clouds of hydrogen gas held in place by magnetic fields arching above the surface. Brighter areas, called plages, are regions of strong magnetic field in which transient flares often erupt.

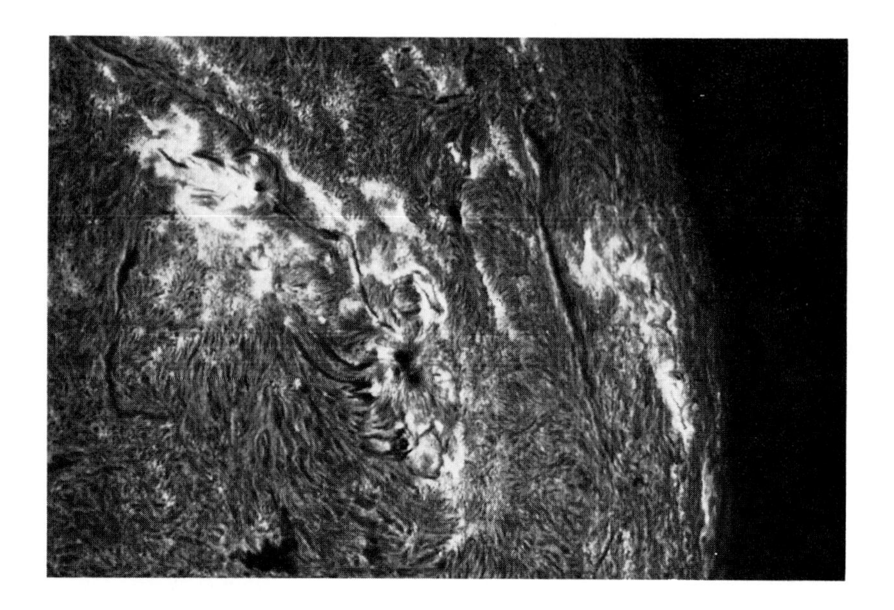

The chromosphere in an active region. Bright plages mark the active sites. Prominences are seen as snaking dark filaments against the disk. Long fibrils spread out from sunspots.

with its characteristic granulation and sunspots. Fraunhofer lines originate in the higher portions of the photosphere, where the darkness of a line depends on the abundance of a particular element and the height of its region of origin. In 1889, George Ellery Hale, then only 21 years old, invented the spectrohelioscope, in essence a sharp filter that gave astronomers monochromatic vision, the ability to narrow down the range of color until only a single Fraunhofer line appears. By choosing a particular line, such as the red line of hydrogen or the violet line of ionized calcium, the viewer can—now more effectively with birefringent filters—view a thin layer of the solar atmosphere in which that line is absorbed. In each line, the face of the sun takes on a remarkably different complexion, and its expression is constantly changing. By analogy with meteorological changes in the earth's atmosphere, changes in the face of the sun reflect patterns of solar weather. Small shifts in the wavelength of a line* reveal strong up-and-down motions,

Motion of a radiating source causes its emitted wavelengths to become shorter if the source is approaching and to become longer if it is receding. In the case of sound waves, this relationship was first noted by the Austrian physicist Christian Doppler in the early nineteenth century, and it is referred to as the Doppler effect. The whistle of a train provides a familiar example: as the train approaches at high speed, the sound has a higher pitch than it does after the train passes and is racing away. By measuring shifts in the wavelengths of visible light, astronomers can determine the speed of a celestial source toward them or away from them.

and the degree to which a line splits indicates the intensity of the magnetic field.

Where the convection zone begins, about nine-tenths of the way to the surface, radiation, which may have been battered about in the interior for millions of years, is quickly absorbed by atoms that have cooled enough to recapture electrons. At this point, radiant energy is transformed into turbulent convection, and large bundles of gas move rapidly toward the surface. The phenomenon is similar to the bubbling of water boiling in a pot: the rising bubbles merge as they zigzag upward and finally burst through the surface. When photographed with a specially designed spectroheliograph that registers dark and light images according to motion toward or away from the observer, bright bubbles represent ascending gas and polygonal, dark boundaries mark the channels of descending gas.

As these bundles, or cells, of gas ascend at a rate of two or three kilometers per second, the surface of the sun takes on a frothy, bubbling appearance called granulation. Each granule has an average lifetime of about eight minutes. Because of atmospheric blurring, these small cells are near the limit of resolution that can be achieved by ground-based telescopes, and some of the best photographs have been obtained from 80,000 feet using a balloon-borne telescope. Granules average about 1000 kilometers in diameter, although some are as small as 300 kilometers. They are separated from each other by about 1500 kilometers on average, and their total number on the sun at any time is about 4 million.

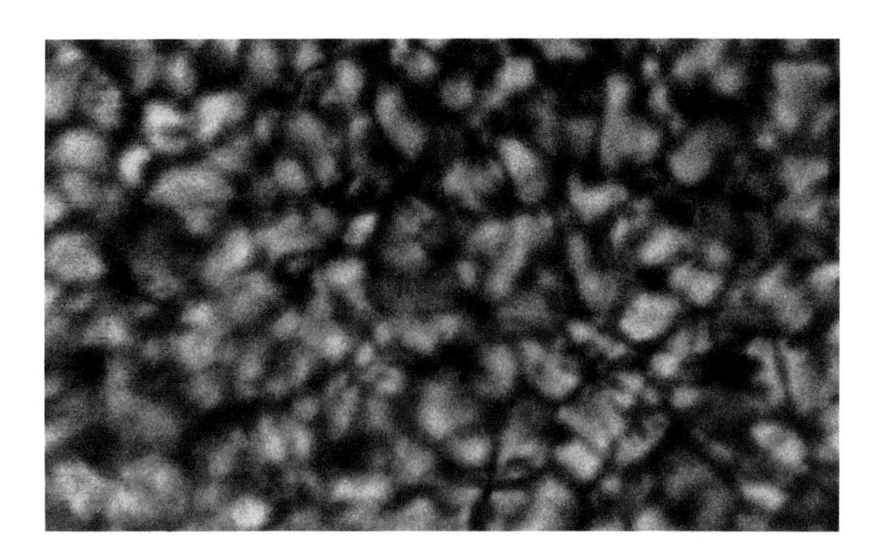

Solar granulation pattern in the photosphere. The granules are typically 2000 kilometers in diameter.

Stratoscope I. At the bottom of the telescope tube is a 12-inch parabolic primary mirror. An enlarging lens produced an image with a scale equivalent to that of a telescope with a 200-foot focal length. Electronic stabilization controlled the steering motors so that the telescope turned less than a fifth of a second of arc during each two-thousandths of a second exposure time. Two flights brought back 16,000 photographs.

Perhaps 2 percent of the granules seem to explode horizontally. Starting out brighter than average, they enlarge around a developing dark center and continue to grow in a ring that finally breaks up into a number of granular fragments. Radial expansion proceeds to about 4000 kilometers at a rate of about 1.5 kilometers per second, and these exploding shapes live about 30 minutes on average.

In theory, the convection zone may have a many-tiered structure, the bottom of the hierarchy being massive convection cells, or supergranules, a few hundred thousand kilometers in diameter, with velocities of about 40 meters per second. As these gigantic cells ascend, a pattern of intermediate cells develops, with each cell about 15,000 kilometers deep and 30,000 kilometers across. Overlying this zone is the pattern of fine cells, or granules, that bubble to the surface. Movement in the lower tiers is much slower than it is near the surface, but the pattern of underlying activity shows up as a widespread heaving of the entire solar surface. Supergranules exhibit a small upflow at their centers and a downflow at their edges, with a lifetime of a day or somewhat longer. Each supergranule is the size of about 300 granules.

The outlines of supergranules form a network, best seen in the violet spectral line of singly ionized calcium, Ca K (the K is the Fraunhofer designation). Hovering near sunspots, self-luminous clouds of calcium are called "flocculi" because in white light they resemble fluffs of sheep's wool. The entire solar surface, seen in the violet of Ca K, takes on a mottled appearance like the skin of an orange. These features mark the separation of the sun's atmosphere from its body.

The Shivering Sun

This model of the sun as shells within shells has implied a rigidity of structure that must now be modified to a certain extent. The sun is a ball of spinning gas, dragging complicated strings and ropes of magnetized plasma in its interior. It is not surprising, therefore, that its surface shows strong differential rotation. Furthermore, the shells inside the convection zone rotate faster than the outer regions. On close examination, the surface exhibits complex oscillations that reflect this interior turmoil.

In recent years, astronomers in the United States, the Soviet Union, and Great Britain have discovered that the sun shivers, quivers, and vibrates in a variety of modes. A remarkable series of observations began with the work of Robert Leighton and his students at the California Institutue of Technology in 1962. The surface of the sun, they

The supergranulation revealed by Doppler shifts of the wavelength of a spectral line. The image is a superposition of two spectroheliograms, one redshifted and the other blueshifted. Light regions are material moving toward the observer and dark regions are material moving away. The pockmarking reveals the horizontal motion of cells about 30,000 kilometers in diameter. From these observations the five-minute solar oscillation was discovered.

found, heaves up and down irregularly at a speed of about a thousand miles an hour. Compared to pulsating variable stars, which swell and contract at speeds of up to 50,000 miles an hour, these solar surges are quite small, and the swelling does not occur over the entire surface simultaneously. Instead, localized portions of the surface, like a choppy sea, seem to pulse independently of each other. Each local rise and fall has a characteristic oscillatory period of about five minutes, although rhythms vary over particular areas from three minutes for a time to six minutes a little later, and a total vertical motion of 700 to 1400 miles.

In theory, the sun can pulsate simultaneously in millions of ways. A mathematical model once made by Robert K. Ulrich implied that oscillations would group into intrinsically distinguishable, observable patterns. In addition to the complex five-minute surges, simpler, longer-period oscillations of smaller amplitudes should be observable on a global scale. Ulrich calculated that such oscillations should characteristically last from 15 minutes to over an hour. And, indeed, oscillations of periods from about 10 to 160 minutes have since been reported by various observers. Much less certain are hints of oscillations lasting up to about a dozen days. Some theorists even propose that periods as long as 10 years should be present.

The phenomenon of solar oscillations calls to mind the seismic vibrations that are excited when the earth is shaken by a powerful earthquake. The entire globe rings like a huge bell, although the sound is below the audible frequencies. Recording the travel time of seismic waves from an earthquake to a network of detector sites reveals certain details of the inner structure of the earth—its chemical composition and physical character, for example. In a similar fashion, it should be possible to probe the inner shells of the sun to construct a kind of "helioseismology." No matter that the sun is a gaseous ball throughout, that its material can slosh back and forth like tides in a bay, and that the entire globe can breathe in and out. Because of the low average atomic weight and the high temperature of the solar interior, the speed of sound is some 200 kilometers per second, far faster than it is in the earth. The velocities of these waves carry information about temperature and pressure at levels from just beneath the surface to deep in the interior.

There are two broad classes of pulsation. One is the familiar sound wave, called the *p* mode for pressure wave, which propagates by alternate compression and rarefaction at the speed of sound. If the sun contracts, increased pressure pushes it back out again past the point of equilibrium where pressure is exactly balanced by gravity.

The sun then falls back inward to repeat the vibration cycle. This fundamental radial mode of oscillation is spherically symmetrical. Such acoustic waves resemble the tones of a musical instrument, and, by analogy, we should expect a wide spectrum of overtones.

The second form of pulsation is the gravity wave, or *g* mode. When a mass of solar matter is displaced, gravity provides a restoring force. Ocean waves offer a familiar terrestrial example. No sharp boundary between the masses is necessary; a variation of density with depth is sufficient to propagate the *g* mode. Unlike sound waves, which are driven by pressure changes, gravity waves are driven by buoyancy. A mass of gas displaced downward finds itself in a denser medium and is buoyed upward. When it passes its point of equilibrium, it becomes heavier than the surrounding medium and falls down again. Gravity waves are not spherically symmetrical, and they have their own set of overtones.

Theoretically, acoustic oscillations of the sun can have periods extending up to about an hour; gravity-wave periodicities may last as long as three hours. Because the internal temperature of the sun ranges from 15 million kelvins in the core to 6000 kelvins at the surface, and because core density is a billion times as great as photospheric density, oscillations develop very complex patterns. Many modes can exist both radially and circumferentially, with vibrations propagating around as well as through the body of the sun.

Other nonradial, asymmetric modes of oscillation distort the shape of the sun. For example, in the quadrupole mode, the poles of the sun may flatten and its spheroid shape become oblate, to be followed by a squeezing of the equator that produces a prolate figure. This push-pull oscillation, which also has overtones, causes the sun to vibrate in somewhat shorter periods.

Henry Hill, who began a study of the shape of the sun at Princeton, transferred his observations to a more advantageous site 8500 feet up in the Santa Catalina Mountains near Tucson in 1973. With a specially designed telescope and better viewing conditions, he and Robin Stebbin failed to detect any deviation from perfect circular symmetry. They did, however, incidentally discover that the edge of the disk vibrated, as though the sun were ringing with a deep bass tone, for a fundamental period of 52 minutes. This observation of a shivering sun opened up a new and fascinating subject for examination.

Teams of Soviet and British scientist soon found evidence of even longer pulsations. Both groups made their discoveries almost simultaneously, and their reports came out in the same issue of *Nature*. The Soviet group, led by G. B. Severny, used a solar magnetograph, a

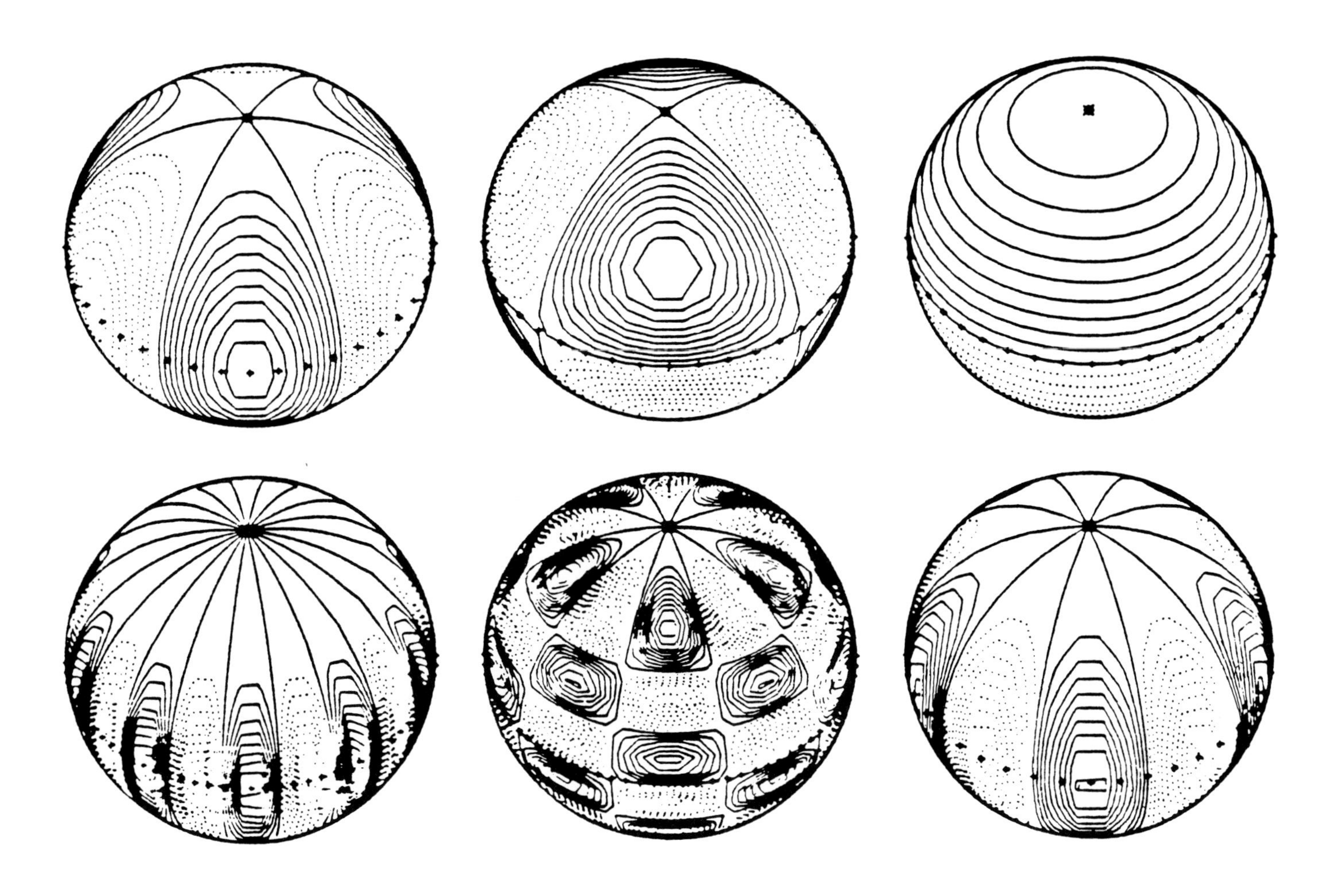

Contour plots of selected modes of oscillation of the sun. The solid lines represent zones of expansion; the dotted lines, contraction. Only six of thousands of possible modes are shown. This selection of modes illustrates progressively complex oscillatory motions from left to right and top to bottom. Oscillations occur with periods ranging from minutes to at least hours.

device normally used to map magnetic fields by the splitting of spectral lines, to measure the velocity of pulsation by observing the Doppler shifts along their sightline. From 122 hours of data, they extracted a period of two hours and 40 minutes, with a precision of one-half minute. The amplitude of oscillation was, they said, about 10 kilometers. The Birmingham University group, led by G. R. Isaak, observed Doppler shifts in the Fraunhofer lines of sodium and potassium from the Pic du Midi observatory in the Pyrenees. They found that pulsations lasted for two hours and 39 minutes, plus or minus two minutes. Great interest attaches to these observations, because the longer the period, the deeper the origin of the vibration in the sun.

Some doubts still remained after several years of attempts to confirm these results, but all uncertainties were finally swept away when a collaboration of French, Soviet, and American scientists set up an observatory at the South Pole. During the Antarctic summer, the sun shines overhead 24 hours a day, and generally good weather made possible a continuous, five-day run of observations. The scientists found clear evidence of a 160-minute oscillation period with an amplitude of about three kilometers.

Solar oscillations will undoubtedly prove to be a rich source of information about the solar interior. Because oscillation amplitudes are small, and because most of the data runs have been short, the extraction of oscillations from background noise has been difficult. As the precision of observations improves, it should become possible not only to probe the depth of the convection zone but also to search for a shear boundary between a rapidly rotating magnetic core and the convection zone.

Solar seismologists have already measured about 1000 frequencies with an accuracy of better than 0.1 percent. This data base compares favorably with the library that earth scientists have to work with. Any present interpretation of solar vibrations is necessarily very limited, but, in general, they seem to match a theoretical spectrum based on the standard solar model. The contemplated transfer of observations to an orbiting spacecraft should greatly enhance our powers of detection. A French attempt to record oscillations from a tethered balloon high above the south pole is also planned, as well as a coordinated effort of ground-based telescopes distributed in longitude so that observations are not interrupted by night.

A Shrinking Sun?

If the sun oscillates globally on a time scale of hours, can it possibly shrink or expand perceptibly over hundreds of years? Lord Kelvin and Hermann von Helmholtz had hypothesized that the sun could generate the energy it radiated by consuming itself, but no observational evidence was available. In 1979, John A. Eddy set off a new controversy by offering archival evidence that the solar diameter of 31 arc minutes 59 seconds, or 1919 arc seconds, appeared to shrink by a meter and a half per hour, or about 0.1 percent per year. (An arc minute is 60,000 kilometers.) So rapid shrinkage would cause the sun to disappear in a hundred thousand years. This, of course, was preposterous, but it was not beyond possibility that the solar diameter does

Solar telescope at the south pole. The inclined platform guides the telescope to follow the sun. In the 1984–1985 austral summer, 450 hours of observations were carried out to determine frequencies of solar oscillations. More than 80 harmonics were measured to an accuracy of a few microhertz with amplitudes above a few centimeters per second. Martin Pomerantz, leader of the program, is shown in the photograph.

vary over epochs, increasing or decreasing after the pattern of known solar activity cycles.

Eddy had stumbled across his evidence while searching through historical data from the Greenwich Observatory, where the local times of centuries of solar meridian crossings are kept on file. He was particularly intrigued by one long run of transit telescope data from 1850 to 1953 that gave solar horizontal diameters.

In analyzing his data, however, Eddy may have ignored two important factors: atmospheric influences and observer idiosyncrasies. Over time, the combined effects of changes in atmospheric pollution and in turbulence have an appreciable effect on the estimation of solar diameter. Pollutants make the sky brighter. Against a clear, dark blue sky, an optical illusion makes the bright solar disk appear larger than it seems when the surrounding sky is whitened by haze. Personal idiosyncrasies on the part of two Greenwich observers were also quickly identified. According to John H. Parkinson, an astronomer at the University College of London:

> The horizontal measurements were made by noting the length of time taken by the sun to drift across a wire in the eyepiece of the telescope. Before 1854, the observers watched the image of the sun while listening to the ticking of a pendulum clock. The precise time at which the sun crossed the meridian was then interpolated mentally between two successive ticks.

Even a casual perusal of the Greenwich records, Parkinson notes, reveals that the two observers "simply couldn't get it right at all." They consistently erred by as much as 2.2 seconds. A difference of even one second in the recording of transit time could lead to an error of 15 arc seconds in the measurement of solar diameter.

In spite of the criticism that quickly developed, the question still intrigued astronomers. Using the Greenwich vertical solar diameter measurements, which were made with a micrometer rather than a clock, a group led by Sabatino Sofia at NASA's Goddard Space Flight Center found a systematic decrease in radius, but only by about 0.2 arc second, roughly 10 to 20 percent of that suggested by the horizontal-transit evidence. If solar-transit observations were the only means of measuring the sun's diameter, the search for historical changes of only a few tenths of an arc second would be hopeless. Fortunately, records of eclipse shadows and of Mercury's transits across the face of the sun make higher accuracy possible.

On the occasion of the total eclipse of 1715, Edmund Halley organized hundreds of observers throughout England to record the exact time of totality at their locations. Halley had no thought of measuring solar diameter, but in 1979, the American astronomer David Dunham undertook just such a project. Using Halley's data from 1715, he compared it with the results of similar observations that he organized for a total solar eclipse over the Pacific Northwest. Dunham found the sun to be 0.34 arc second smaller in 1979 than it was in 1715.

Sabatino Sofia has reservations about this result. He suspects that records of the exact station of each of Halley's observers may not have been precise to better than one or two hundred meters. A more accurate record, he thought, might have been kept of the eclipse of 1925, the path of which was followed as the shadow swept across Manhattan. For that eclipse, an employee of the Consolidated Gas Company of New York organized a network of observers at rooftop stations on every city block along the length of Riverside Drive. Perhaps the shadow could be timed block by block. Sofia made a diligent effort to resurrect the records, but all his searches failed. After comparing shadow data from a 1976 eclipse over Australia and the 1979 eclipse with the 1715 data, Sophia and his colleagues did conclude that the solar radius had decreased by about 0.34 arc second.

About 13 times a century, Mercury transits the face of the sun. Records of these events go back to the seventeenth century. Some 50 observations have been made since then, and possibly two dozen offer sufficiently precise timing to gauge solar diameter in a useful fashion. It takes more than five hours for the tiny planet to drift across the solar disk—a far more leisurely pace than the quick, high-noon transits of the earth across the sun. Still, the measurement of this passage has its own intrinsic difficulties. A phenomenon known as "black drop" impairs observations. When the transit begins, Mercury is a dark object against a dark sky. As the planet impinges on the edge of the disk, it appears to be accompanied by an afterimage—the "black drop"—that trails along for tens of seconds or even minutes and obscures the instant of contact. Even forewarned of this effect, an inexperienced observer has difficulty in accommodating for it. Irwin Shapiro of the Smithsonian Astrophysical Observatory has been collecting historical records of Mercury's transits to determine a correction for the earth's motion in connection with his radio studies of quasars. When the controversy arose over Eddy's claims, Shapiro analyzed his Mercury transit records for an independent measure of the solar diameter. He concluded that the sun may be shrinking, but by no more than 0.15 arc second per century, if at all.

Clearly any claims about a shrinking sun must be regarded skeptically, and more precise observational techniques must be devised. At the High Altitude Observatory in Colorado, a new Meridan transit telescope has been installed, which times the noon transits electronically. Several banks of over 500 photosensitive diodes each measure both vertical and horizontal diameters 30 times a second. Each noontime crossing provides thousands of bits of data. But it will still take perhaps five or 10 years to provide the observations necessary for a credible decision as to whether or not the sun is shrinking.

3

The Solar Atmosphere

The Sun, whose rays
Are all ablaze
With ever-living glory
Does not deny
His majesty—
He scorns to tell a story!

Gilbert and Sullivan, *The Mikado*

Solar eclipse photograph obtained on February 16, 1980. A neutral density filter with a strong radial gradient suppressed the extreme brightness differential between the inner corona and its outer regions. Wispy streamers of luminous gas reach out hundreds of thousands of kilometers. This corona is typical of solar maximum, with its streamers symmetrically surrounding the sun like the petals of a dahlia. At solar minimum the streamers concentrate toward the equatorial plane.

The spotted, wrinkled face of the sun does not scorn to tell a story. Careful scrutiny of its features reveals its innermost secrets. As an eclipse masks the blaze of light, the chromosphere and corona shine forth, a crowning glory to behold.

Sunspots: Magnetic Footprints on the Sun

Aristotle taught that the sun was a globe of pure fire without blemish, and this belief held sway for many centuries. Even in Aristotle's time, however, his pupil Theophrastus mentioned spots seen on the sun with the naked eye. The first written record of a sunspot appears in the Chinese *Book of Changes* before 800 B.C., and Chinese annals from A.D. 188 make reference to "flying birds" on the face of the sun. Subsequent records from the Orient, however, mention sunspots fewer than a hundred times in the next 18 centuries.

Under normal conditions, the bright sun blinds the eye, but when fog and haze reduce its glare, especially near sunrise or sunset, sunspots three times larger than the earth can be discerned. Soot from the great forest fires in Russia in 1871 created ideal conditions for detecting sunspots, and the Russian chronicles of the fourteenth century described "dark spots on the sun as if nails were driven into the body of the disk."

With the invention of the telescope in 1609, four men began seeing a spotted sun. Johannes Fabricious in Holland, Christopher Scheiner, S.J., in Germany, Galileo Galilei in Italy, and Thomas Harriot in England all claimed the discovery of sunspots between 1610 and 1612. The first publication of the news in 1611 is attributed to Fabricious, who may have made his discovery as early as December, 1610. Father Scheiner first saw sunspots in March, 1611. Rebuffed for contradicting Aristotle—as his ecclesiastical superior informed him, "I have read Aristotle's writings from beginning to end, and I can assure you that I have nowhere found in them anything similar to what you mention; . . . be assured that what you take for spots in the Sun are the faults of your glasses or your eyes"—he published his discovery under a pseudonym. When Galileo demonstrated a 9-power spyglass to the Venetian Senate in 1609, he pointedly emphasized its wartime uses, but soon he constructed a 20-power instrument and aimed it at the moon, the stars, and the planets. Immediately, he saw four moons of Jupiter and the mountains and craters of the moon and the Milky Way resolve into swarms of countless stars. It does appear that Galileo may have been the last of the four to point his telescope at

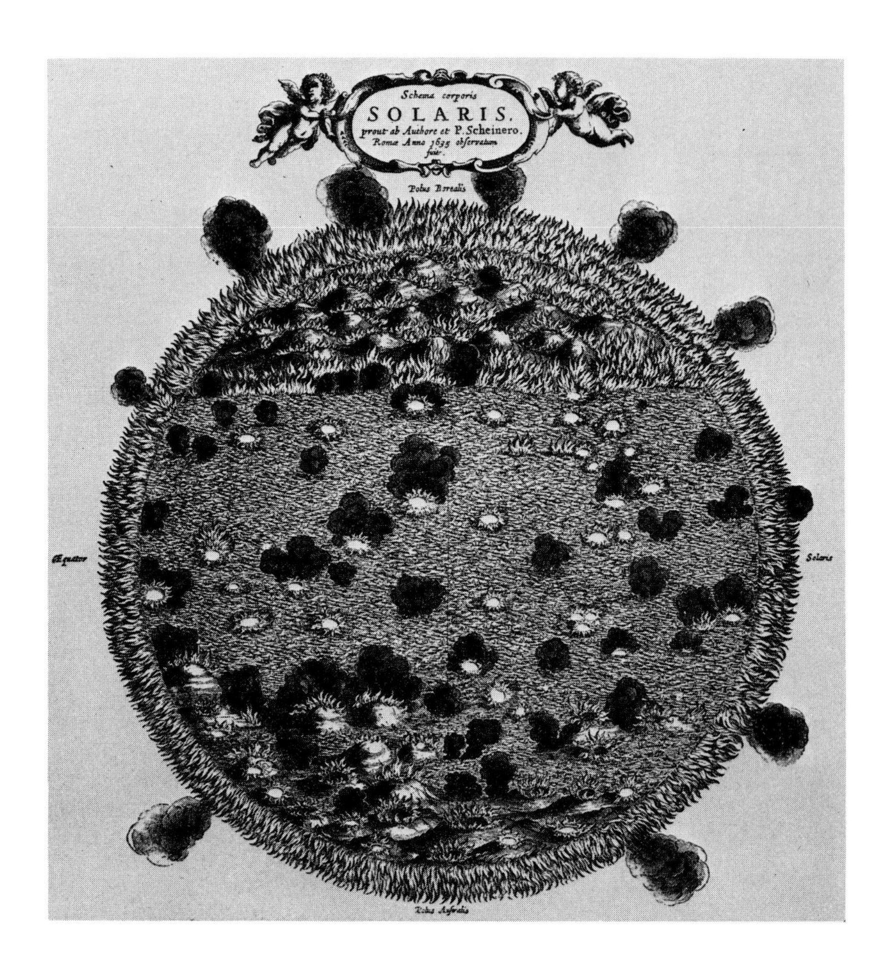

Drawing of sunspots by A. Kircher in *Physica Sacra,* 1665. Kircher drawings can be used to determine the differential rotation of the sun 375 years ago. No difference is discernible from the present rotation.

sunspots; his first known mention of them appears in a letter dated October 1, 1611.

Many philosophers and churchmen of the time refused to look through Galileo's telescope, protesting that it gave them headaches. Rather than contradict Aristotle, they preferred to consider any spots to be bodies outside the sun. To assuage their theological concerns about blemishes on the handiwork of God, they argued that such spots must be clouds floating over a perfect sun. In response, we have Galileo's own words:

> I therefore repeat and more positively confirm . . . that the dark spots seen in the solar disk by means of the telescope are not at all distant from its surface, but are either contiguous to it or separated by an interval so

> small as to be quite imperceptible. Nor are they stars or rather permanent bodies, but some are always produced and others dissolved. They vary in duration from one or two days to thirty or forty. For the most part they are of irregular shape, and their shapes continually change, some quickly and violently, others more slowly and moderately. They also vary in darkness, appearing sometimes to condense and sometimes to spread out and rarefy. In addition to changing shape some of them divide into three or four, and often several unite into one.

The description can hardly be improved upon today.

Two hundred years ago, some astronomers thought the spots must be solid mountaintops protruding above an ocean of flowing lava that, they reasoned, had high and low tides. As the tide ebbed, the higher mountaintops emerged as dark bodies. In 1774, however, Alexander Wilson, a Scottish astronomer, observed that the spots appeared saucer-shaped, with inclined edges, like the slopes of a crater leading to a dark interior inside a brilliant shell. Sir William Herschel proposed that they were portions of the surface of a cold, solid crust shielded by two cloud layers, the outer layer being the brilliant, incandescent, hot photosphere and the inner being a cool, protective shield that shaded the crust. When the clouds parted to reveal the underlying cool crust, a spot would appear. Herschel went so far as to suggest that the dark surface beneath the clouds might support life.

Incredibly enough, the idea of a dark solar interior persisted through most of the first half of the nineteenth century. Even in the 1850s, David Brewster, in his popular book *More Worlds than One,* insisted that "we approach the question of the habitability of the Sun with the certain knowledge that the sun is not a red-hot globe, but that its nucleus is a solid opaque mass receiving very little light and heat from its luminous atmosphere." Such a sun could not continue to shine for more than a day or two.

Modern Observations

The entire surface of the sun appears granular, as if it were paved with cobblestones. Typically, a sunspot begins as a dark pore in the midst of this grainy pattern, and soon several pores coalesce with one another to form a spot. Sometimes the spot lasts only a few hours, but very large spots occasionally grow and last for weeks or months. Every spot has a dark inner umbra, surrounded by a half-shaded penumbra that appears to be embroidered with a filigree of threads combed outward from the center of the spot. These threads may be fine convection

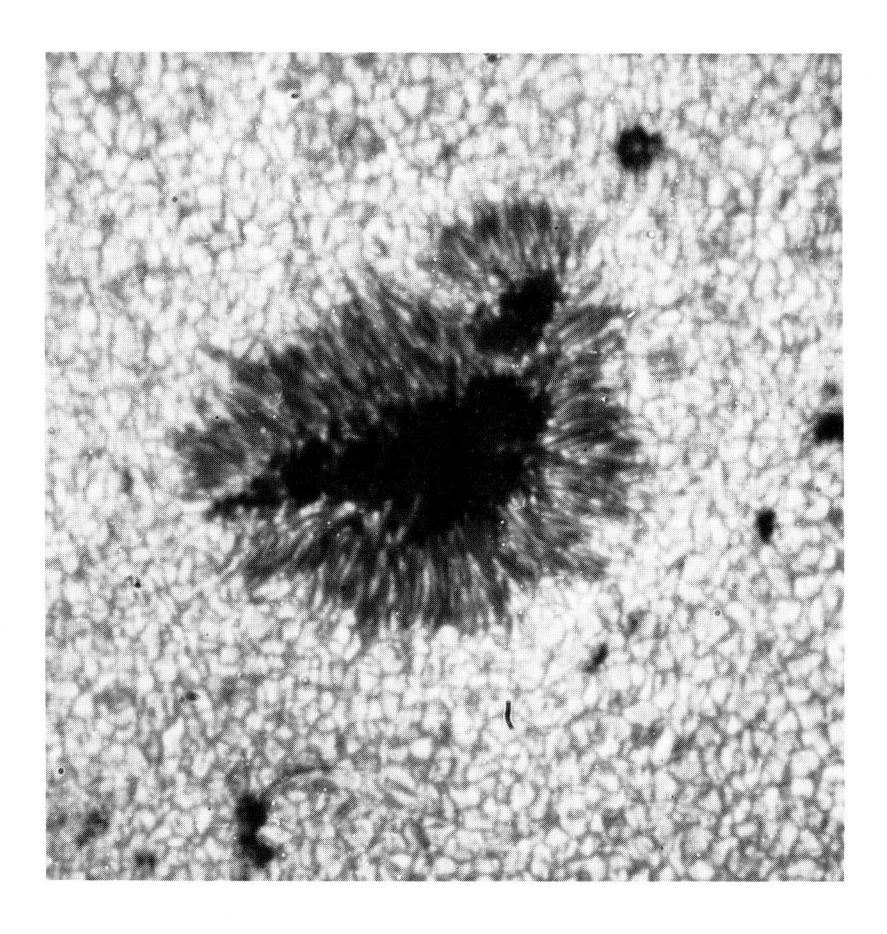

A portion of the solar surface photographed with Stratoscope I. The dark sunspot umbra is surrounded by a penumbra of narrow, long filaments. The entire sun, outside sunspots, is covered by a granulation pattern of hot convective gases rising from the interior.

channels carrying hot gas to the periphery. In shape, the spot most often resembles a funneled cavity 400 or 500 miles deep. In the dark umbra, the temperature is about 3800 K, hotter than the hottest blast furnace on earth. An isolated spot of medium size is brighter than the moon or a powerful arc lamp, yet in contrast to the 5875-K temperature of the surrounding photosphere, the spot appears dark.

A relatively small spot has a diameter of a few thousand miles, roughly the size of the earth. One of the largest spot groups on record started to form on February 2, 1947. It grew to a size greater than 7 billion square miles in April, and after five solar rotations, its last vestiges were still visible in September.

As markers on the clear disk, sunspots reveal that the sun rotates from east to west, but in a very peculiar way. Unlike the solid earth, the gaseous sun does not rotate with the same angular speed at all

Convection cells in altocumulus clouds, seen from above, closely resemble solar granulation patterns.

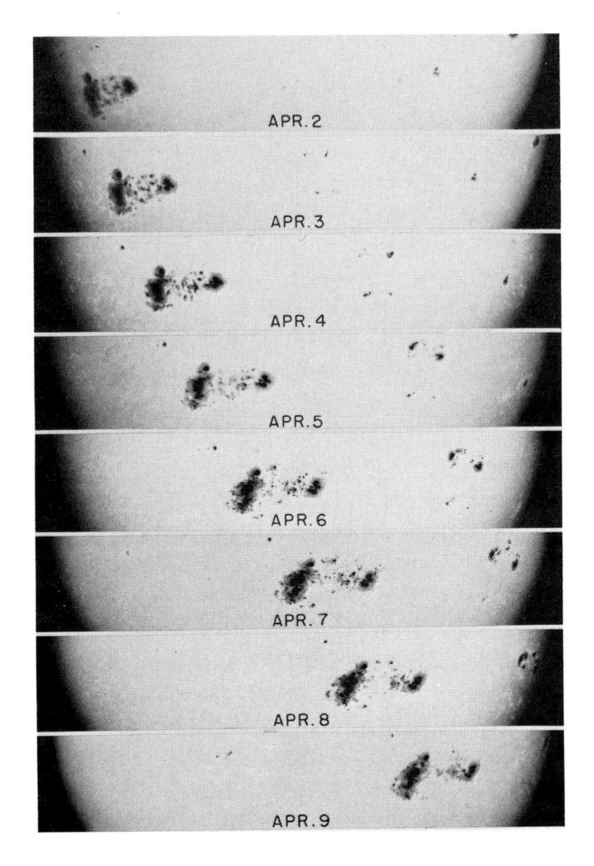

Sunspots are markers on the surface of the sun that reveal its rotation. The period varies from about 27 days near the equator to as long as 34 days near the poles.

latitudes. It twists on its axis, so that, contrary to expectation, the equatorial regions rotate faster than the polar caps. A spot close to the equator completes a rotation in 26.9 days, a spot at 30° latitude takes 28.3 days, and the polar zone may take as long as 34 days. Most of the changing features that we observe on the surface of the sun must relate in some way to the contortions it undergoes.

For reasons hard to understand, astronomers neglected the study of the sun for nearly two centuries after the discovery of sunspots between 1610 and 1612. For 25 years, beginning in 1826, Heinrich Schwabe, an apothecary in Dessau, Germany, and an amateur astronomer, kept careful records of the sunspots he observed through a two-inch telescope mounted on the roof of his house. Actually, he was hoping to see the hypothetical planet Vulcan pass across the face of the sun. When the significance of his discovery that sunspots came and

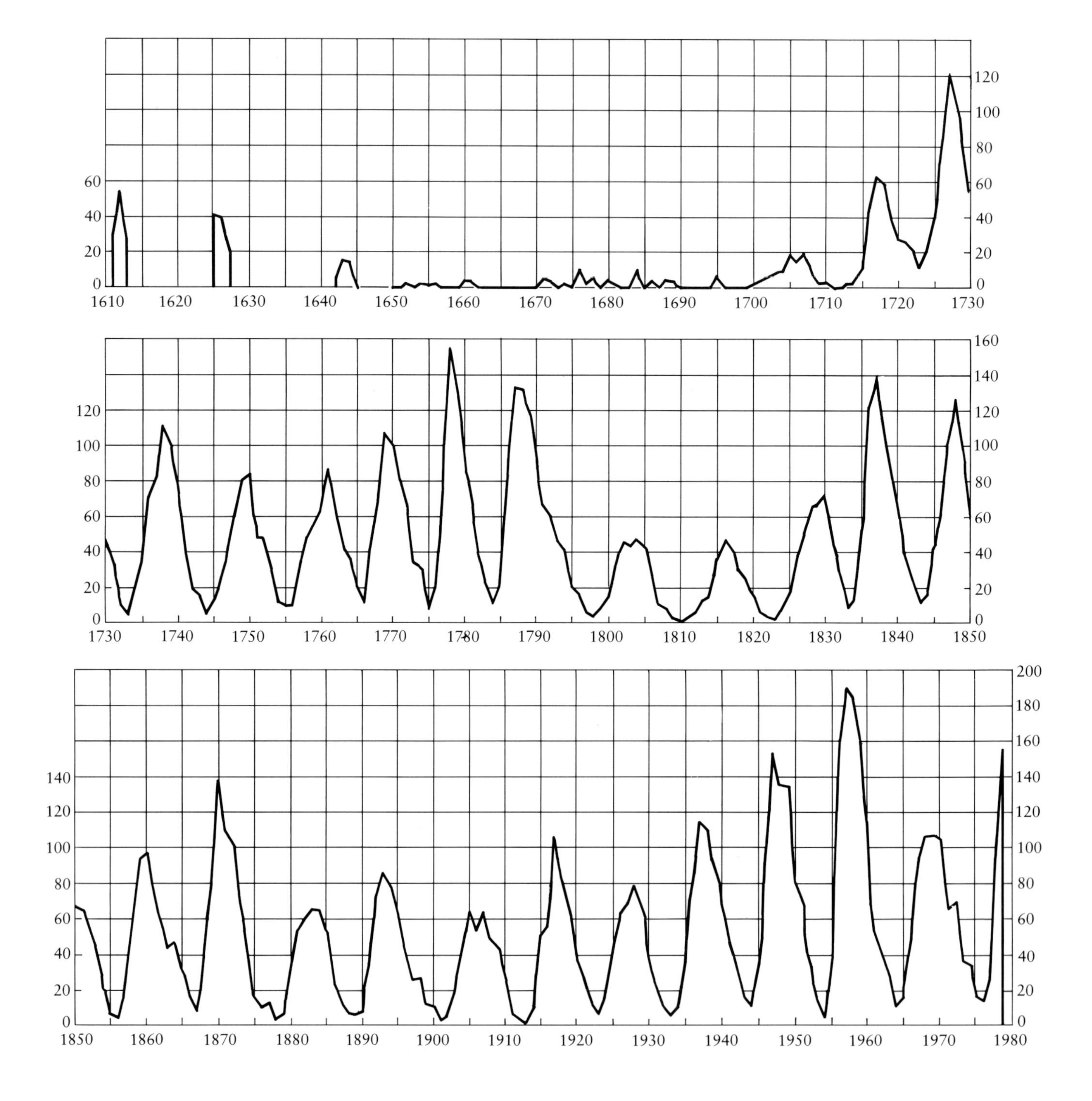

0
20
40
60
80
100
120
1610
1620
1630
1640
1650
1660
1670
1680
1690
1700
1710
1720
1730
140
160
1740
1750
1760
1770
1780
1790
1800
1810
1820
1830
1840
1850
180
200
1860
1870
1880
1890
1900
1910
1920
1930
1940
1950
1960
1970
1980

The solar sunspot cycle from the time of Galileo to the present. Coverage length of a cycle is 11 years. The shortest cycle was seven years, the longest 17 years.

went with a period of about 11 years dawned on him, he wrote to the British astronomer Carrington, "I may compare myself to Saul who went to seek his father's ass and found a Kingdom." In 1857, Schwabe received the Gold Medal of the Royal Astronomical Society for his discovery of the periodicity of sunspots. The president's address summed up Schwabe's accomplishment: "Twelve years he spent to satisfy himself, six more years to satisfy, and still thirteen more to convince mankind of his discovery; . . . the energy of one man has revealed a phenomenon that had eluded the suspicion of astronomers for 200 years."

Using records going back as far as 1729, we can now trace 23 sunspot cycles, of an average length of 11.1 years, in which the number of spots varies from minimum to maximum and back to minimum again. One cycle was as short as seven years and one as long as 17. In the most intense cycles, those that maximized in 1870 and 1957, the sunspot numbers were three times as high as in the weakest maxima, and at times as many as 300 spots could be observed. During the periods of sunspot minima, months may pass without any visible spots.

At one time it was thought that the sunspot cycle was a truly periodic phenomenon associated with the motions of planets, especially Jupiter, which has an 11-year orbital period. The evidence today weighs more heavily in favor of some form of repeated eruption. Each new cycle begins with the appearance of spots in a belt about 40° north or south solar latitude. As the cycle progresses, new spots appear at lower and lower latitudes down to about 5° from the equator at the end of the cycle. High-latitude spots of the new cycle often begin to appear before the spots of the old cycle have vanished. Sometimes this overlap may last as long as four years. Such erratic behavior, coupled with departures from the mean value of 11 years, favors an independent eruption as the source of each outbreak of new spots. The mechanism of this eruption lies hidden deep within the sun, and theories about its source are still highly speculative.

In successive 11-year cycles of growth and decay, the polarity of the solar magnetic field reverses. A full cycle of variation from maximum to maximum with a return to the same polarity is, therefore, more fundamentally 22 years, rather than 11 years. To complicate matters, the process of field reversal every 11 years is not perfectly symmetrical in both hemispheres. At the minimum of the cycle in progress during 1953, the polarity was positive in the north and negative in the south. Between March and July, 1957, the south polar field reversed while the positive polarity in the north persisted until late 1958, when it finally reversed.

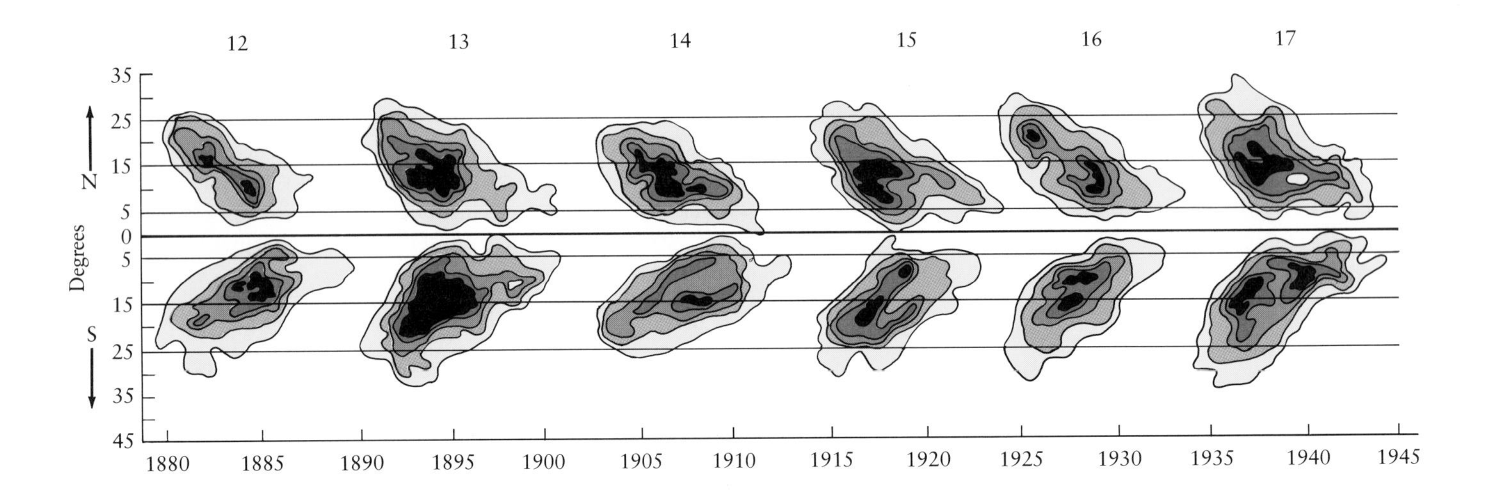

The Maunder butterfly pattern shows the early appearance of sunspots near latitudes of 30° north or 30° south at the beginning of a sunspot cycle and their subsequent drift toward the equator. Darker colors correspond to higher concentration of sunspots.

Sunspots and Magnetic Fields

Early in the nineteenth century, the British scientist Michael Faraday laid a sheet of paper over a magnet and covered it with a scattering of iron filings. When he tapped the paper, the iron filings arranged themselves in a sweeping pattern of graceful curves emanating from the poles of the magnet. Generations of schoolchildren have repeated this simple, beautiful experiment. Faraday theorized that the arrangement of filings revealed lines of magnetic force in the space around the magnet. Along each line of force, the strength of the magnetic field is constant. Closer to the poles of the magnet, the magnetic force becomes greater, as evidenced by the clustering of the filings in denser lines. With increasing distance and weakening field strength, the lines thin out. If a compass is brought near the magnet, its needle aligns with the lines of force.

Faraday's concept of a field of force defined by magnetic lines of force was later most elegantly expressed by James Clerk Maxwell in a mathematical theory that unified all the phenomena of electricity and magnetism in just four simple equations and became a landmark of modern physics. According to Maxwell, any motion of electric charge generates a magnetic field in its surrounding medium and, conversely, any motion of a magnetic field induces the flow of an electric current.

The strength of any magnetic field is described in units of gauss; a small, steel bar magnet may have a strength of some tens of gauss close to each pole. The lines of force emerge from the north pole and loop back to reenter the magnet at the south pole. The earth's magnetic

Faraday map of magnetic field strength made by sprinkling iron filings around a bar magnet. The steel key, placed in the magnetic field, becomes magnetized.

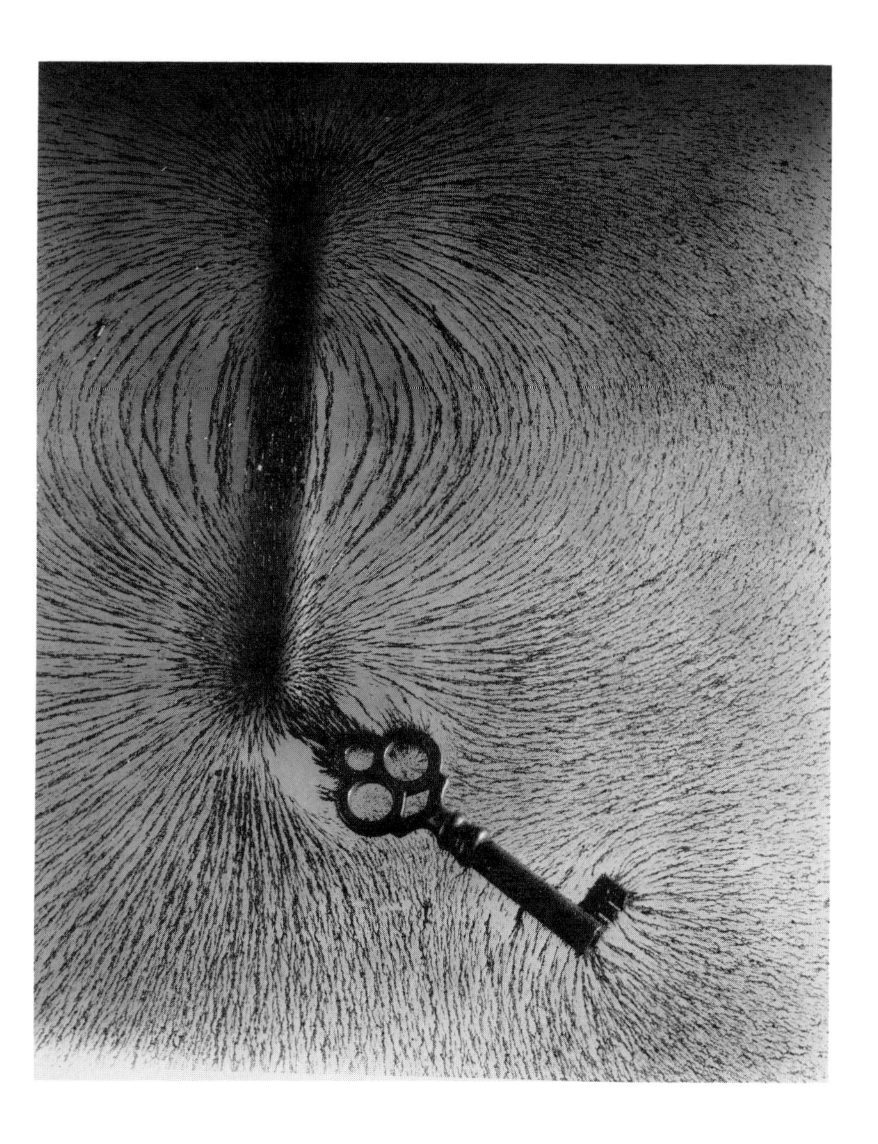

field can be described quite well in terms of a simple bar magnet, or "dipole," model, with a north magnetic pole in Prince of Wales Island (100°W, 73°N) and a south magnetic pole, nearly opposite, in Antarctica. Its field strength in middle latitudes is a few tenths of a gauss.

With superior modern magnetic materials, a small horseshoe magnet may have a field strength as high as 1000 gauss between its poles. To create such a strong field over a very large area becomes impractical with permanent magnets, and we make use instead of large coils carrying electric currents. Current passing through a long helical coil produces an axial magnetic field within the coil that emerges from one end and loops back to the opposite end, just as the field of a magnetic bar does. Very large coils and strong currents are needed to supply the magnetic fields of modern particle accelerators, and fields of a few thousand gauss may be produced over areas of a few square feet. All man-sized efforts are puny, however, compared to the magnetic force produced in a sunspot.

The riddle of sunspots was largely solved early in this century by George Ellery Hale. He surmised that the spots were magnetized regions, and he devised a spectroscopic method of measuring their field strengths. Magnetism splits the spectral emission lines of hot gases. Where the sun is unspotted, spectral lines are single, but when the slit of the spectrograph is aimed at a sunspot, spectral lines are doubled or tripled. Hale found that magnetic fields over spots were often a thousand times stronger than the fields over undisturbed neighboring areas. Each spot acts as a north or south magnetic pole, often covering an area tens of times as great as the earth's equatorial cross section. These immense fields must derive from great currents in the range of a thousand billion amperes circulating within the sun. (A 100-watt lamp carries one ampere of current.)

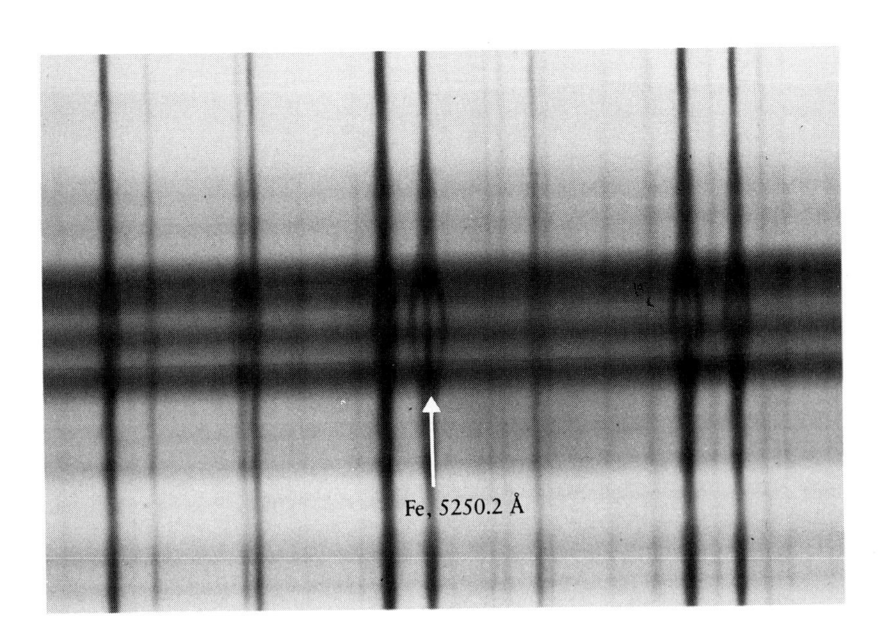

Spectrum lines over a sunspot are divided into triplets by the strong magnetic field. The separation of the components is proportional to the strength of the magnetic field, here about 4000 gauss.

Magnetic lines of force emerge nearly radially from a north polarity spot, and they may loop thousands of miles through the solar atmosphere to reenter the surface at a spot of south polarity. Sometimes a unipolar spot appears from which the lines emerge without any apparent return path, but, later, a spot of opposite polarity usually develops, as though the magnetic field precedes the cooling that makes the spot visibly dark. In effect, it seems that the magnetic field of a spot is a giant refrigeration mechanism, suppressing violent turbulent flows of gas within the spot and forcing hot bright gas to well up to the surface by some roundabout path.

It has been proposed that strong magnetic sunspot fields originate in the highly convective gas surrounding the inner nuclear furnace. Hot-gas streams carry burned nuclear fuel outward, and cooler gases carry fresh nuclear fuel inward toward the center. Because of the solar rotation, these streams get twisted into whorls that detach like smoke rings from the convective region of the core and rise through the quiet radiation zone of the solar interior toward the surface. The motion of turbulent ionized gas or plasma in the outer convection zone twists and kinks the magnetic fields embedded in the plasma, stretching the field lines like rubber bands and squeezing them closer together to create intense local fields. Under the influence of their own buoyancy and the upwelling of convecting gas, the intensified "ropes" of magnetic field break through the photospheric surface, where their footprints appear as sunspots.

In the interior of the sun, the interaction between differential rotation and convection wraps an initially north-south (poloidal) magnetic field (*a*) around the sun into an amplified azimuthal (toroidal) field (*b*). The toroidal field is driven to the surface by convection or floated to the surface by magnetic buoyancy. Columns of rising or sinking convection twist as a result of Coriolis acceleration and impart the twist to the magnetic field. These twists develop into small poloidal components (*c*). The combined effects of many rising convective cells eventually reestablishes the original poloidal field.

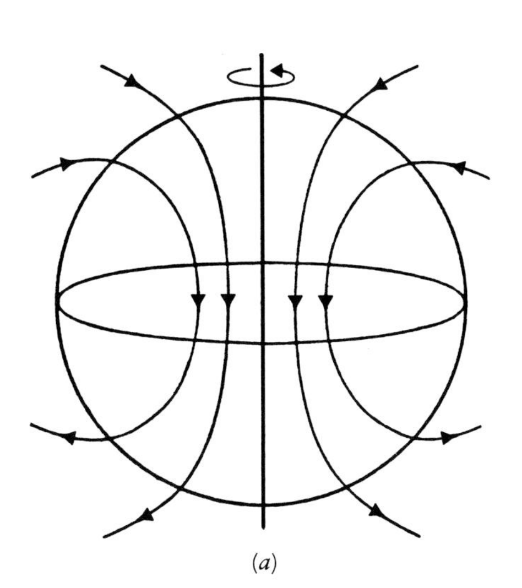

(*a*)

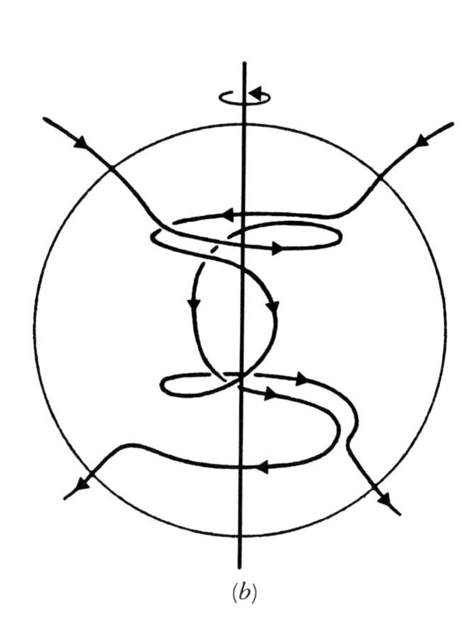

(*b*)

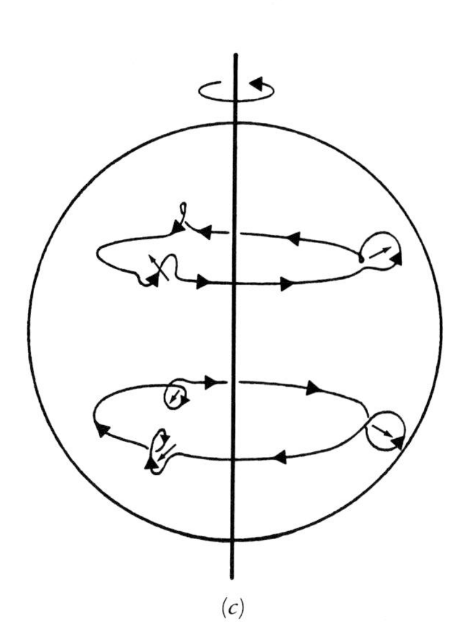

(*c*)

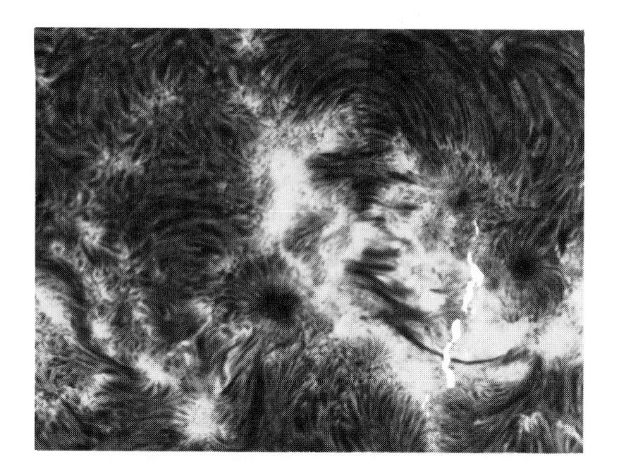

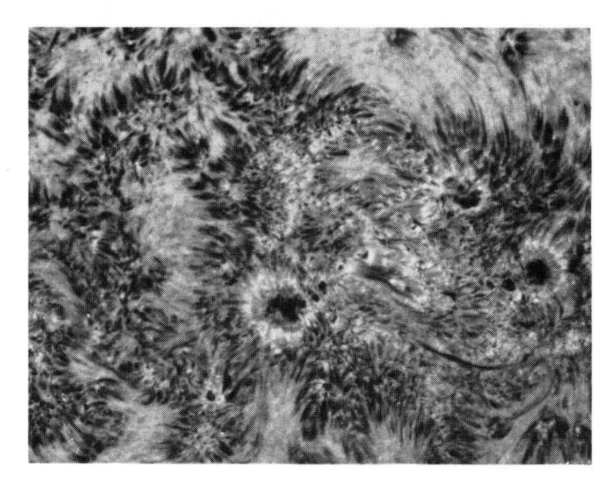

Spectroheliograms photographed in the wavelength at the center of hydrogen-alpha and slightly displaced from center. Active sunspot regions are marked by bright plages, bordered by long fibrils.

Viewed from afar, the strong localized magnetic fields on the surface average out to a weak, general, axial north-south field with a strength of about one gauss—not very different from that of the earth. As a magnet, the sun pales in comparison to "magnetic" stars. White dwarfs often have fields of thousands of gauss, and neutron stars have fields of trillions of gauss.

A closer view reveals magnetic fields that vary from a few gauss to 100 or 200 gauss scattered over the solar surface. The pattern of strong magnetic fields is duplicated by regions of brighter-than-normal gas, called plages, which are best seen in the monochromatic light of hydrogen or ionized calcium. All major forms of solar activity are associated with these magnetized plage areas that surround the sunspots.

In recent years, high-resolution observations have demonstrated that outside of sunspots the magnetic field over most of the solar surface consists of very small features that have spatial scales of less than one arc second (1000 kilometers) and very large field strengths of 1000 to 2000 gauss. Various names have been applied to these regions—gaps, magnetic knots, invisible sunspots—but they are now commonly referred to simply as magnetic elements.

Sunspots and Earthly Activities

Very clear connections exist between sunspot cycles and phenomena in the earth's atmosphere such as auroras, magnetic storms, and radio communications. The effects of sunspots on tropospheric weather and social phenomena, however, are highly speculative. Sunspot numbers have been compared with patterns of annual growth rings in trees, levels of lakes, layering of sediments and deposits in lakebeds, numbers of icebergs, abundance of the annual Atlantic salmon catch, rates of Australian rabbit reproduction, amounts of tropical rainfall, admissions to psychiatric hospitals, and, of course, the behavior of the stock market. Those who suggest market correlations claim that exposure to more ultraviolet radiation during sunspot maxima has a therapeutic effect that makes people feel bullish; insufficient ultraviolet leads to malaise and the mood to sell. Magnetic storms have been supposed to have subtle effects on electrical currents in the nervous system, especially the brain. An epidemiologist has claimed that a surprisingly close correlation exists between the last six sunspot peaks and worldwide outbreaks of influenza. None of these speculations appears to be founded on any credible statistical evidence.

Although the sun supplies the heat that drives the atmosphere, the circulation of the atmosphere is caused by unequal distribution of that heat—more solar energy arriving at the equator than at the poles—rather than by any variation in total input. A small variation in solar activity could perhaps affect the apportionment of water between sea and air and, therefore, the production of snow and ice. This in turn might influence the severity of winters and the levels of rivers, lakes, and seas.

In 1958, during a strong sunspot maximum, an unusual menu of freakish weather was reported. Rain in the Congo persisted more than three months beyond its normal cutoff date. It rained in Libya all through the middle of what should have been a long, dry summer, and Southern Rhodesia was wet during its normally dry winter. Everywhere in Africa the weather story told of floods and unseasonal rain, but that continent was not alone. Flooding occurred in China, Europe, India, and Argentina. In Australia, New South Wales had snow in June for the first time in 60 years, a snowstorm hit Portugal in May, Czechoslovakia had its worst heat wave in 183 years, and a local thunderstorm in England produced over 2000 flashes of lightning in the space of two hours. Although many people speculate that such abnormal weather has some connection with solar activity, as yet we have no persuasive evidence of any positive correlation between sunspot cycles and meteorology. (The broad subject of sun, weather, and climate is discussed more thoroughly in Chapter 6.)

Sunspots and Auroras

Another proxy indicator of solar activity is the eerie natural phenomenon known as the aurora borealis—the northern lights. Unlike magnetic storms, which are revealed only by delicate and sensitive instruments, the strange lights of the aurora sometimes emblazon the sky all around the world. Most frequently, however, they appear over two doughnut-shaped zones centered around the geomagnetic north and south poles. The band of the northern zone circles through lower Hudson's Bay in Canada, Labrador, Iceland, Scandinavia, Siberia, and Alaska. Seeing auroral displays in those skies, ancient peoples were awestruck and often terrified by the flaming, pulsating, red-and-green glows. Historians of science have been able to trace solar activity through accounts of auroral phenomena as far back as several hundred years B.C.

Aristotle refers to auroras during the fourth century B.C., and in the reign of Tiberius (A.D. 14–37), Seneca describes how, one night, a blood-red light glowed in the west. Believing that the seaport of Ostia at the mouth of the Tiber River was on fire, the Emperor dispatched his legions to fight a blaze they could never find. The vivid apparitions

Early woodcut of an aurora seen from Kuttenberg in the kingdom of Bohemia on January 12, 1570. The text with the print reads:

"In the year 1570, the 12th of January, for four hours in the night between midnight and sunrise, the portent appeared in the heavens after this fashion. At first a very black cloud went forth like a great mountain, in which several stars showed themselves, and over the black cloud was a very bright streak of light, burning like sulfur and in the shape of a ship; standing up from this were many burning torches, like tapers, and among these stood two great pillars, one toward the east and the other due north, so that the town appeared illuminated as if it were ablaze, the fire running down the two pillars from the clouds above like drops of blood. And in order that this miraculous sign from God might be seen by the people, the night watchman on the tower sounded the alarm bells; and when the people saw it they were horrified and said that no such gruesome spectacle had been seen or heard of within living memory. Wherefore, dear Christians, take such terrible portents to heart and diligently pray to God that he will soften his punishments and bring us back into his favor so that we may await with calm the future of our souls and salvation. Amen."

in the sky described during biblical times most certainly were auroras. A typical account in 2 Maccabees dates from the second century B.C.:

> About this time Antiochus sent his second expedition into Egypt. It then happened that all over the city, for nearly forty days, there appeared horsemen charging in midair, clad in garments interwoven with gold—companies fully armed with lances and drawn swords. . . .

During the Middle Ages, frequent descriptions were recorded of "fire beams" and "burning spears" in the sky. One of the most spectacular auroras was seen over most of Europe on January 12, 1570. Such manifestations were often, and quite understandably, taken as portents of evil—a coming of the end of the world.

After an active period in the 1620s, few auroras were seen for a century, and the memory of such events almost vanished. The absence of auroras through most of the seventeenth century coincided with the period of missing sunspots known as the "Maunder Minimum." But sunspots and auroras returned. In 1716, Edmund Halley, at the Royal Society, carefully described the arches and rays of a startling aurora as they were observed from widely separated places. Rejecting a popular suggestion that these displays were caused by sulfurous vapors escaping from the earth's interior, Halley proposed the essentially modern view that they were "magnetic effluvia," constrained to move along the lines of force of the earth's magnetic field. (As a phenomenon resulting from solar particle radiation, auroras are fully described in Chapter 5.)

Sunspots and Comets

Chinese annals refer to comets as "broom stars" or "guest stars." Ancient peoples reacted to these apparitions violently as they did to auroras, and cometary events are therefore well documented. Among other things, these records provide extremely interesting evidence of periodicities in the appearances of comets that fit the sunspot cycle. Comets are certainly not outbursts of solar material, and they travel through the solar system in vastly elongated orbits. Why, then, should they have any relationship to the sunspot cycle?

The answer must lie in the variability of astronomical weather. Comets are seen best when the background sky is blackest and airglow is weakest. Airglow arises from the action of solar ultraviolet rays and X rays on atmospheric molecules during the day; energy stored in daylight is slowly dissipated as fluorescence during the night. Even though the airglow provides no more light than a candle at a distance of about 300 feet, it makes the high atmosphere bright enough to hide the dimmer comets. At times of maximum sunspots, ultraviolet and X-ray input is greater, airglow brighter, and the number of reported comets lower. In this indirect way, astronomical climatology proves to be more useful than meteorology in tracing the sunspot cycle. Fairly good airglow cycles can be deduced from cometary observations to 500 B.C., and crude evidence goes back to 2300 B.C.

Solar Eclipses: Nightfall in Midday

A total solar eclipse provides the most spectacular and awe-inspiring display in nature's repertory, and it is therefore not surprising that anecdotal material abounds in ancient, classical, and medieval literature. The first written account of a solar eclipse dates back more than 4000 years to October 22, 2137 B.C. The Chinese classic *Shu Ching* tells the story. Regarding the accuracy of eclipse predictions, Chung-Kang, the fourth emperor of the Hsia dynasty, had warned that "being before the time, the astronomers are to be killed without respite; and being behind the time, they are to be slain without reprieve." Although aware that the eclipse was imminent, the two royal astronomers, Hi and Ho, were "so sunk in wine and excess that they neglected the ordering of the seasons and allowed the days to get into confusion." They failed to warn the populace, and crowds raced wildly through the streets, beating drums and shooting fireworks to frighten away the dragon that was swallowing the sun. For their sins, the astronomers lost their heads.

Allusions to eclipses appear in the Holy Scriptures. Most specific is the reference in Amos 8:9, written about 787 B.C.: "I will cause the sun to go down at noon and I will darken the earth in the clear day." Two centuries later, the eclipse of 585 B.C. interrupted a six-year war between the Medes and Lydians. So awed were the warring armies by the midday darkness that they threw down their arms and made lasting peace, cemented, in the fashion of the time, by a double marriage. During the eclipse of May 5, A.D. 840, Louis Le Debonnaire, son of Charlemagne and heir to his great empire, is said to have died of fright, whereupon his three sons divided their inherited lands into France, Germany, and Italy.

Observers on a rooftop in upper Manhattan photographed the eclipse on January 24, 1925. The edge of totality was at 96th Street.

On average, there are 237 solar eclipses per century, of which about one-fourth are total, for approximately two totals every three years. The longest eclipses last for about seven minutes, but most take only about three minutes. An astronomer dedicated to observing every eclipse over an active lifetime of 50 years and lucky enough to have clear skies on every occasion would have about 150 minutes of total observing time. In any typical community on earth, the local inhabitants are unlikely to see more than one total eclipse in their lives, and ignorance and fear often deprive them of even that small prospect. Warned by the news service of the 1970 eclipse, farmers in North Carolina turned off their television sets and hid inside their houses to escape "dangerous radiation" from the eclipsed sun. For scientific observers, the unveiling of the spectacle often brings on an almost paralyzing hypnosis that upsets all carefully laid plans for photography and instrumental measurements. As Princeton Professor C. A. Young commented after the 1869 eclipse in Iowa, "I cannot describe the sensation of surprise and mortification, of personal imbecility and wasted opportunity that overwhelmed me when the sunlight flashed out."

Stories abound of animals confused and their biological rhythms upset by the unexpected darkness. Thomas Edison carefully set himself up to observe the eclipse of 1878 in a Wyoming chicken coop. As the sun dimmed before totality, the chickens came in to roost. Edison was so occupied fighting the chickens that he missed almost the full three minutes of totality and caught only the briefest glimpse of the spectacle.

The coincidence of eclipse observation with high adventure in remote parts of the earth has seduced scientists into chancing unpredictable weather and enduring the hazards and discomforts of travel

Eclipse photographed from a supersonic jet aircraft at 40,000 feet on February 26, 1979. Active chromospheric forms appear in red. Numerous streamers reach out to distances of 4 million kilometers.

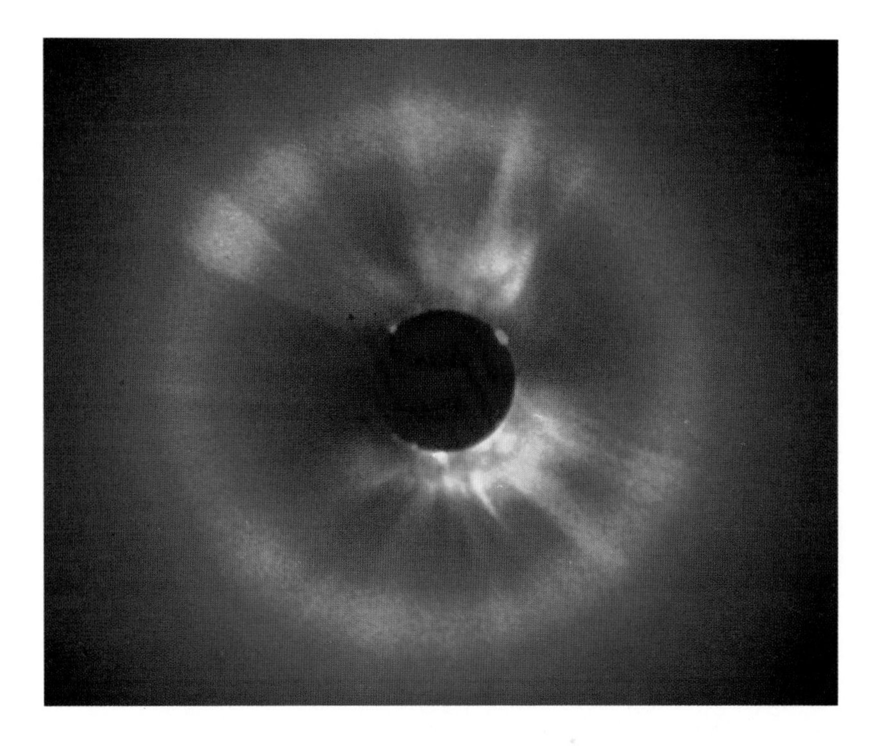

by every conceivable conveyance—stagecoach, sailing ship, balloon, and, most recently, supersonic jet* and orbiting space station. In the midst of the Revolutionary War, a truce was arranged so that an expedition led by Professor Samuel Williams of Harvard College could observe the eclipse of October 27, 1780. The Board of War provided a galley to carry Professor Williams' party to Penobscot Bay. On eclipse day, they were dismayed to find that they had positioned themselves just outside the edge of totality.

An eclipse on July 18, 1880, was especially exciting for European and American astronomers because it offered the first opportunity to use photography. Totality covered a band that stretched all the way

In the last decade and a half, chasing an eclipse by supersonic jet has become a highly perfected form of eclipsemanship. When the fiftieth total eclipse of this century occurred over Africa in 1973, Los Alamos scientists aboard a French-British Concord raced the shadow of the moon over the Sahara Desert at 1300 miles per hour (Mach 2) and observed an almost incredible 74 minutes of totality. At the jet's altitude, the earth's horizon appeared clearly curved against a purple sky. Keeping pace with the moon's shadow made it possible to follow the eclipse in slow motion from start to finish.

from Hudson's Bay in Canada to Spain and Africa in the opposite hemisphere. European astronomers in Spain enjoyed clear skies, but only one of three American expeditions had any success. The most traumatic experience befell a group sponsored by the Navy's Nautical Almanac Office, which included 25-year-old Simon Newcomb, later to become Director of the Naval Observatory and the most distinguished American astronomer of his time.

To observe the eclipse of July 18, 1880, Simon Newcomb led a party on a wilderness trek of 3000 miles into northern Canada. On eclipse day, the site was flooded and the sky overcast. This sketch by Samuel Scudder, who kept the time for the group, shows Newcomb's telescope set up in the crotch of a tree.

Newcomb and his partners left St. Paul, Minnesota, on June 14. After several days by stagecoach across the muddy roads of Minnesota, they transferred first to a steamer on the Red River and then to a barge on Lake Winnipeg. Finally, they boarded a birch-bark vessel with five paddlers to complete their journey. Surrounded by desolate wilderness, they paddled and poled against the current in almost constant rain while swarms of vicious mosquitoes attacked them. Time slipped away, and it took an almost superhuman effort, paddling nonstop for the last 16 days, to reach the edge of the eclipse path the night before the event. The plains were flooded, and only a boggy ridge offered footing. Newcomb set up his telescope in the crotch of a tree. It rained all night, and the morning of eclipse day was heavily overcast. As sadly recorded by one of the party: "This, then, is our success. Three thousand miles of constant travel occupying five weeks, to reach by heroic endeavor the outer edge of the belt of totality, to sit in a marsh and view the eclipse through clouds!"

Is tragedy the rule in scientific eclipse expeditions? One success makes up for many failures. C. A. Young's description of the so-called flash spectrum of the chromosphere during the eclipse of 1870 in Spain cannot be improved upon:

> As the moon advances, making narrower the remaining sickle of the solar disk, the dark lines of the spectrum for the most part remain sensibly unchanged, though becoming somewhat more intense. A few, however, begin to fade out and some even turn palely bright a minute or two before totality begins. But the moment the sun is hidden, through the whole length of the spectrum in the red, the green, the violet, the bright lines flash out by hundreds and thousands, almost startlingly, as suddenly as stars from a bursting rocket head and as evanescent, for the whole thing is over in 2 or 3 seconds. The layer seems to be only something under a thousand miles in thickness and the moon's motion covers it quickly.

Temperature Inversion in the Chromosphere

The flash spectrum of the chromosphere reveals a very inhomogeneous structure. Regions of high and low temperature exist in close

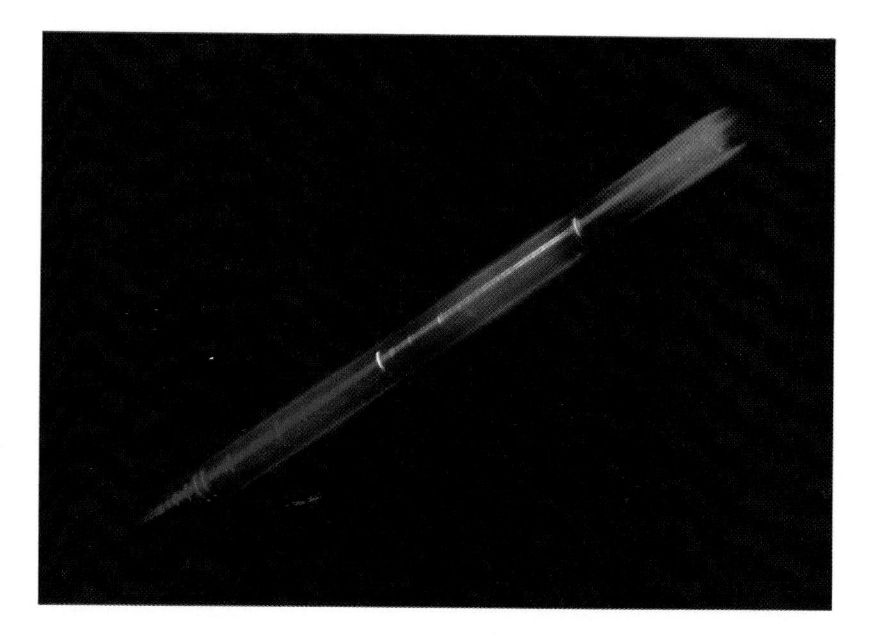

A flash spectrum of the chromosphere photographed against the totally eclipsed disk.

proximity, but most surprising is a steep temperature inversion immediately above the photosphere. From the thermonuclear furnace in the core, the temperature decreases steadily from about 15,000,000 K to about 4700 K at the top of the photosphere. In the higher solar atmosphere we would expect even cooler gas, but, instead, the temperature shoots above 100,000 K in less than 10,000 miles in the chromosphere and above 1,000,000 K in the corona.

Normally, heat flows from hot to cold. How, then, can the chromosphere and corona derive very high temperatures from a much cooler photosphere? Obviously, the photosphere does not heat the corona by conduction or radiation, as a stove heats a kettle. Some mysterious form of heat transfer feeds energy directly into the high atmosphere from far down in the photosphere.

According to one hypothesis the bubbling granules break like ocean waves against the bottom of the chromosphere. This churning of the ocean of solar gas creates a din of unimaginable violence. Its noise is far below the threshold of bass notes to which the ear is sensitive, about 20 cycles per second. Infrasound in the solar atmosphere has a wavelength of perhaps 500 kilometers—about one cycle per minute. Although subsonic to the ear, the vibration pressure would shatter the eardrums of an imaginary listener. As these sound waves rush upward into more rarified gas, they increase in speed until

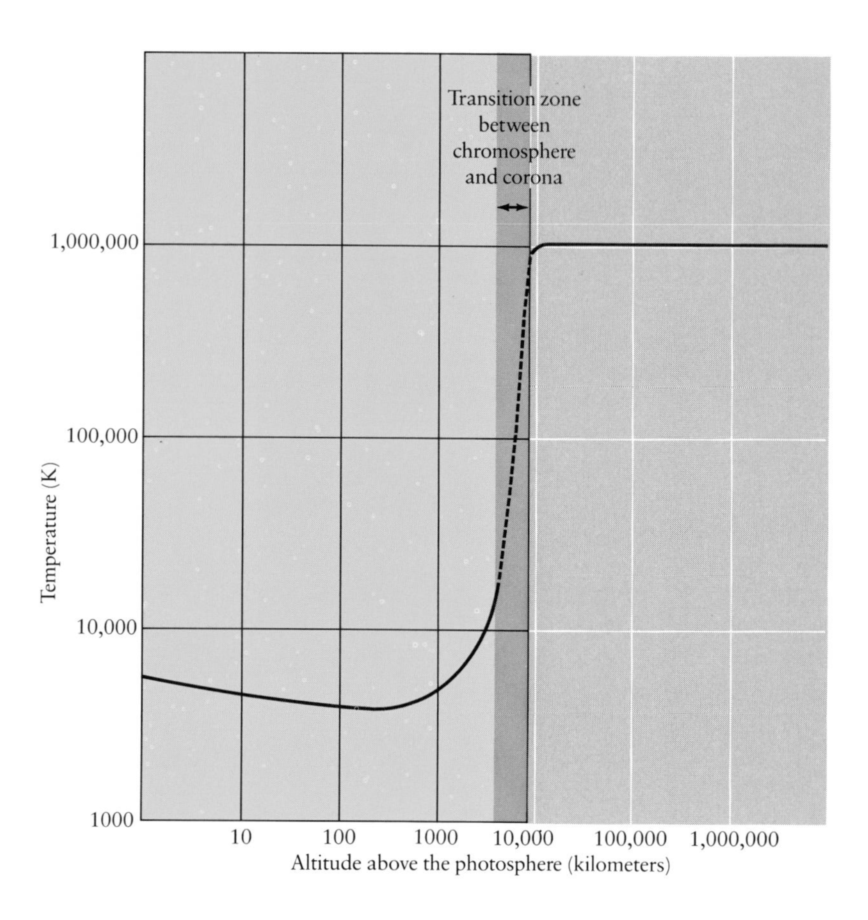

Abrupt rise in temperature across the chromospheric transition region from the photosphere to the corona.

they propagate supersonic shocks like thunderclaps or the sonic booms of jet aircraft. The shock-wave energy dissipates in friction that would heat the coronal gas to high temperature.

Unfortunately, observations designed to detect sound waves propagating upward into the corona have not produced any supporting evidence for this concept of coronal heating. Shock waves can be observed at and near the surface, but their energy dissipates before reaching any significant height in the corona. Solar physicists have therefore turned to models based on varying magnetic fields. Surface magnetic fields are always moving as new fields bubble up from inside and turbulence twists them about. These moving magnetic lines of force in an electrically conducting plasma provide a dynamo mechanism for generating electric current, a current that heats the resistive plasma much as current heats the filament of a light bulb.

The energy necessary to heat the corona is estimated to be about the energy dissipated in a 100-watt lamp for every square meter of the solar surface—about one-millionth of the energy that passes through the corona in radiant heat and light from the photosphere. Yet such a relatively weak power input can heat the gas to a million degrees because the corona is too thin to radiate the energy away quickly. If the corona were as dense as the photosphere, at a temperature of a million degrees it would be bright enough to vaporize the earth.

Eclipse pictures of the rim of the sun show a hairy structure, picturesquely compared to prairie grass, hedgerows, and flames. Tongues of gas, called spicules, spring like frothy jets above the bursting granules. They surge up from the top of the photosphere and fall back again in five to 10 minutes, rising with speeds of 10 to 15 miles a second to heights as great as five or six thousand miles. An average

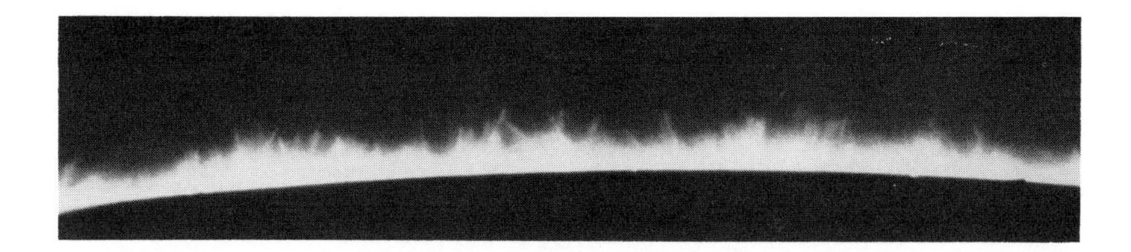

The chromosphere at the limb, showing spicules, which jet above the surface at speeds of 30 kilometers per second or more, to heights as great as 10,000 kilometers.

Spicule bushes outline supergranule boundaries. The individual fibrils appear as spicules at the limb, extending upward 5000 kilometers and higher.

spicule weighs a million tons, and its internal temperature may rise to over a hundred thousand degrees. At any given instant, as many as a hundred thousand spicules cover the face of the sun and some seem to vanish into the corona.

The chromosphere has thus been called the "spray of the photosphere." It is estimated that the spicules cover about 1 percent of the solar surface, and they appear to be grouped in bushlike structures, with five to 15 spicules in a bush. Noticeable changes occur in the shapes of these bushes every hour or so. At the polar caps, giant spicules lasting as long as an hour rise to 40,000 kilometers and spread from 15,000 to 20,000 kilometers at the base over areas more than twice as large as the earth.

Where the chromosphere blends into the corona, solar gas suddenly becomes about a thousand times thinner. So transparent is the corona that stars can easily be seen through it at the time of an eclipse. If the eclipse occurs at sunspot maximum, coronal streamers surround the black moon very symmetrically. At sunspot minimum, great equatorial streamers that distort this symmetry can be seen stretching millions of miles into space. Observers in airplanes flying in the shadow of an eclipse high above most of the dust-filled terrestrial atmosphere have used photoelectric cells to detect the light of the corona as far out as 20 million miles (about 40 solar radii).

Radio Probing of the Far Reaches of the Corona

Radio astronomy provides the most distant measurements of the corona. As the celestial sky passes overhead each year, the extended corona intercepts radio stars, and the effects of these interruptions can be observed. More than a dozen radio sources come sufficiently close to the sun to have some of their emission scattered.

Occultation—that is, masking by the sun—can first be detected at distances of more than 100 solar radii, and it is useful as a probe of the very extended range of the corona. The phenomenon of occultation occurs when plasma density fluctuations in the corona scatter radio emissions from nearby sources, just as stars scatter, or twinkle optically, when seen through the irregularities of the terrestrial atmosphere. Both magnitude of electron density in the plasma and fluctuations in that density affect the degree of scattering. By proper interpretation, then, it becomes possible to draw conclusions about the density and shape of the corona.

The Crab Nebula, the remnant of a supernova explosion seen in A.D. 1054 and one of the brightest radio sources in the sky, comes within five degrees of the sun. At close range, its radiation probes the polar regions of the corona. At the greatest angular distance from the sun, its radiation probes the corona closer to its equator. By applying radio-scattering techniques to several celestial sources over the course of the year, radio astronomers can construct two-dimensional maps of the corona.

Artificial Eclipses by Means of the Coronagraph

At the time of the eclipse of 1882, the new photographic dry plate revolutionized the application of photography to science. William Huggins, a leading astronomical spectroscopist, was so greatly impressed with the clarity of the corona seen at eclipse that he believed he could photograph it in full sunlight. Like other distinguished observers, he was totally unsuccessful in attempts from Pikes Peak and Mt. Etna. Almost 50 years after Huggins gave up, Bernard Lyot, a French astronomer, finally succeeded where so many had failed.

Near the edge of the solar disk, the brightness of the corona equals that of the full moon. If we can see the moon in daylight, why is it so difficult to see the corona? Scattering of sunlight by air molecules and dust makes the sky bright, and this brightness, along with the dazzling light from the sun, overwhelms the faint corona. Even if the sky were clear, we would find it very difficult to mask out the light of the disk, even with a sharp-edged shield; the edge simply shines too brightly.

About the time that most astronomers had given up on observations of the corona without an eclipse, Lyot succeeded in developing the coronagraph. No radically new principles were involved in its design; it succeeded, rather, by reducing stray light in every conceivable way. Basically, a coronagraph employs two lenses in series. The first, or objective, lens images the sun on a metal disk of exactly the right size to eclipse the photospheric disk while allowing coronal light to pass through. A second, or camera, lens focuses the eclipsed image onto photographic film. Light scattered on internal surfaces is shut out by baffles, and the objective lens is made of only one, flawless glass element, free of bubbles, scratches, and other imperfections. All surfaces must be scrupulously cleaned. Finally, the instrument must be set up on a mountaintop. Because rising air currents tend to carry dust upward, especially over warm land, snow-covered terrain is best. Precipitation is also very effective in scavenging dust from the air. Lyot selected as his site the Pic du Midi Observatory at 2870 meters (9415 feet) in the snow-capped Pyrenees.

Donald Menzel of Harvard College went to view Lyot's telescope and its results first-hand. He came home determined to duplicate it in the United States. Recalling the clear blue skies of his childhood in

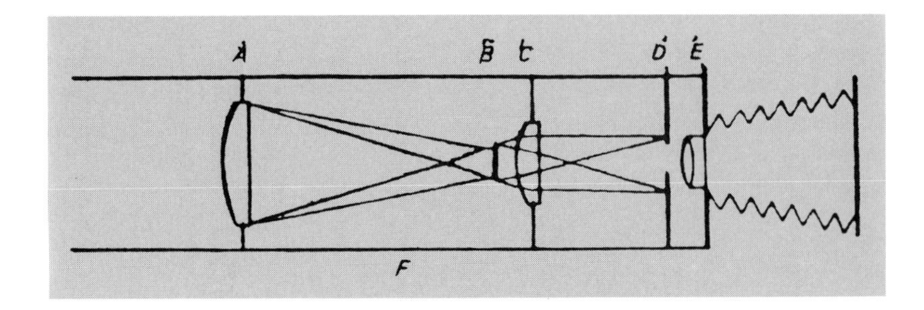

In a coronagraph, light passing through the objective A, an optically pure simple lens, forms a solar image on the stop. This obstacle is introduced in order to cut out the light coming from the solar disk and passes the light of the corona 10 to 20 seconds of arc beyond the solar limb. The final image is formed by the objective lens E on either a photographic plate or a spectrograph slit.

Leadville, Colorado, he chose a site at Fremont Pass, just 15 miles north of Leadville, at a height of 3512 meters (11,520 feet). His earliest financial support for the coronagraph came from the Department of Agriculture under Secretary Henry Wallace. At that time, shortly after the great dust bowls of the mid-1930s, a number of scientists, including Menzel and Charles Greeley Abbott of the Smithsonian, thought that the sun's cyclic activity might be a factor in the comings and goings of the great droughts in the high plains.

Today, coronagraphs are operating around the world, atop Haleakala crater on the island of Maui, Sacramento Peak Observatory in New Mexico, and on the Jungfrau in Switzerland, as well as on the Pic du Midi. Observations can be made on any clear day out to about 20 arc minutes from the edge of the disk.

The Pic du Midi Observatory sits atop the French Pyrenees at a height of 9450 feet. The thin, dry air and the suppression of dust by the snow cover that normally covers the slopes make the setting ideal for coronagraphic photography of the sun. It was here that Bernard Lyot made the first successful coronagraphic observations in 1931.

The coronagraph and a filter that passes red hydrogen light reveal huge streamers of bright gas looping sometimes as high as a hundred thousand miles into the corona. These prominences stretch even farther across the surface of the sun, often achieving a length five times as great as their height, and they may contain a billion tons of material. Time-lapse motion pictures reveal continuous changes in their overall shapes, along with complicated internal streaming.

Seen against the disk, prominences appear as dark "filaments" snaking across the vicinity of sunspot groups. Actually, both prominences and filaments are equally bright. Against the dark background sky beyond the sun's edge, or limb, a prominence appears bright, but against the bright disk, it looks like an obscuring ribbon of smoke. Prominences usually appear to be rooted in sunspot groups, and from their very beginning they show the typical arched structure so indicative of magnetic fields. Where the streamers are anchored to the photosphere, violent convection twists and shifts the magnetic field lines about, simultaneously causing the material in the high arches to react in spectacular whipping, streaming, and eruptive patterns. A filament usually points toward the spot of greatest magnetic field strength, and it often keeps growing for months, long after the spots have faded. Prominences may also break up explosively and fly off into space.

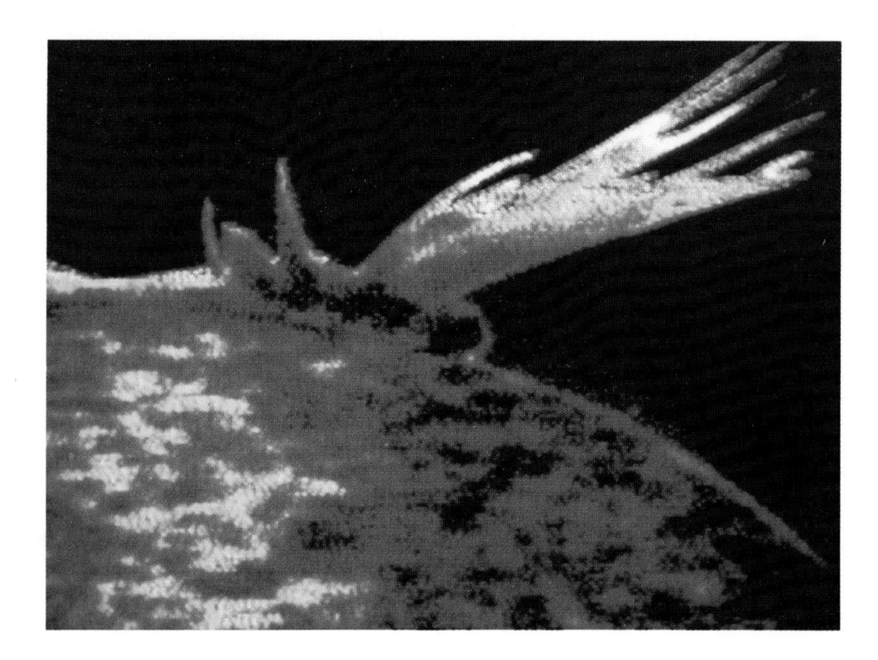

A surging spray photographed in singly ionized helium and colored by computer graphics according to intensity. White is brightest, red is coolest.

A twisting eruptive prominence arches 400,000 miles above the sun. The photograph, taken in the light of singly ionized helium at temperatures from 30,000 to 40,000 K, was obtained with the spectroheliograph aboard Skylab. The eruption began with a low-lying prominence, which suddenly soared to its maximum height in about 20 minutes.

A massive quiescent prominence hanging over the solar limb. Arrays of fine, almost vertical ropes cluster like fiery pillars supporting the roof of the prominence. Gases rain down at speeds of about one kilometer per second. Random whipping motions are faster.

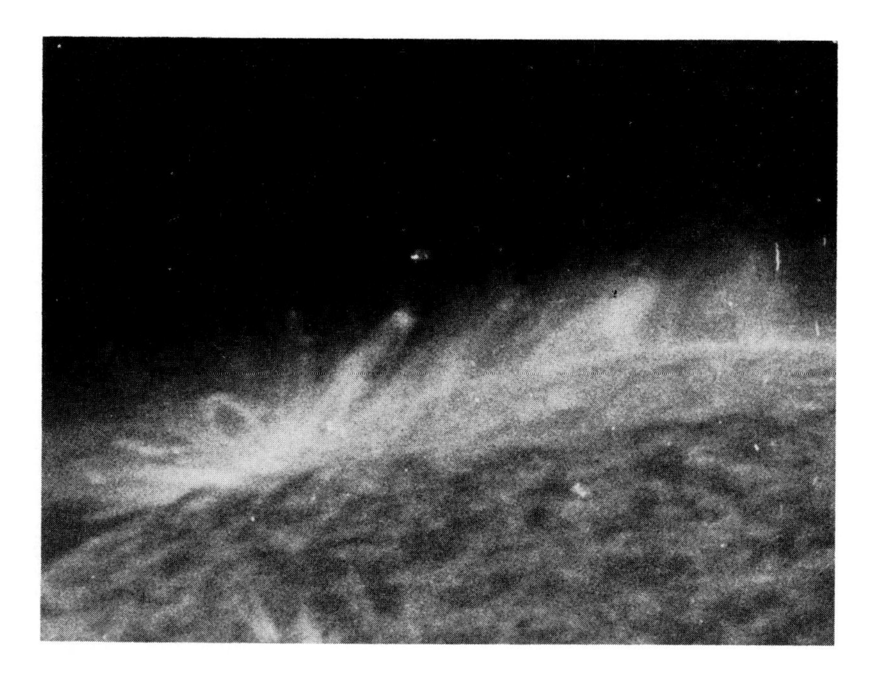

Erupting prominence in the light of six times ionized neon (at 600,000 K). The active limb is covered by looping magnetic field lines that guide the soaring prominence material.

Close to the chromosphere and photosphere, the loops become shorter and tighter until the photosphere itself takes on the appearance of a closely knit sweater. At the poles of the sun, the surface is covered with a brush of fine plumes, standing straight up from the surface.

The corona is a hot, gaseous plasma, a fourth state of matter distinguishable from the more familiar gaseous, liquid, and solid conditions. Plasma forms when electrons are stripped from gaseous atoms, leaving an assemblage of ions and electrons to move about in random thermal motion. A candle flame is a plasma, as is the gas discharge in a fluorescent light or a neon sign. When meteors enter the atmosphere and burn up, their vaporized remains are plasmas. The entire earth is blanketed by a plasma sheath called the ionosphere, which is produced by solar ultraviolet and X rays that tear electrons loose from atmospheric molecules. Radio waves bounce back from such a plasma unless their frequency is above a certain critical point at which they can seep through. (We shall discuss the ionosphere more fully in Chapter 4.)

Some plasmas, like the candle flame, are relatively cool, while others, like the core of the sun, are very hot. When we speak of heat, we mean the total energy of motion of gaseous particles, but when we speak of temperature, we refer to the average speed of those particles.

This chromospheric pattern, seen in hydrogen-alpha, shows tightly knit loops enmeshing the entire surface of the sun.

The corona is a million-degree plasma, and its ions and electrons race about at great speeds compared, for example, with the thousand-degree particles of the ionosphere. But the coronal plasma is so tenuous that its heat content is relatively small. It would impart very little sensible heat to a body in contact with it.

Ninety-nine percent of the light of the corona is white, exactly like the color of the photosphere. In fact, the corona is visible because it scatters the light that emerges from the photosphere. To draw an analogy to the halo surrounding a streetlamp on a foggy night, the lamp is the sun and the fog droplets that scatter its light are the free electrons in the corona, which have the property that they scatter light with no average change in color.

We infer the high temperature of the corona from a variety of evidence based on the quality of its light, the radio noise that it emits, and its great extent. For example, in comparison to the terrestrial atmosphere, the corona seems to defy gravity. Our atmosphere is relatively cool. Most of it is at a temperature of only a few hundred kelvins, and the pull of gravity binds it to the earth in a thin shell. The 6000 K temperature of the photosphere would not be nearly hot enough to support the corona. In the sun's strong gravitational field, a corona of that temperature would reach up only a few dozen miles. It is easy to compute that the coronal plasma must be at a temperature of about a million kelvins to expand outward a million miles against the strong pull of the sun's gravity.

Even near the base of the corona, Fraunhofer lines are missing. Actually, the lines are undetectable because they are so smeared out by the Doppler effect; light reflected from hot, speeding electrons is shifted in wavelength in proportion to the speed of the electrons. At a "smear" temperature of 6000 K in the photosphere, electrons are too slow to cause much blurring, but at million-degree temperatures, the lines spread into each other.

Working in Spain during the eclipse of 1870, C. A. Young discovered in the faint white light of the corona a narrow, green emission line. He was not able to match it to any known spectral line ever observed in the laboratory, and it was presumed to be an unknown solar element, "coronium." After the turn of the century, many more lines were detected in the red and violet wavelengths, as well as in the green ones, but none could be identified. Roughly 1 percent of all the visible coronal light was contained in about 20 of these emission lines. Finally, in 1938, the Swedish physicist Bengt Edlén proved that the brightest line was the green wavelength of an iron atom that had been stripped of half its normal 26 electrons. Other lines came from iron atoms minus nine and 10 electrons; from calcium, minus 11 and 12 electrons; from nickel, minus 11 and 15 electrons; and so on. Such heavily ionized atoms could only be produced by the violent collisions of particles racing about in a million-degree plasma.

A million-degree plasma radiates most intensely in the X-ray and ultraviolet wavelengths. X-ray emission from the solar corona is about a hundred times as strong as the visible green and other colored lines from the highly stripped ions. Once the nature of the coronal plasma became clear from Edlén's explanation of "coronium," it was possible to predict X-ray emission. To observe solar X-rays then became one of the first challenges to the new rocket astronomy and one of its earliest successes.

The Solar Optical Telescope (SOT)

Space observatories thus far have been enormously successful in revealing the distribution of far ultraviolet and X rays on the face of the sun. They have not, however, been able to refine spatial details any better than observatories on the ground. The smallest features on the solar disk that can be distinguished from the ground measure about 700 kilometers, the size of the state of Texas. Sunspot drawings by Scheiner and Galileo at the turn of the seventeenth century and solar granulation photographs by Janssen in 1890 are nearly as good as those obtained with the best solar telescopes today. Atmospheric turbulence still permits optimum seeing—that is, a resolution of about 0.3 arc second—only a few hours a year, and even then for only 10 or 15 minutes at a time. In longer exposures, the images dance all over the field of view. Most of the time, the best resolution achieved approaches the theoretical performance of a 12-inch telescope. We can infer that the fine structure of magnetic fields is well below this limit, but what we see is an averaged pattern that is very deceptive.

The prospect that perfect skies in space will yield the ultimate sharpness in telescopic images has been one of the strongest motivations for establishing space observatories. A Solar Optical Telescope (SOT), weighing 4000 kilograms and seven meters long, to be carried aboard a space shuttle is expected to offer a major advance in resolution (0.1 arc second) of images on the sun. It will have a primary mirror diameter of 50 inches and features especially designed to compensate for thermal distortions. To maintain ultraviolet reflectivity, for example, protection of the mirror surface from accumulation of contaminant molecules will require continuous heating. The astronomers have chosen a Gregorian telescope design that passes the image through a small aperture in a baffle at the primary focus and rejects all the incident heat flux outside this small image spot. Active coolant will circulate to protect the most vital elements. Optical alignment of primary and secondary mirrors must be held to 50 micrometers. Jitter must be controlled to less than 0.1 arc second, and pointing accuracy to a couple of arc seconds. Rated against such rigid specifications, most conventional telescope structures would appear to be rubbery. Because of the folded Gregorian optics, the SOT is much less susceptible to bending and warping.

The unequal heating of the shuttle when the telescope is pointed at the sun for long exposures will compound existing thermal imbalances—the sunward side will get very hot, the back side very cold. Observations will need to be regularly interrupted for rotations of the shuttle that will barbecue it uniformly on all sides.

A seven-day mission is expected to produce some 50,000 photographs of the sun. In addition, the telecommunications relay satellite will handle a vast flow of data to ground. These electronic transmissions will come from charge coupled devices (CCDs) that serve as sensors for an ultranarrow band filter and spectrograph.

A successful SOT will bring the perception of fine details of the solar surface into sharper focus than ever before.

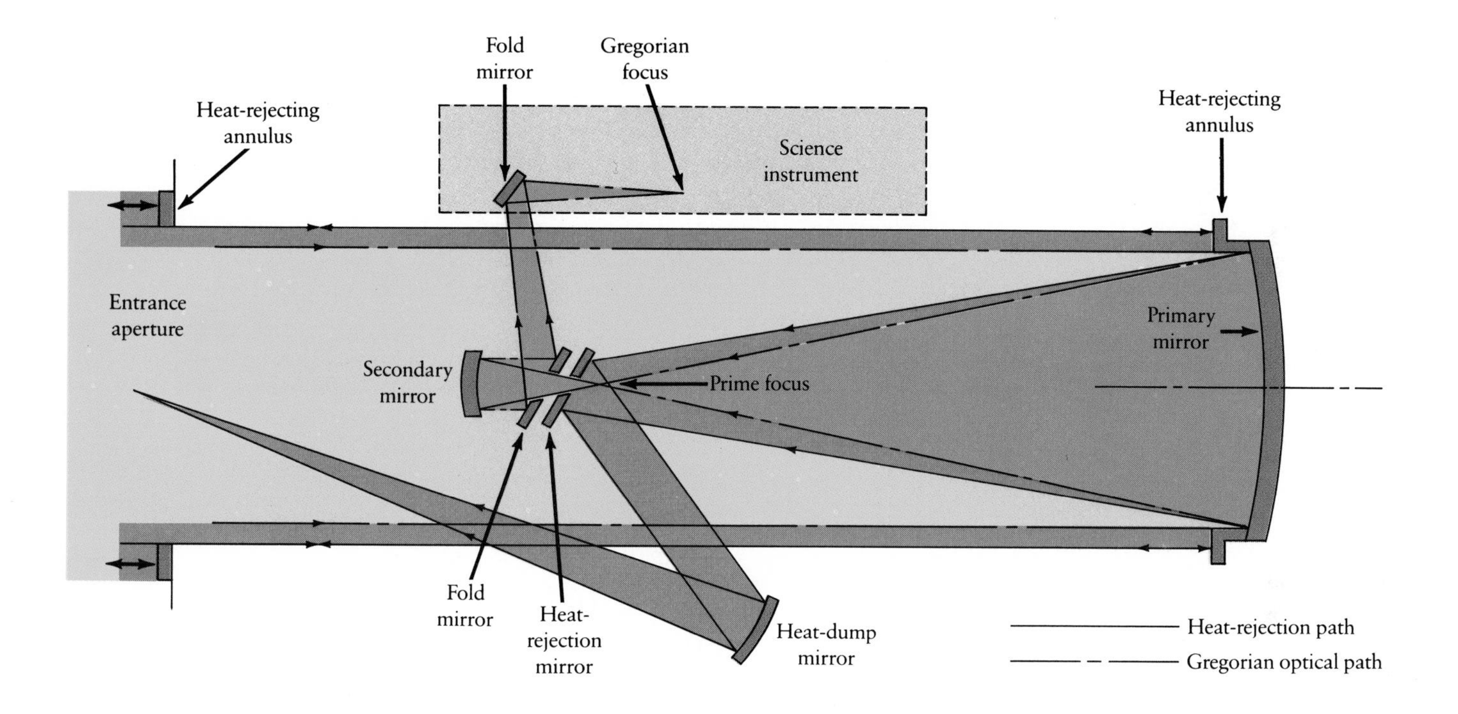
Fold mirror
Gregorian focus
Heat-rejecting annulus
Science instrument
Heat-rejecting annulus
Entrance aperture
Primary mirror
Secondary mirror
Prime focus
Fold mirror
Heat-rejection mirror
Heat-dump mirror
Heat-rejection path
Gregorian optical path

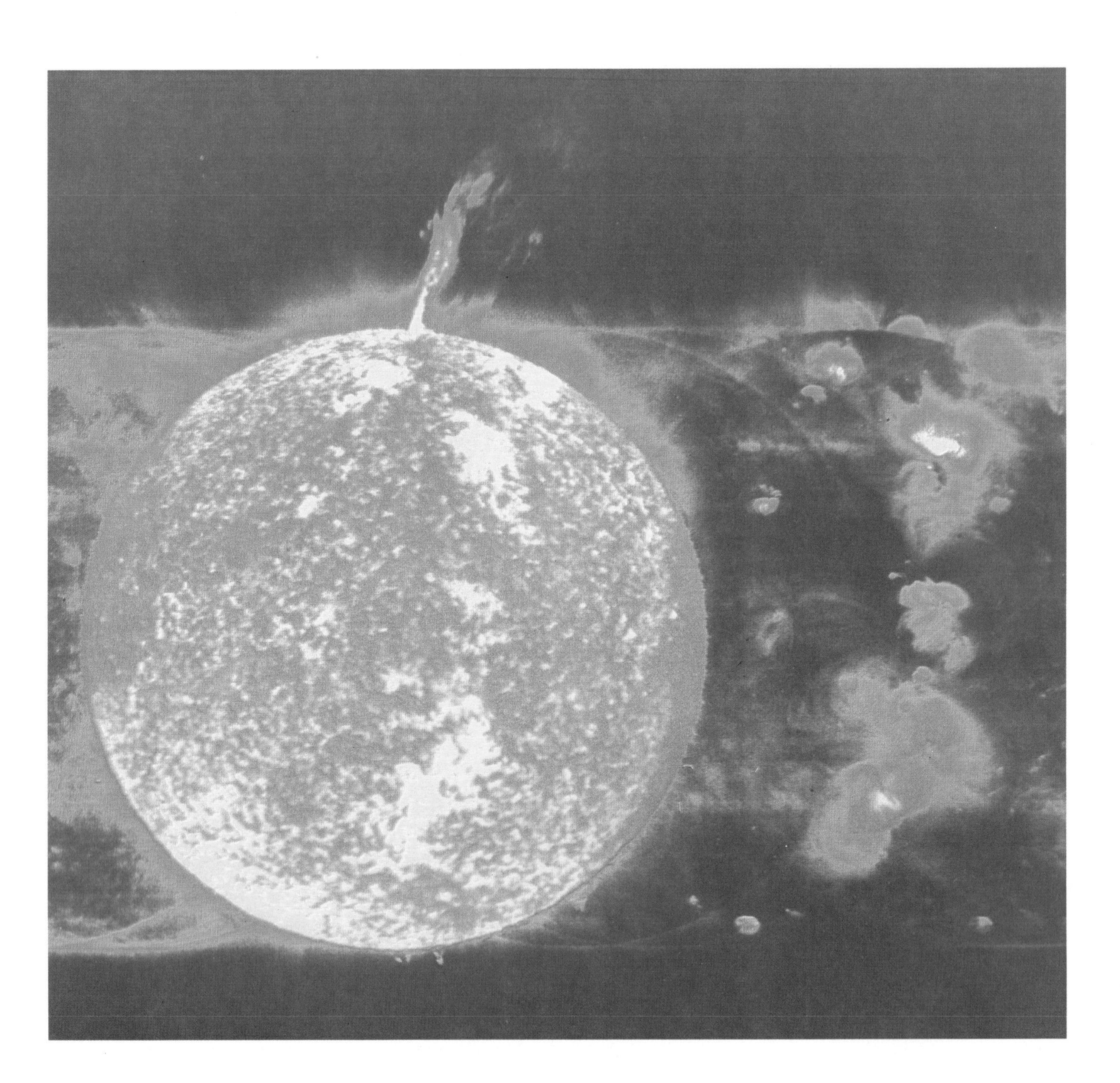

4

An Inconstant Sun: Electromagnetic Radiation

Bright star! Would I were steadfast as thou art— . . .

John Keats, "Sonnet XX"

Spectrally dispersed images of the sun obtained with an objective grating spectrograph aboard Skylab. With the spectroscope slit removed, overlapping images of the sun in every wavelength appear at the spectral line positions. Here the full disk appears in the 304-Å wavelength of ionized helium. Immediately to the right is an image in 14 times ionized iron, which is emitted only from localized coronal condensations.

The radiant flux of energy from the sun at the earth is called the solar constant. But is it really constant? Visible light and heat waves, which represent more than 99 percent of the sun's radiation, vary by less than 1 percent. Climatologists, however, believe that a change of even a fraction of 1 percent, sustained over a span of years, could have a significant impact on the earth's climate, and dedicated observers have sought evidence of any trace of variability in the solar constant for the past hundred years. By contrast, the tiny fraction of the solar constant delivered in short-wavelength ultraviolet and X rays, which is less than a few hundredths of 1 percent, varies enormously on every time scale. Because only the very short wavelengths can ionize the air and transform its chemistry, its significance for the upper atmosphere is out of all proportion to its energy content. But the nature of extreme ultraviolet and X rays was largely unknown before the advent of rockets and satellites that could carry detectors above the opaque, intervening air.

Charles Greely Abbot, secretary of the Smithsonian Institution from 1919 to 1944, dedicated almost his entire career to measuring the solar constant from the ground. In retrospect, all of his work can be summed up in the statement that no evidence exists of a change in the solar constant by as much as 1 percent, but his experiments were not sensitive enough to detect smaller changes. With far more sensitive instruments now carried aboard satellites, it has been confirmed that the solar constant does indeed vary, at times by more than 0.2 percent.

A successful NASA effort to monitor the solar constant was conducted on the Solar Maximum Mission (SMM), beginning February 14, 1980. The SMM carries the most advanced sensor developed thus far, the active-cavity radiometer designed by Richard C. Willson and his Jet Propulsion Laboratory colleagues, which has an absolute-measurement capability sensitive up to 0.001 percent for rapid variations and very good long-term absolute stability.

Results published through the end of 1981, summarizing nearly 23 months of continuous record, showed frequent excursions of 0.05 percent from average and several dips of up to 0.1 percent. Over the 18 months from February, 1980, to August, 1981, there was an average decrease of 0.1 percent, with the largest drop yet measured, 0.23 percent, in July, 1981. The correlation of dips with the blocking effect of sunspots persists throughout the observing period. In fact, the greatest dip coincided with the passage of a very large sunspot group that covered a thousandth of the sun's disk—a major blackout as sunspots go.

The new technology of Willson's radiometer does indeed seem to represent the long-dreamed-of capability to monitor the solar constant

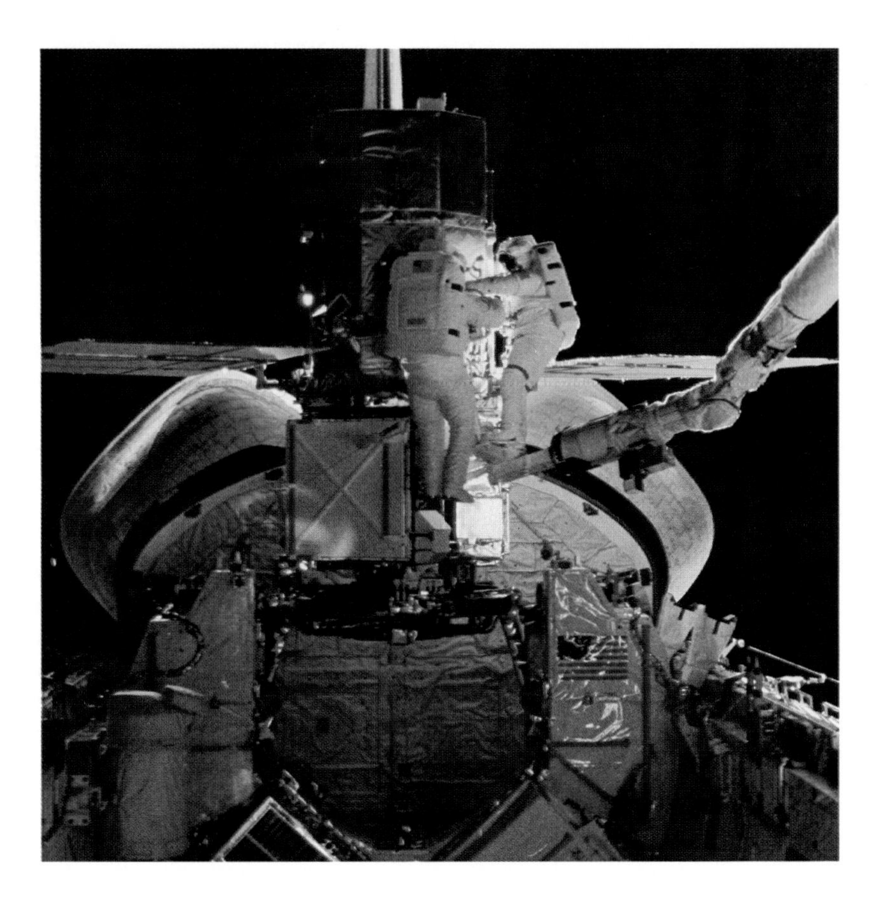

The Solar Maximum Mission spacecraft was designed to study solar flares and coronal transients. Ten months after launch, the attitude control system that pointed the satellite instruments to the sun failed. In 1984, NASA astronauts retrieved the SMM with the space shuttle, repaired it, and released it to resume its scientific program once again.

continuously at a level significant for climate effect. Stephen H. Schneider, of the National Center for Atmospheric Research, reminds us just how complex this connection is. Any effect of solar irradiance on climate must be delayed by the enormous heat capacity of the oceans, the different mixing rates in different parts of the oceans, and a host of land-sea and local influences. Schneider's computer models suggest that a 0.1 percent drop in solar irradiance could bring about a 0.1°C drop in average global temperature. In such an event, the terrestrial response would take five years to reach half that effect, and then perhaps another 100 years to complete the change. Schneider does concede, however, that if a change of 0.05 percent persisted for as long as 10 years, a significant climate signal could occur.

The hundred-year quest for solar-variability data has now entered the space age. Dedicated monitoring over the next few decades may

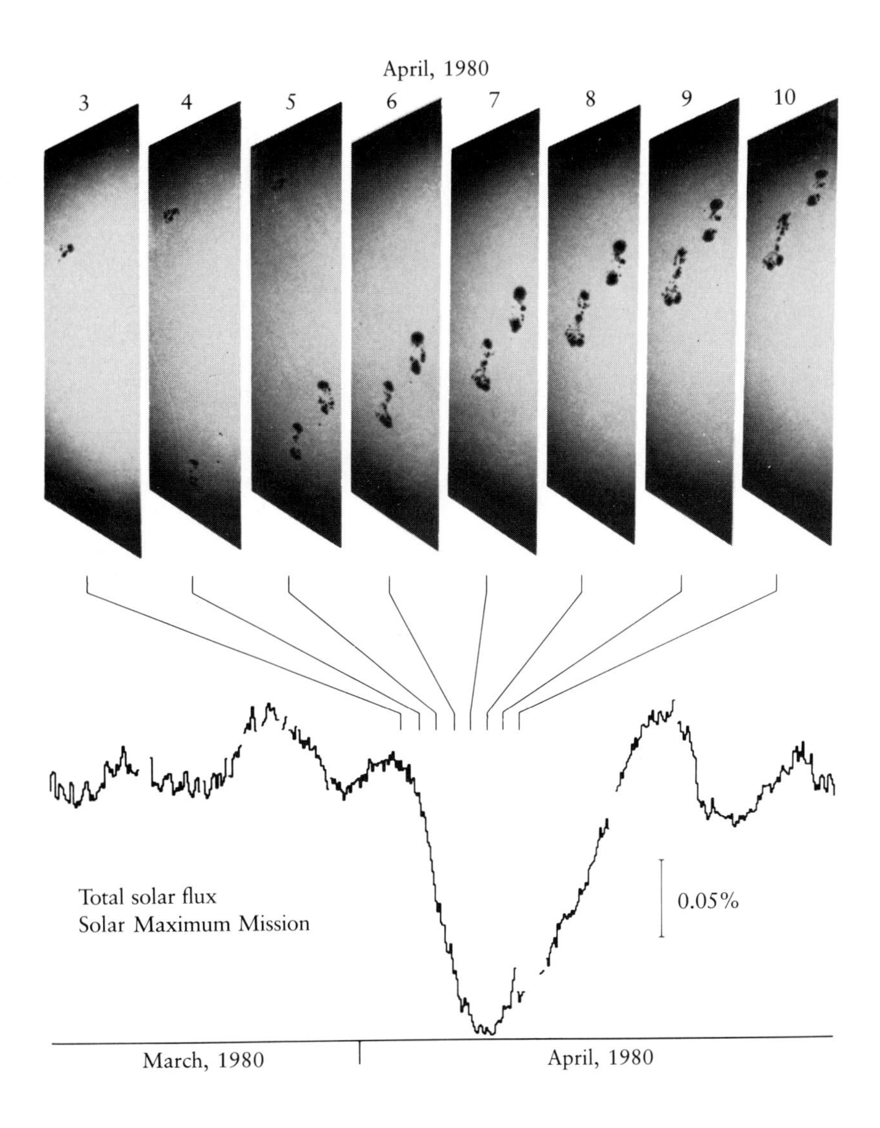

The variation of solar flux during March and April, 1980, measured from space by means of the cavity radiometer, developed by R. C. Willson at the Jet Propulsion Laboratory. The deep drop in early April coincided with the passage of a large sunspot group across the central meridian of the sun.

uncover some significant sun-climate relationships. Chapter 6 describes evidence that prolonged changes in solar insolation of only tenths of a percent have triggered catastrophic climate changes over the past million years.

The serenely steady visible sun gives little hint of its high variability and its explosive outpourings of invisible ultraviolet and X rays. These energetic radiations produce the ionosphere, and their stormy nature is reflected in dramatic effects on radio communications.

Early Probing of the Ionosphere

High overhead, a blanket of electrified plasma, the ionosphere, wraps around the earth. From dawn to nightfall, it waxes and wanes as the sun passes from horizon to horizon. When radio engineers discovered the ionosphere in the 1920s, the visible sun gave no clue as to the nature of its ionizing radiation. For 25 years, inferences about the sun's invisible ultraviolet and X rays were drawn primarily from observations of ionospheric behavior. Not until captured German V-2 rockets were brought to the United States after World War II was the mystery of how the sun produces and controls the ionosphere finally solved.

Radio Communication Beyond the Horizon

The early history of the ionosphere is closely linked to the work of Guglielmo Marconi. In 1895, when he was only 21, he built a demonstration wireless telegraph on his father's estate near Bologna, Italy. In 1899, after many experiments with short-range transmissions, he established the American Marconi Company, which later evolved into

Guilielmo Marconi, who received the Nobel Prize for his pioneering accomplishments in radio communication.

the Radio Corporation of America. On December 12, 1901, Marconi radiotelegraphed three Morse code dots for the letter *s* from Poldu, near Land's End, England, to a flimsy wireless station in Newfoundland, 1800 miles across the Atlantic. How did the waves get over the bulge of the ocean? The curvature of the earth amounted to a barrier 100 miles high, and radio waves, like all other forms of electromagnetic radiation, were expected to travel in straight lines, except for a small amount of bending, or diffraction, that was theoretically far too minute to curve the radio signals around the earth. A popular supposition at the time was that the ocean salt water acted as an electrical transmission line.

The successful transmission was wildly celebrated, and Marconi quickly became an international hero. Although engineers put little stock in the practical value of the demonstration, it stimulated many of the best scientific minds of the time to rush into print with theoretical models.

By March of 1902, a plausible explanation was advanced by Arthur E. Kennelly, a self-taught engineer and onetime assistant to Thomas Edison. His reputation in British electrical engineering established, he had moved to the Harvard School of Engineering about the time of Marconi's experiments. Kennelly conjectured that radio waves got around the curvature of the earth by reflection from an electrically conducting layer at a height of 50 miles. About a decade later, Lee de Forest, the inventor of the vacuum-tube triode amplifier, deduced a good measurement of the height of the reflecting layer—62 miles.

Almost simultaneously with Kennelly, Oliver Heaviside made a similar proposal in England. Kennelly's hypothesis was fleshed out in some detail; Heaviside, described by his colleagues as a brilliant misanthrope, put forward his idea in just three brief sentences:

> There may possibly be a sufficiently conducting layer in the upper air. If so, the waves will, so to speak, catch onto it more or less. Then the guidance will be by the sea on one side and the upper layer on the other.

Heaviside also wrote a detailed paper supporting his theory of an electrified high-altitude layer, but a referee for *The Electrician* rejected it. In 1912, W. H. Eccles, who was familiar with Heaviside's unpublished paper, went out of his way to credit it in his own sophisticated paper on the propagation of radio waves in a plasma. Eccles' acknowledgment served to attach Heaviside's name to the layer, while Kennelly's name is often forgotten.

The system of circulating currents in the high atmosphere that causes daily changes in the magnetic field. Currents flow along the contours indicated in the direction of the arrows. Ten thousand amperes flow between the contour lines so that, by day, the total current between the center and the edges of the current system is 60,000 amperes. The map is drawn as seen from the sun.

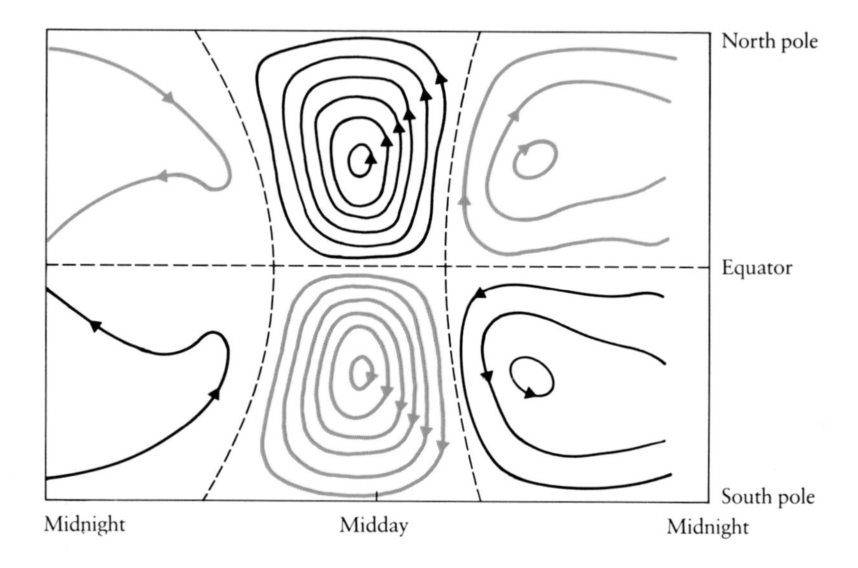

The compass needle oscillates as the earth rotates under the ionospheric current system.

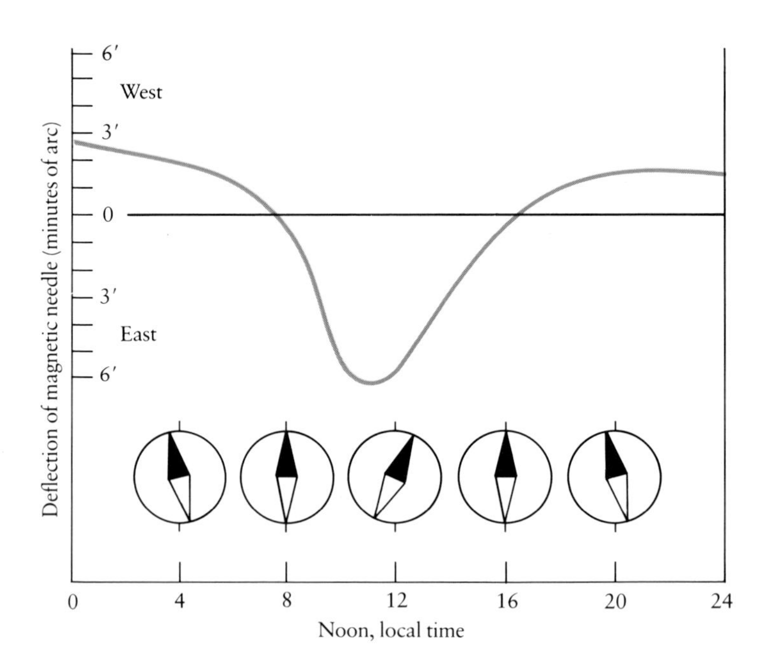

Rapid expansion of radio broadcasting in the early 1920s called attention to an erratic fading of signals over distances of about 100 kilometers. The phenomenon was quickly attributed to interference between two wave trains, one traveling along the ground (the ground wave) and the other reflected from the Kennelly-Heaviside layer (the sky wave). By 1924, Edward Appleton and his student Miles Barnett in England, Edward O. Hulburt and E. Hoyt Taylor of the U.S. Naval Research Laboratory, and Gregory Breit and Merle Tuve of the Carnegie Institution had each independently begun a series of investigations that led to a remarkably clear description of the nature of the ionosphere and its control by solar ionizing radiation.

To determine the height of reflection of a continuous wave, Appleton prevailed on the British Broadcasting Company to send a gradually varying wavelength from its London transmitter after the end of its broadcast day, while Appleton received the interference pattern of ground and sky waves at Oxford. With Barnett, he observed the elapsed time between emission and reception of the same frequency as the broadcast frequency was oscillated back and forth, and the two quickly published their demonstration of the "reflection of wireless waves from the upper atmosphere." Early experiments used frequencies that penetrated only the lowest portions of the reflecting layer. Later, as the frequency was increased, Appleton discovered a new reflecting layer above the Kennelly-Heaviside layer. By 1927, his experiments confirmed a lower layer as well as an upper layer bracketing the Heaviside layer. He labeled these three layers D, E, and F, in order of increasing height.

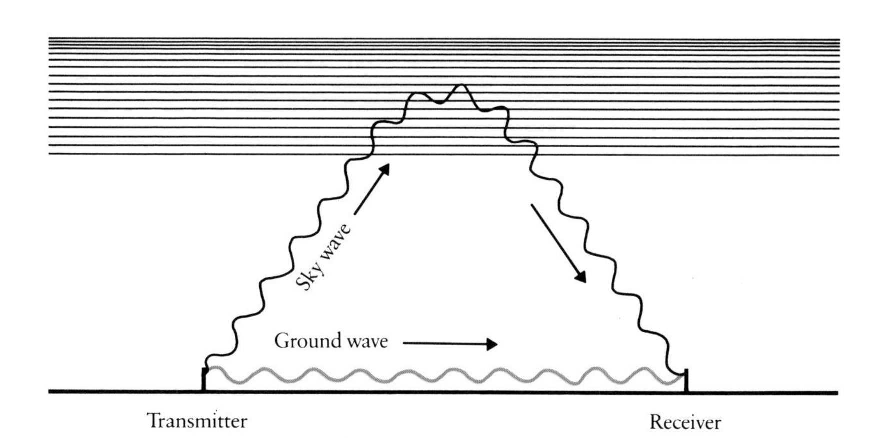

Shortwave radio signals travel a straight path along the ground and reflected paths via the ionosphere. Fading occurs when the ground wave and sky wave are out of step at the receiver.

Skip Distance

Much of the original research in the United States was based on the experiences of radio amateurs, who truly pioneered round-the-world shortwave (less than 50 meters) reception. When vacuum tubes became generally available after World War I, radio hams began to communicate around the world. Most amateurs could get tubes with power outputs of only two to 50 watts, and their transmissions were limited to the very short waves, less than 200 meters, which were thought to be almost useless for long-distance transmission. Still, the radio hams were having astonishing success with five-watt transmissions, communicating from hemisphere to hemisphere. They found that the short waves skipped over regions near the transmitter and returned to ground at greater distances. The close-in region, 20 or 30 miles in diameter around the transmitter, defined a zone of silence, the skip region. Beyond the skip distance, a zone of reception extended hundreds of miles.

In the early 1920s, Hulburt and Taylor, at the newly established Naval Research Laboratory (NRL) in Washington, D.C., organized a systematic study of radio communications with the help of radio amateurs. The skip distances for wavelengths of 16, 21, 32, and 40 meters were found to be about 1300, 700, 400, and 175 miles respectively, during daytime. Skip distances were greater at night than during the day, and greater in winter than in summer in mid-latitudes. The electrification of the Heaviside layer, expressed in number densities of

Radio waves are repeatedly reflected from the ionosphere as they propagate around the earth.

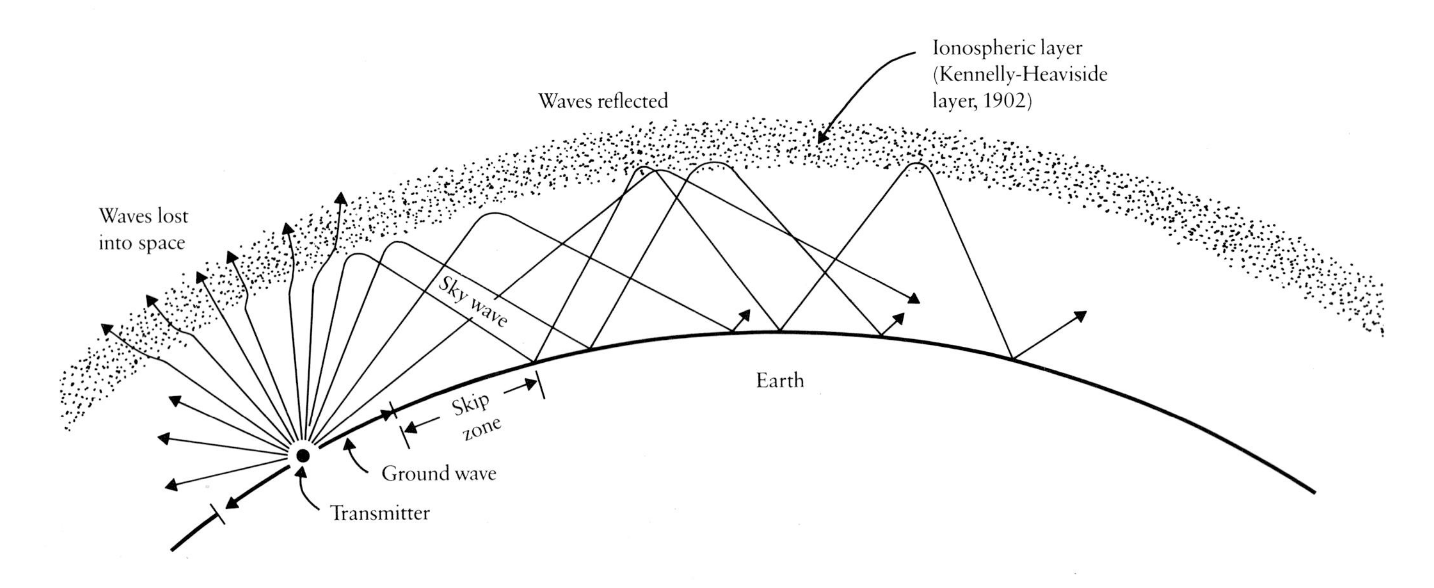

At the Naval Research Laboratory, Edward O. Hulburt pioneered studies of the height of the ionosphere and its diurnal variations by the method of skip distances of shortwave radio signals. His theoretical modeling of ionospheric production led to his suggestion of the importance of solar X rays in the formation of the E region.

Merle Tuve and his colleagues at the Carnegie Institution in Washington, D.C., developed the techniques of radio-pulse echo sounding of the height of the ionosphere.

electrons and ions, was calculated to be about 1 million per cubic inch at maximum, and the height of reflection was about 100 miles. From daytime maximum around 2 p.m., the ionization fell to about one-tenth that value in the small hours of the night. By 1926, Hulburt and Taylor had enough data to publish a remarkably accurate account of the diurnal ionospheric behavior as it relates to solar elevation. Hulburt was even led to speculate about the possibility of solar X rays as the source of ionization.

Although such study had fundamental scientific value, skip distances had no obvious value for communication until their function as a mode of long-distance communication at surprisingly low power was revealed. One hundred watts powered transmissions of several thousand miles; raised to kilowatts, radio signals could be picked up even after traveling twice around the earth. Navy researchers prepared charts of skip distances that could be used to select wavelengths for specific paths of communication. In preparing for his flight over the south pole on November 29, 1929, Richard E. Byrd had to find a way to maintain radio contact with the base at Little America. His plane had very little power available for communication, and a special program of shortwaves matched to skip distances over the course of the flight was planned. For the first 200 miles, the wavelength was 69 meters; over the next 200 miles, it was switched to 45 meters; the final 380 miles to the pole were covered on 34 meters. Except for a loss of contact at the 400-mile mark, radio communication was continuous.

The possibility of pulse-reflection experiments had certainly occurred to scientific investigators during the early 1920s, but the difficulties of measuring time between transmission and reception in milliseconds, with only delicate string galvanometers that had comparatively long restoring times, loomed large indeed.

In the winter of 1924–1925, Breit and Tuve, at Carnegie, collaborated with Hulburt and Taylor, who sent pulses at 4.2 megahertz by means of a 10-kilowatt shortwave NRL transmitter to the Carnegie laboratory seven miles away in northwest Washington. The Carnegie scientists devised a multivibrator keying circuit to pulse at full power for 10^{-4} second at 80 cycles per second. The receiver was next moved to NRL and placed close to the transmitter. There precise phase measurements, with a range sensitivity of about 20 meters in path length, were begun. This "ionosonde" was the forerunner of the basic instrumentation still used for the study of the ionosphere.

Flight operations from adjacent Bolling Field, however, continually harassed the experimenters. Detectable effects on echo phasing were observed long before the approach of an incoming plane could be heard, and the perturbations grew larger and larger until the plane landed nearby. Tuve described how, in 1929, the scientists aired their exasperation with the airplane problem frequently over lunch at NRL. A young Navy lieutenant, William S. Parsons (later Admiral Parsons, one of the key figures in the development of the atomic bombs that were dropped on Japan), happened to overhear their complaints. He subsequently suggested that these very aircraft problems might have useful applications in detecting airplanes approaching Navy ships or other ships approaching in fog or darkness. His proposal to the Bureau of Ships was classified "Secret," and thus the Navy began its development of radar.

With rapid advances in radio-pulse techniques in the 1920s, radio science came of age. Probes of the structure and diurnal variability of the ionosphere were accomplished with precision, more from a need to improve the reliability and efficiency of radio communication than to serve any basic understanding of natural processes. Scientific modeling of sun-ionosphere relationships had to await the rocket years that followed World War II.

Robert H. Goddard with his first successful rocket. Propelled by gasoline and liquid oxygen, the rocket flew on March 16, 1926, to a height of 41 feet and a distance of 184 feet.

The Sun's Ionizing Radiation

Early Rocket Astronomy

Scientists and engineers had envisioned rockets and satellites for probing the near reaches of space long before World War II spurred the great technological surge that made them a reality. In 1929, Robert H. Goddard pioneered upper-atmosphere research when he sent aloft a rocket carrying a barometer, a thermometer, and a camera to photograph their readings. His rockets may have been toys compared to those soon to follow, but Goddard established the principles of design for all future liquid-fueled rockets. How difficult Goddard's accomplishment was to comprehend is evident from the following quote from a 1921 *New York Times* editorial:

> That Professor Goddard, with his "chair" in Clark College and the countenancing of the Smithsonian Institution does not *know* the relation of action to reaction and of the need to have something better than a vacuum against which to react—to say that would be absurd. Of course he only seems to lack the knowledge ladled out daily in high schools. . . .

The German rocket-development program at Peenemünde was established in 1936. By 1939, Wernher von Braun and his team had perfected the V-1 rocket, and, before the war's end, England was blitzed by V-2s. When American forces entered Germany in 1945, they raced the Russians to Nordhausen, the huge underground V-2 rocket factory.

Nordhausen was a hellish place a mile underground, run by the slave labor of thousands of prisoners from France, Poland, Holland, and the Soviet Union. At the time of its capture, it had produced more than 3000 rockets, about 1000 of which had been launched toward Britain. The military goal was 50 rockets a day into the heart of London; Moscow was to receive similar treatment next. Although Nordhausen lay within the Soviet occupation zone, the U.S. Army got there first and made off with about a hundred rockets. Lashed to the decks of Liberty ships, the V-2s crossed the Atlantic and continued on flatbed railroad cars to the White Sands Missile Range in New Mexico, where they lay neglected on the desert, barely protected from sandstorms. It is altogether remarkable that more than 60 were restored to flight condition between 1946 and 1952. While the Army studied the V-2's propulsion technology, its warhead space was given over to the scientific community for high-altitude research. From the very beginning, studies of the sun carried a high priority.

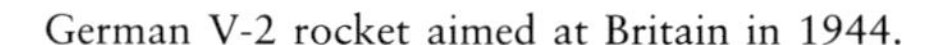
German V-2 rocket aimed at Britain in 1944.

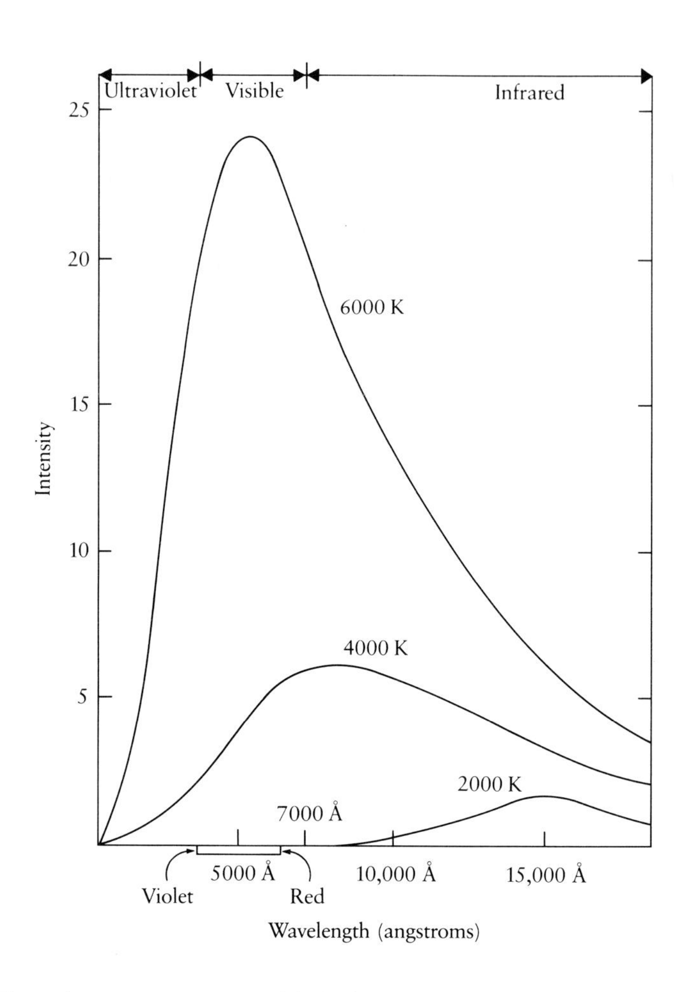

Black-body curves showing the spectral distribution of intensity at different temperatures. At 5800 K the sun radiates most of its energy in the visible range.

The solar spectrum in visible light resembles a 6000-K black body. (Black-body radiation has the same energy density in each wavelength range as the radiation from a perfectly absorbing heated body. Radiation in a state of thermal equilibrium is black body.) Its spectral intensity peaks in the yellow-green and falls off toward the long-wavelength infrared and the short-wavelength ultraviolet. No detectable X-ray emission would be expected. Early in the century, before the recognition of million-degree temperatures in the corona and before the advent of rockets for direct observations above the atmosphere, inferences about the sun's short-wavelength radiation had to be drawn from indirect evidence based on ionospheric behavior.

Nitrogen and oxygen molecules in the air can be ionized by extreme ultraviolet radiation or X rays, but not by visible light. Because the ionosphere develops as the sun rises and decays as the sun sets, it

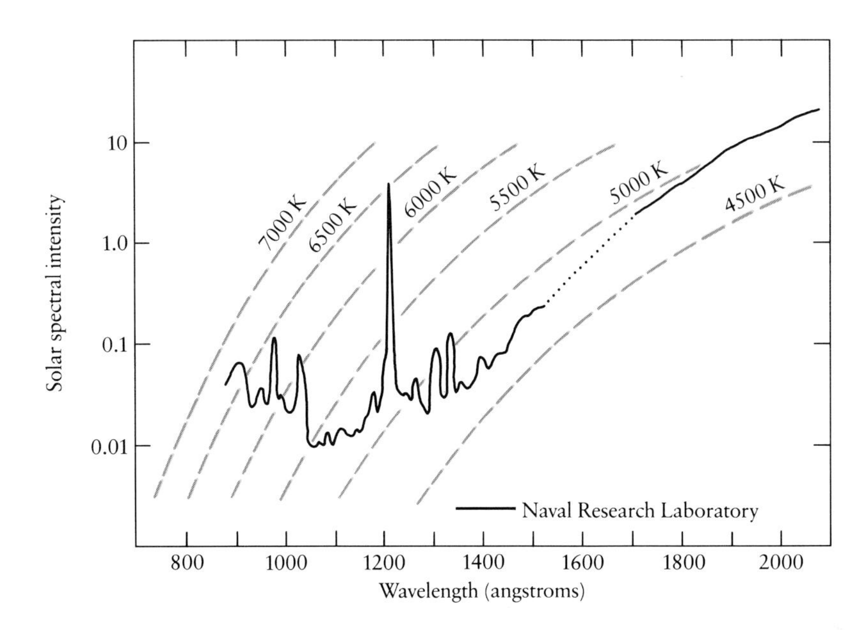

In the far ultraviolet, the solar continuum is weaker than that of a 6000-K black body. Dashed curves show black-body spectra at various temperatures. Strong line emission appears below 1300 Å and rises above the 6000-K curve.

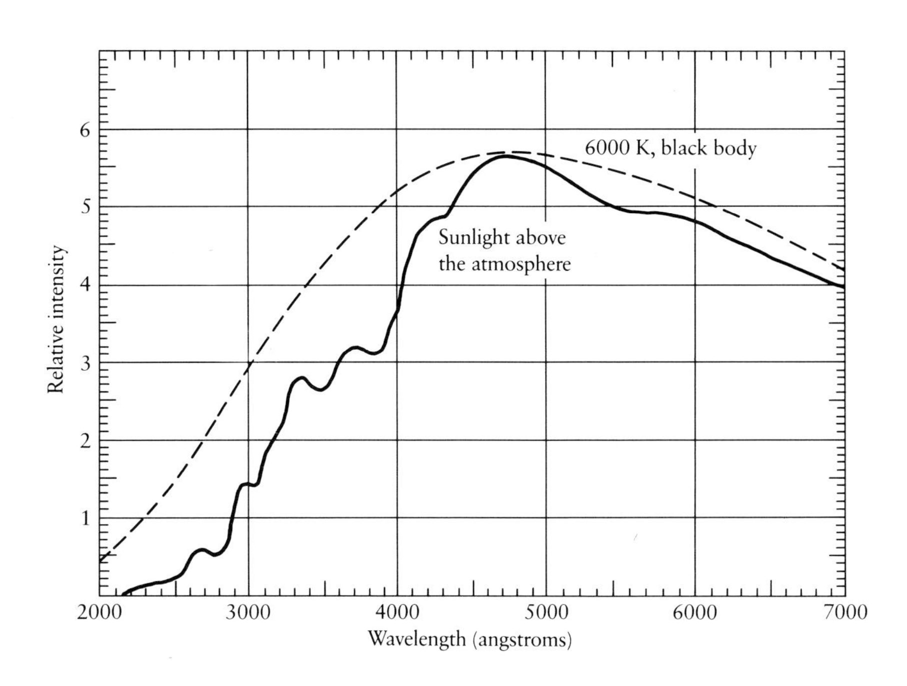

Energy distribution of the solar spectrum at 2000 to 7000 Å compared to that of a 6000-K black body. At wavelengths shorter than those of green light, the solar spectrum falls below the 6000-K black body.

must be produced by solar ultraviolet or X rays. Ultraviolet cannot ionize the principal constituents of the atmosphere unless its wavelength is less than about one-fifth that of visible light. If the solar spectrum retained a 6000-K black-body characteristic into the far ultraviolet, it could deliver enough ionizing radiation to produce a substantial ionosphere. But if the effective solar temperature were significantly lower in the ultraviolet, its flux at earth would be insufficient. Scientists in the 1940s knew that the corona could produce X rays, but it appeared so dilute that its radiant flux did not seem adequate to have a major effect on the ionosphere.

Before high-altitude rockets, scientists could observe the solar spectrum from the ground only to about 2900 angstroms, where ozone in the stratosphere begins to absorb strongly (an angstrom, Å, is 10^{-10} meter). A spectrograph carried on the balloon flight of Explorer II in 1935 to 22,000 feet failed to show any detectable solar ultraviolet below 2900 Å. An atmospheric transparency, or "window," sets in at about 2150 Å, the point at which ozone begins to transmit but before the atmosphere has become opaque because molecular oxygen (O_2) does not transmit. Attempts to observe in this band from 3500 meters high on the Jungfrau in Switzerland had also been inconclusive. It remained, then, for the V-2 rockets brought to White Sands in 1946 to make possible a breakthrough in knowledge of the invisible range of the solar spectrum.

Forty-five feet tall and five feet in diameter, the V-2 could carry a 1000-pound payload to about 90 miles. In those early days, we learned by hard experience how to design equipment that could survive rocket launches and reentry. The rocket carrying the first spectrograph above the ozone layer came screaming down in streamlined

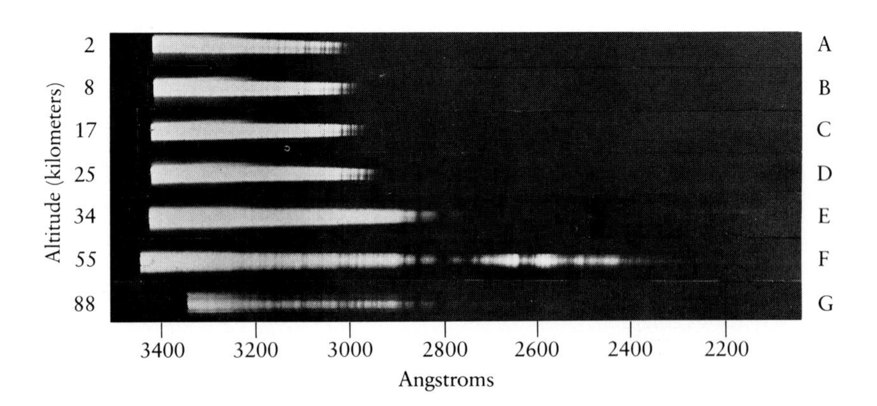

The solar spectrum photographed from a V-2 rocket at White Sands on October 10, 1946. With increasing altitude, ozone absorption decreased, and the spectrum extended farther into the ultraviolet. Strip G was too poorly exposed to give a good spectrum image.

A V-2 rocket ready for launch at White Sands, New Mexico, in 1948. A spectrograph is mounted in its nose cone.

flight and buried itself out of sight in the New Mexico desert. From the enormous crater, some 80 feet in diameter and 30 feet deep, only a bucketful of identifiable debris was recovered, as though the rocket had vanished almost completely on impact.

On the next attempt, our team from the Naval Research Laboratory mounted the spectrograph in the tail section, which was severed from the main body by an explosive charge on reentry. Like a falling kite, the tail fluttered down relatively gently at less than 300 feet per second to the floor of the desert, and the film cassettes were recovered undamaged. Success was finally achieved with a spectrograph mounted in the nose cone and separated by dynamite at 50 kilometers. Below 3000 Å, the spectral intensity rapidly declined until it resembled the emission of a 5000-K black body at 2000 Å. This temperature was so low that it ruled out the possibility that a thermal ultraviolet continuum of solar radiation could produce the ionosphere.

V-2 rocket after landing on the desert floor.

A V-2 rocket flight usually lasted about five minutes. To stabilize the craft in flight, its fins were canted, thereby inducing a slow roll about the long axis. A slit spectrograph mounted so that it looked out through the side of the spinning rocket would get very brief glimpses of the sun from this merry-go-round platform. Because the intensity of solar radiation was expected to fall rapidly with decreasing wavelength in the ultraviolet, a great deal of ingenuity went into the design of the spectrograph in order to maximize exposure to the sun. The conventional slits were replaced by ultraviolet transparent spheres of lithium fluoride, two millimeters in diameter. These beady eyes formed tiny solar images immediately behind them, and these served as source points of the radiation to be analyzed by a concave diffraction grating. Now the useful field of view became a wide cone, 140 degrees in diameter. A spectrograph of this design yielded the first extension of the solar spectrum below 3000 Å on October 10, 1946.

After the first flush of success, the invention of a stabilized platform took highest priority. The earliest versions took out the roll, but did not compensate for pitch and yaw. Even so, development of solar pointing systems progressed slowly, and spectrograph payloads proved disappointing for one reason or another—poor rocket performance, misbehavior of the pointing control, damage to film cassettes on impact, and so on. After this series of failures, it seemed wise to attempt some simpler, broadband measurements using newly developed detectors so sensitive to selected narrow wavelength intervals of

Diagram of the Naval Research Laboratory spectrograph flown in a V-2 rocket on October 10, 1946. Sunlight reaching the two lithium fluoride beads was focused to a point immediately behind each bead. The ray paths are shown from beads to mirror, to grating, and to film. As the rocket spun, the sun was alternately seen by one or the other bead.

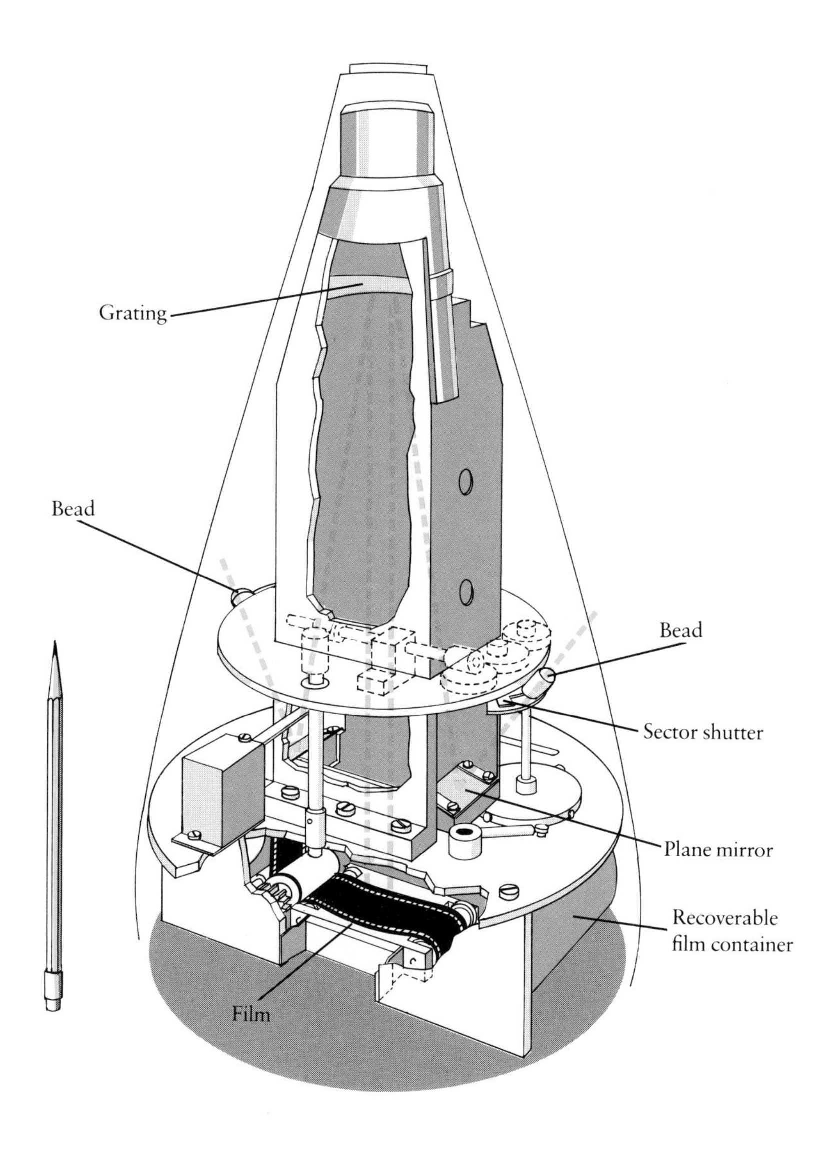

ultraviolet and X rays that no pointing control was necessary. A flight in 1949, carrying spectrally selective X-ray and ultraviolet Geiger counters was entirely successful, and the first quantitative indications were obtained of the intensity of the solar spectrum in the far ultraviolet and X-ray regions, along with the range of altitudes in which the various wavelengths are absorbed.

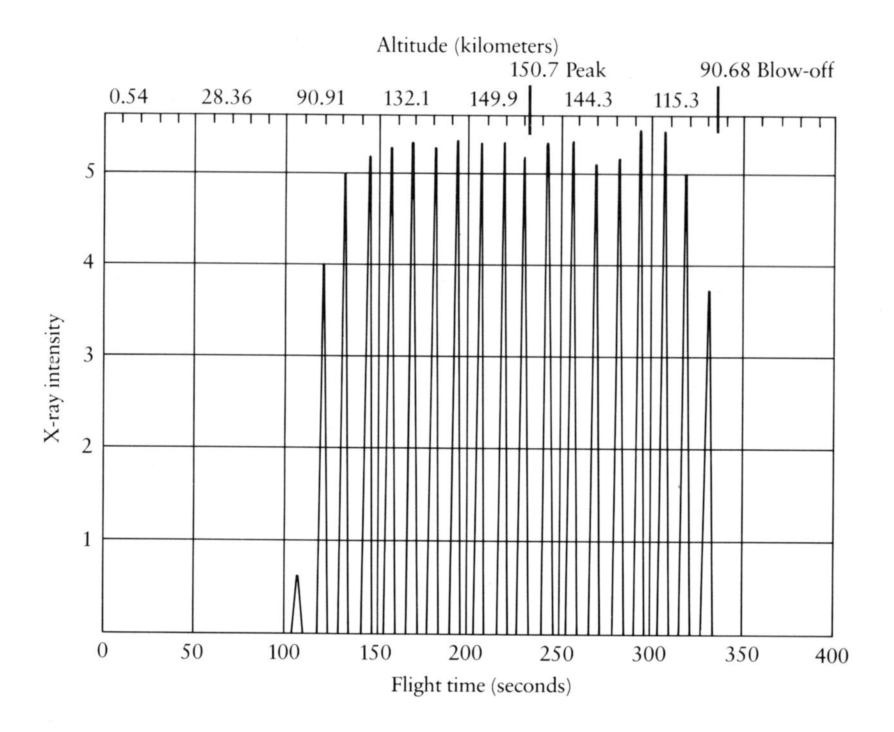

Telemetry record of solar X-ray flux obtained from a spinning V-2 rocket on September 29, 1949. Each signal spike is a measure of X-ray intensity at the indicated altitude.

We mounted an array of detectors so that they looked out from the skin of the rocket. The rocket climbed and spun silently until it reached 45 miles. Then the radio telemeter registered signals from a detector sensitive to the ultraviolet resonance line of hydrogen (1216 Å) known as Lyman-alpha. By far the strongest line in the solar spectrum, it is undoubtedly the most pervasive wavelength radiated throughout the universe. Although Lyman-alpha cannot ionize the oxygen and nitrogen that make up the bulk of our atmosphere, it can ionize a trace of nitric oxide smog that exists above 45 miles. So potent is the interaction between Lyman-alpha and nitric oxide that it alone can produce nearly all of the lowest region of the ionosphere. Ten miles higher, the rocket radioed back another band of ultraviolet, known as the Schumann region. This ultraviolet can dissociate oxygen molecules into their constituent atoms, and the breakup has a profound effect on the altitude profile of the ionosphere. At almost the same height, strong X-ray signals were detected. It thus became clear that X rays play an important role in producing the portion of the ionosphere that reflects broadcast-band radio waves.

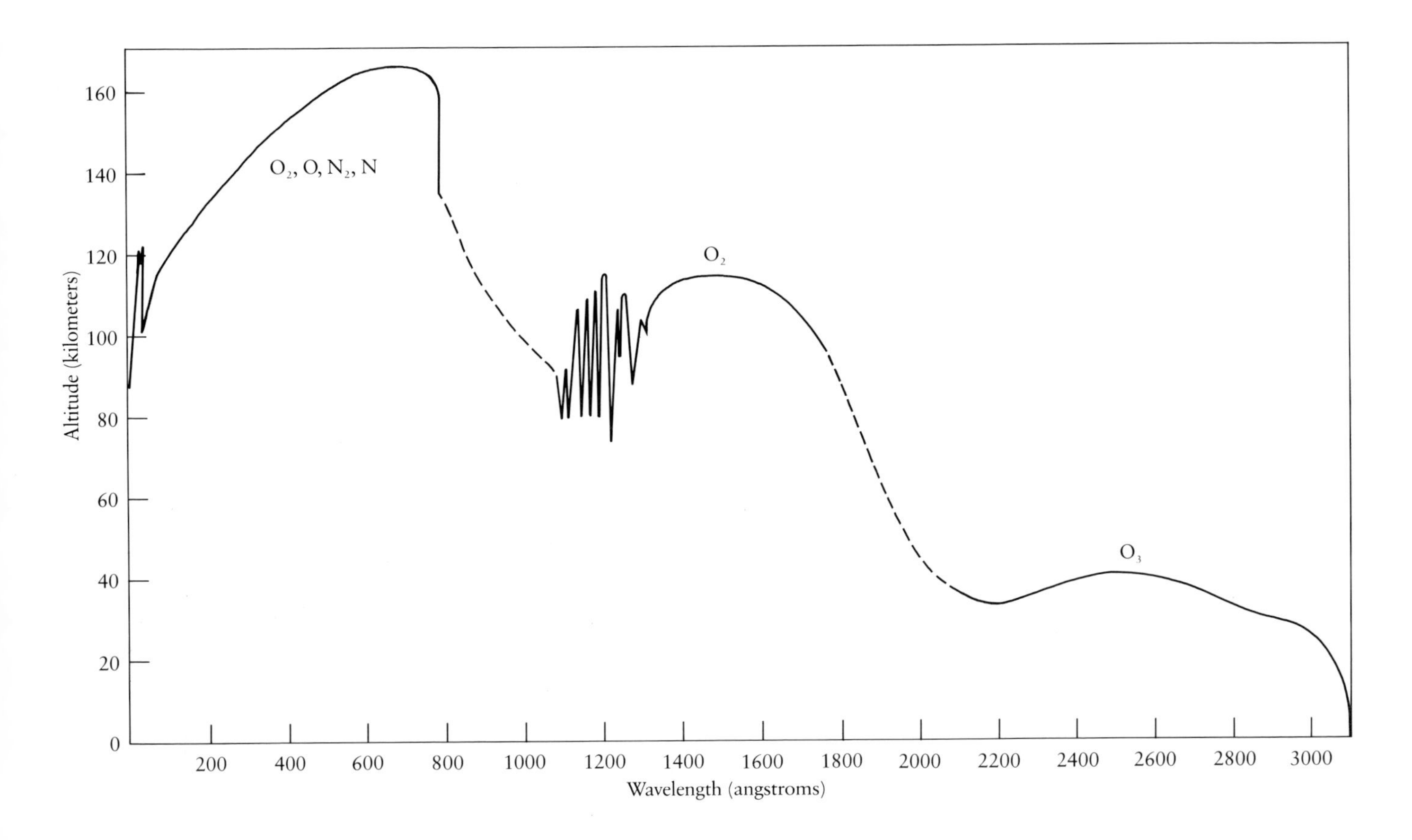

Altitudes at which solar radiation is attenuated to 1/*e* of its intensity above the atmosphere. (The logarithmic base *e* equals 2.7.) Ozone (O_3) becomes opaque below 3000 Å. Transmission improves slightly near 2100 Å, after which molecular oxygen is the principal absorber. Molecular and atomic nitrogen and oxygen are strong absorbers below 1000 Å.

This beginners' luck was immediately followed by a series of ignominious failures. Spurred by the success of the first photometry observations, we planned an elaborate array of detectors that would follow the interaction of a wider sample of solar X-ray and ultraviolet radiations with the upper atmosphere. The V-2 rocket rose about one mile, laid over horizontally, and headed for an inglorious crash. A year later, we tried again. This time, the V-2 climbed only inches and "walked" toward the blockhouse. Directly in front of our window slits, the huge rocket fell over and exploded. When the flames died out and the smoke cleared, nothing remained but a charred carcass. The V-2 rockets had been an unpredictable gift. In addition to those that exploded on ignition, some somersaulted end over end; one landed on the edge of Juarez, just over the Mexican border, and another inside White Sands National Park.

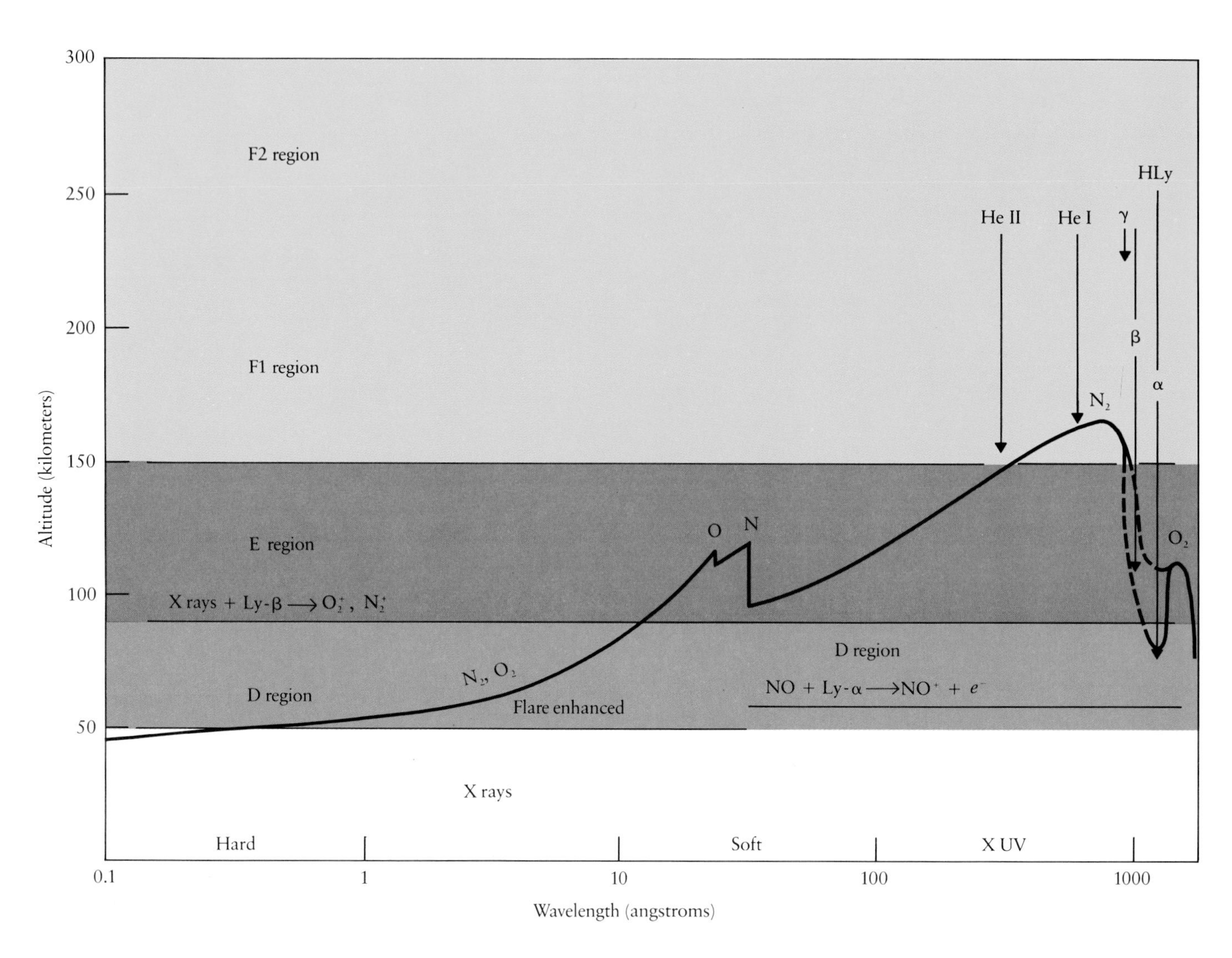

Penetration of solar ionizing radiation into the earth's atmosphere. The curve shows the level of attenuation to $1/e$. Molecular and atomic constituents are indicated in their spectral absorption ranges. The normal D region is produced by hydrogen Lyman-alpha ionization of a trace amount of nitric oxide. Flare X rays ionize N_2 and O_2 down to the base of the D region. The E region is produced primarily by the action of X rays and ultraviolet on oxygen and nitrogen. The F1 and F2 regions result from ionization of nitrogen and oxygen by extreme ultraviolet rays.

Second-Generation Rockets of the Fifties

In 1952, the Viking rocket, with many technical innovations such as a gimbaled motor that could fishtail to guide the rocket and spin jets that stabilized it against roll, was groomed to replace the V-2. We were assigned the precious space in the nose cone for our radiation measurements, and we lined up 60 detectors that were to be pointed at the sun steadily throughout flight. Immediately after launch, something went wrong—the rocket turned 90 degrees away and flew to 70 miles with all our detectors looking at nothing but empty sky. Then it started to roll so rapidly that the sun passed through the field of view in a blur and centrifugal force ripped the instrumentation from its mounting. Recuperating from surgery at flight time, I opened a copy of the *New York Times* the following day to read the headline "Viking Rocket Sets New Altitude Record"—not a word in the account about our dreadful disappointment.

While such traumatic experiences were more frequent than we care to recall, the technology developed rapidly. It was clear that we needed a rocket more specifically tailored for scientific research, and the Aerobee became that rocket. While at the Johns Hopkins University Applied Physics Laboratory, James Van Allen had specified its basic characteristics. Developed with Navy support, the small rocket was inexpensive enough to assign it completely to a single experimenter and powerful enough to carry a 150-pound payload to ionospheric heights.

To aim spectrographs at the sun with high accuracy from a spinning rocket, stabilized platforms now had to be developed. With support from the Air Force Cambridge Research Laboratories, the University of Colorado designed a sophisticated biaxial pointing control that corrected for pitch and yaw as well as spin. This sun-seeker was one of the most complex robots developed in the 1950s. No matter how the rocket heeled and spun, it held the spectrograph pointed at

The recovery of an Aerobee nose cone and instrumentation section that parachuted to the ground after severance from the propulsion section.

Herbert Friedman adjusts the sun-seeking "eye" of a solar camera in the nose of an Aerobee rocket. American scientists conducted most of their solar research with such rockets throughout the 1950s.

the sun. At launch, the optical platform and its instrument were locked inside the rocket's nose cone. Once above the sensible atmosphere, the spar carrying the spectrograph popped out of the rocket nose, and its photoelectric eyes searched for the sun. With the sun captured in its sights, the control system never relinquished its target. As Newman Bumstead put it, "Even with his horse bucking 60 degrees back and forth while pinwheeling around 90 times a minute, the sharpshooter holds his gun on target and scores seven hits in 10 shots at a four-inch bull's eye 100 yards away." This device made it possible to increase exposure to the sun greatly and to observe the shortest wavelengths of the solar spectrum in spite of rapid intensity falloff. Below 1750 Å in the far ultraviolet, Fraunhofer lines disappeared, to be replaced by strong emission lines all the way to X rays against a fading continuum.

The many rocket measurements made over the course of a full sunspot cycle showed a clear connection between the intensity of the more energetic X rays, the number of sunspots, and the amount of electricity in the ionosphere. In 1953 and 1954, when very few spots were present on the sun, X-ray intensity decreased. The corona appeared to have cooled to somewhat less than a million degrees. At the same time, the ionosphere became weaker. In 1957, when the spot numbers approached a maximum, X-ray emission increased sevenfold in intensity; at the more energetic short wavelengths, emission grew 50 to 500 times brighter. In these shortest wavelengths, the corona appeared to be hotter than 2 million degrees. Indeed, the steady sun, as seen by the rocket's X-ray eyes, behaved like a highly variable star. As X-ray and ultraviolet intensity increased with sunspot numbers, so did the density of the ionospheric plasma. In October of 1957, the ionospheric mirror was so strong that BBC television was received in the United States; in November, it was picked up as far away as Australia.

X-Ray Sources in the Corona

Radio scientists, who probe the ionosphere constantly with pulsed radio signals, find a 27-day periodicity in the effects they observe. From this we can infer that the solar regions producing X rays and ultraviolet radiation remain fixed in the solar atmosphere and turn with it through more than one solar rotation.

Observations taken during solar eclipses strongly suggest that ionizing radiation is produced above sunspots and plages. If such radiation were emitted uniformly over the whole face of the sun, an eclipse would cause a smooth decline in the ionospheric electron density to minimum value at totality, and a steady recovery to normal, just as visible light first smoothly fails and then smoothly returns. Instead, careful ionospheric measurements usually indicate rather abrupt deviations, as though much ionizing radiation comes from small, local regions that are suddenly covered and equally suddenly uncovered.

Back in 1958, when wide-field-of-view detectors were the only instruments available, a total solar eclipse offered the best opportunity to observe the distribution of X-ray sources on the sun. As part of the International Geophysical Year (IGY) program, a team of ground-based astronomers, led by John Evans, director of the Sacramento Peak Observatory, and a group of rocket astronomers from the Naval Research Laboratory, under my direction, combined to observe the October 12 eclipse in the South Pacific. On January 24, 1925, when I was nine years old, I had made my first expedition to witness a solar eclipse, traveling 13 miles by subway from Brooklyn to upper Manhatten to see the edge of totality cross 96th Street. My second eclipse expedition, 33 years later, took me 9000 miles by air and naval vessel.

The eclipse was to begin at sunrise on the equator near New Guinea, race across the Pacific Ocean for about 8500 miles to the coast of Chile near Valparaiso, then leave the earth at sunset. It would take three hours and 10 minutes to cross the Pacific at an average speed of 2670 miles per hour. In all its long path, which was only 150 miles wide at its widest, the eclipse would be observable on land only from a small number of tiny coral atolls. Evans' group chose the Danger Islands—three small islets, Puka Puka, Motu Ko, and Motu Katava, and a few sandbanks around a barrier reef enclosing a lagoon—on which to set up their spectrographs.

Nike-Asp rocket launched from the deck of the U.S.S. *Point Defiance* at the onset of totality during the eclipse of October 10, 1958, at the Danger Islands.

To support the whole expedition, the Navy provided the services of the U.S.S. *Point Defiance*, an LSD (landing ship dock), which made a remarkable base of operations. The helicopter flight deck at the stern provided a platform for the six rockets to be fired in sequence as the eclipse progressed and still left room for two helicopters that shuttled back and forth between ship and shore. The montage of South Sea Islands, Polynesian natives, naval warship, and modern rockets made for a rather new look in the traditional high adventure of eclipse expeditions.

When our huge vessel arrived off Puka Puka, it could not put in close to shore—the coral barrier completely ringed the atoll. The entire population of 750 natives, who had never seen anything more mechanical than a bicycle or larger than a copra schooner, thronged on the shore, gaping first at the great leviathan and then at the spectacle of the "whirlybird" bringing in the first visitors.

On the eclipse day, everybody aboard ship was up at dawn. The sky had been clear for 10 days, but this morning was gray and overcast. Although clouds would not threaten the mission of the high-flying rockets, prospects looked dim for the land-based astronomers on Puka Puka. Out at sea, our slim 27-foot rockets shot like arrows to the sky in sequence. When the smoke had cleared, the campaign was clearly a success. As our thoughts turned to our colleagues on Puka Puka, we picked up their sad story on radio. Clouds had completely obscured their observations. A year's preparation before embarking, a journey of 6000 miles from San Diego, 50 days of effort on Puka Puka—all had come to naught. Forty miles at sea, we had enjoyed a clear view of the spectacle, a bonus not at all essential to our mission.

Quickly we read the results of the rocket observations from the telemetry records. X rays came from the corona; even at the midpoint

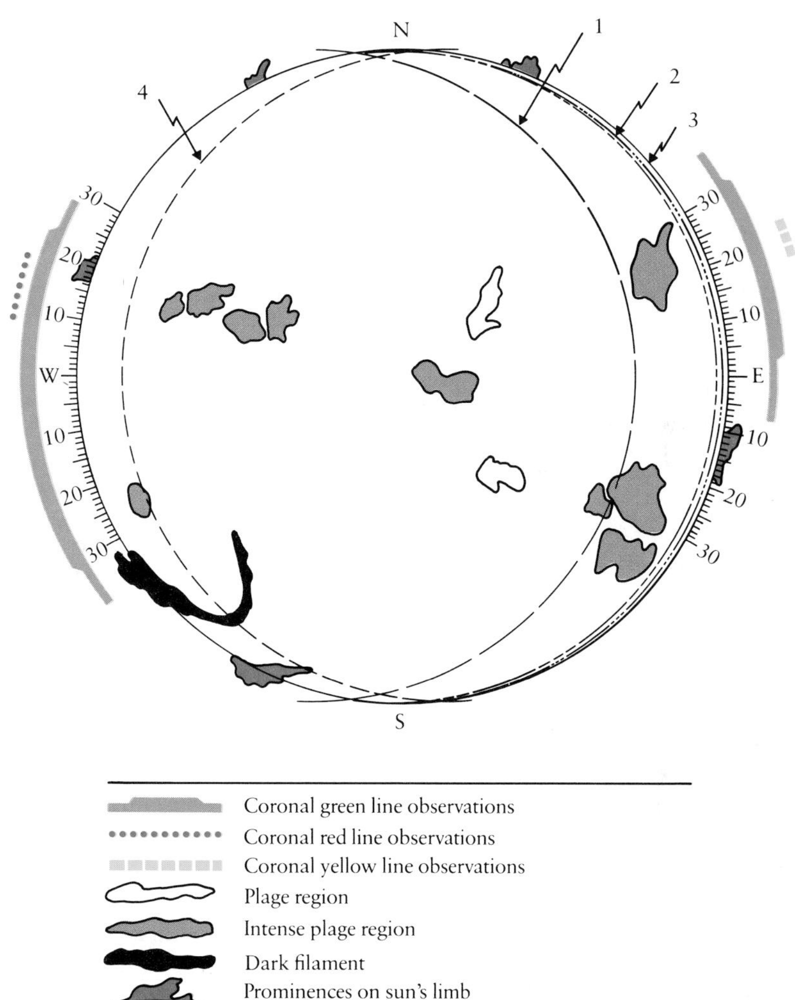

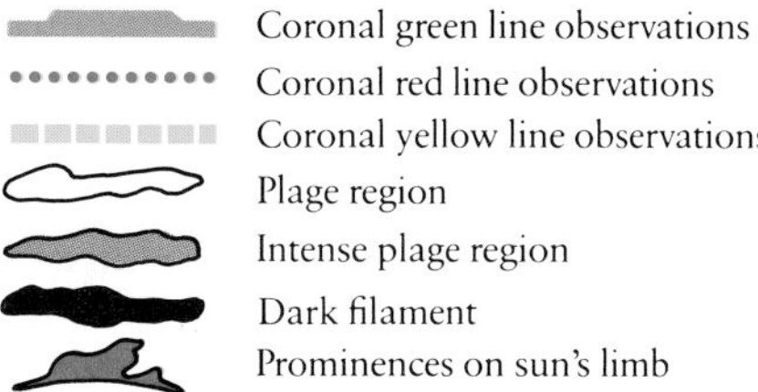

Solar activity map on October 12, 1958. Rockets 1, 2, and 3 were fired at different times to observe the masking of plage areas on the east limb. Rocket 4 was fired just after exposure of the west limb. The east-limb area containing the concentration of plages was six times as bright in X rays as the west limb.

of totality, 13 percent remained unobscured. In contrast, the ultraviolet rays were completely eclipsed, proving that they originated in the chromosphere. Furthermore, as the moon intercepted individual sunspot areas, the X-ray flux diminished abruptly, establishing that the X-ray sources were concentrated over sunspot groups. Because X-ray emission correlated exactly with hot condensations of the corona directly above the sunspots, it became quite clear why such emissions did follow the sunspot cycle.

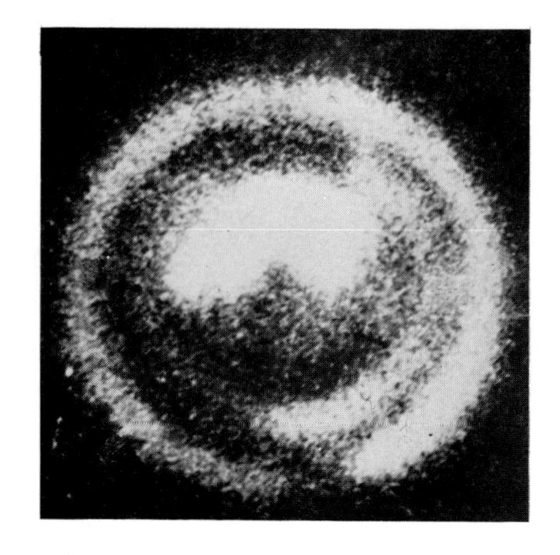

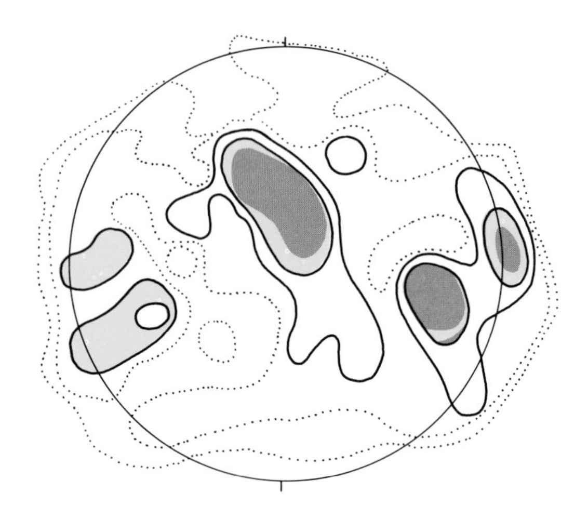

The X-ray and radio sun on April 19, 1960. The X-ray photo (top left) was taken with a pinhole camera aboard an Aerobee rocket. Visible light was excluded by an aluminum filter. The two-axis, stabilized camera mount did not remove rotation about the pointing direction to the sun, and localized sources are smeared into arcs by a 60-degree rotation during exposure. The 9.1-centimeter radioheliograph recorded at Stanford University (bottom left) shows strong concentrated regions of emission. The rotated radioheliograph (right) gives an image that closely resembles the X-ray photo.

The next few years brought rapid progress in X-ray studies of the sun. In 1960, we obtained a crude X-ray photograph of the sun with a simple pinhole camera carried on an Aerobee rocket. Under the headline "Sun Photo Taken by Shoe Box," the Paris edition of the *Herald Tribune* reminded readers of their childhood cameras made with a pinhole in one end of a shoe box and a piece of tissue paper across the other that showed the image. The two-axis pointing control could not correct for the rotation of the camera and the image was blurred, but its significance was clear. At least 80 percent of the X-ray emission came from no more than 5 percent of the area of the disk. After 1960, high-resolution X-ray telescopes were developed and flown on rockets, and the X-ray photographs taken by Skylab during the 1970s provided as much detail as optical images. X-ray telescopes today compare favorably in aperture and resolution with optical telescopes flown on space observatories.

Explosive Flaring

All solar variability related to plages and sunspots concerns what we may call a quiet sun. Such phenomena are insignificant compared to the explosive outburst and terrestrial impact of a solar flare. Even relatively frequent small flares expend an amount of energy equivalent to millions of hydrogen bombs. The largest and rarest flares have the power of over a billion hydrogen bombs exploded during the course of a few hours. This tremendous energy is exported by a brilliant burst of

Ionospheric Layering

The newly acquired knowledge of the solar X-ray and ultraviolet spectrum made it possible to create models of the structure and solar control of the ionosphere. Appleton's D, E, and F layers could now be understood in terms of the wavelength distribution of incoming radiation, the spectral absorption characteristics of atmospheric gas, and recombination processes that consume free electrons.

Ionizing radiation from the sun in the ultraviolet and X-ray parts of the spectrum penetrates the top of the atmosphere, where the gas is very thin, with very little attenuation. As gas density increases toward earth, the rate of ionization increases. Eventually, the rate of absorption of incoming energy is matched by the rate of increase of gas density. At this level, the production of electrons maximizes. Further down, the penetrating remnant

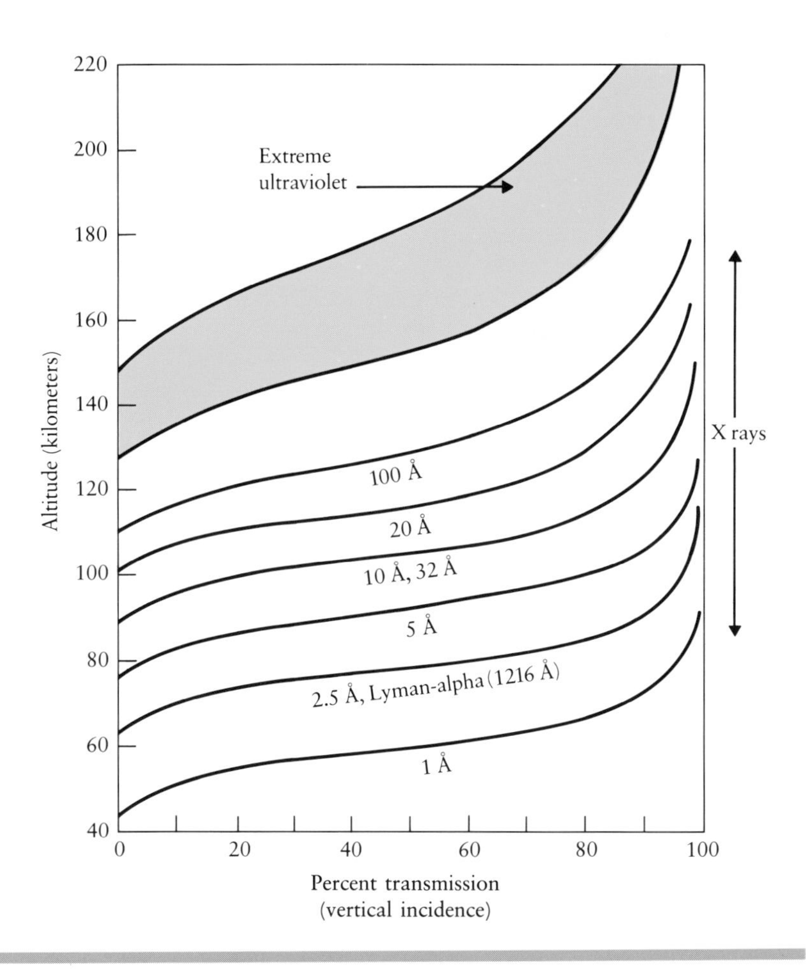

Penetration of the atmosphere by solar X rays and ultraviolet radiation. Hydrogen Lyman-alpha has the same characteristic absorption as 2.5-angstrom X rays.

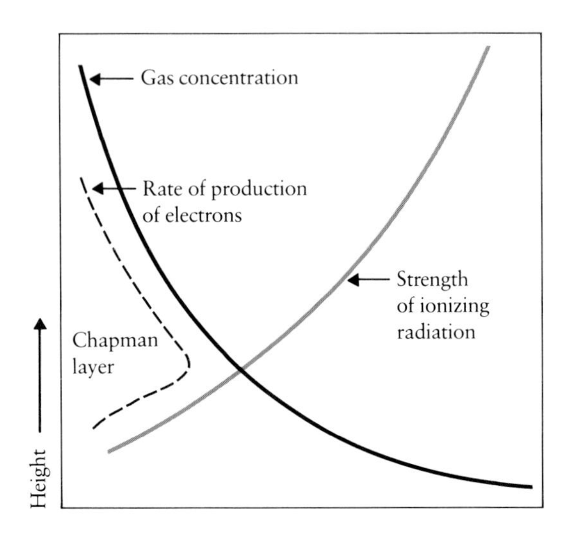

The production of a Chapman layer of electrons when ionizing radiation enters the atmosphere.

of incoming radiation quickly vanishes. This "layering" was first theoretically described in detail by Sidney Chapman in 1931, and such structures have since been called Chapman layers. They form deepest when the sun is overhead, and decrease in density as the angle between the direction of the radiation and the vertical increases.

Light of any given wavelength has a unique absorption coefficient in atmospheric gas and produces its unique Chapman layer. A broad spectrum of continuous radiation is absorbed continuously over a correspondingly broad height, and each Chapman layer spreads out over the same range. Characteristic D, E, and F regions overlap but retain their identities. Except for the resonance line of hydrogen, Lyman-alpha, the spectrum of ionizing radiation begins at about 1000 Å. Lyman-alpha penetrates to the D region, where it can ionize the trace constituent, nitric oxide. The rate of electron loss at each altitude is equally important in establishing the ionization distribution with height. The loss processes may involve the recombination of electrons and ions and the attachment of electrons to neutral atoms or molecules to form negative ions.

Free electrons in the ionosphere act as radio relay stations. They oscillate in tune to the radio frequency of passing waves and rebroadcast the same frequency. If an electron collides with

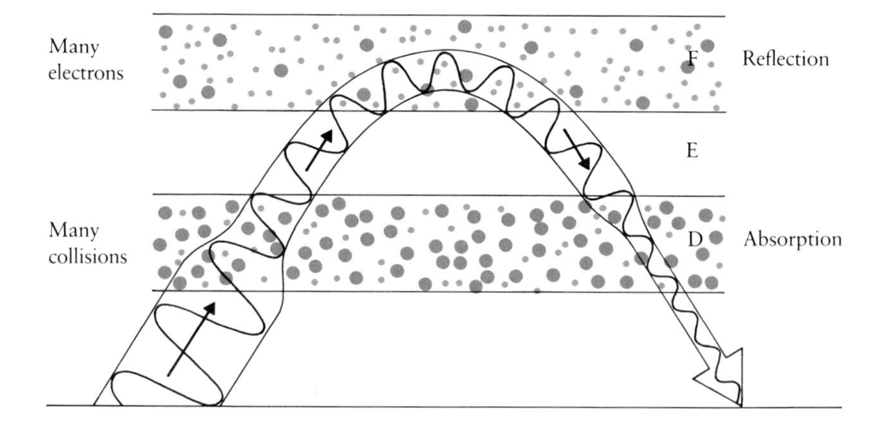

When there are sufficient electrons high up, radio waves are reflected; when there are too many electrons low down, they are absorbed. Red dots represent atoms; green dots represent electrons. The radio frequency up and down remains unchanged.

a neutral gas particle, its oscillation is damped and the energy that was picked up from the radio wave passes over to the gas. In the E and F regions where the gas density is lower, frequency of collision and absorption of radio-wave energy are lower. Although there are fewer and fewer electrons at deeper levels, the concentration of neutral particles increases very rapidly. Hence the collision rate and radiowave absorption maximize in the D region.

Competition between their increasing reflection at high levels and their increasing absorption in the D region as the sun rises overhead governs the strength of radio reflections from the E and F regions. As radio-wave frequency increases, the concentration of electrons necessary for reflection increases and absorption decreases. If the frequency is too low, absorption dominates. All that remains for effective radio communication is a relatively narrow band of frequencies between the limits of absorption and reflection.

Day-to-day analysis of the reflecting and absorbing characteristics of the ionosphere is carried out with ionosondes, devices that transmit a short pulse upward and receive its reflection at the same place. The frequency of the radio pulse is continuously varied over a few minutes, and the time delay of the echo is automatically recorded on an oscilloscope photograph, an ionogram.

Ionograms reveal what appears to be a layered structure of ionization. For the maximum electron density in a layer, there is a corresponding vertical frequency at which reflection disappears. At medium frequencies (0.3 to three megahertz), reflection is from the E region at heights of 100 to 150 kilometers. At higher frequencies (three to 30 megahertz), reflection is from the F region up to heights of 250 kilometers or more. Waves in the range 30 to 300 megahertz undergo little reflection and escape into space. Waves in the low-frequency (30 to 300 kilohertz) and medium-frequency bands may be totally absorbed in the lower

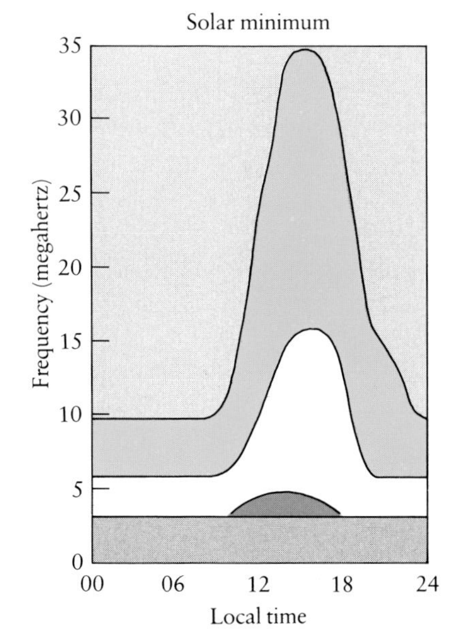

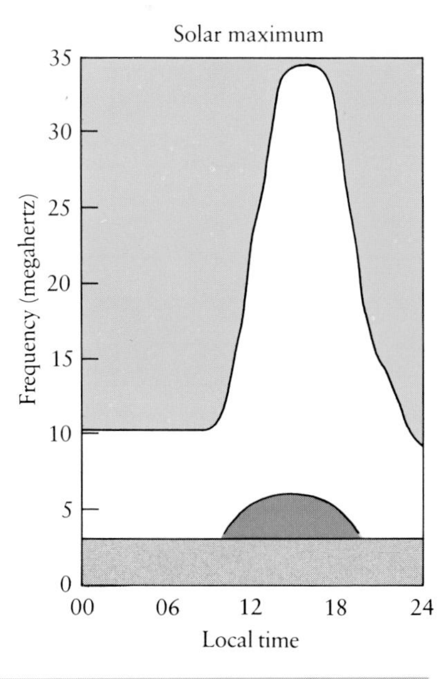

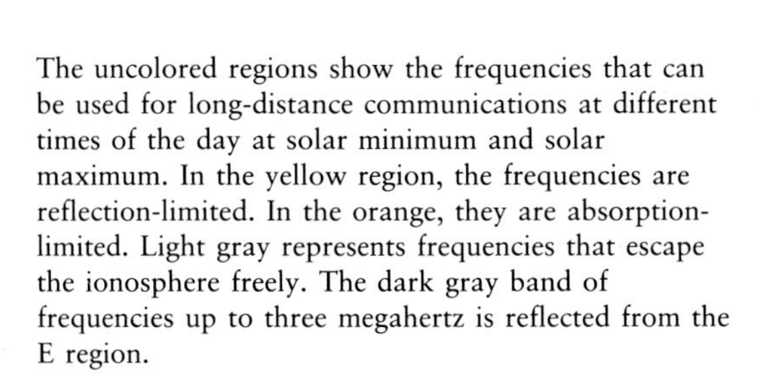
The uncolored regions show the frequencies that can be used for long-distance communications at different times of the day at solar minimum and solar maximum. In the yellow region, the frequencies are reflection-limited. In the orange, they are absorption-limited. Light gray represents frequencies that escape the ionosphere freely. The dark gray band of frequencies up to three megahertz is reflected from the E region.

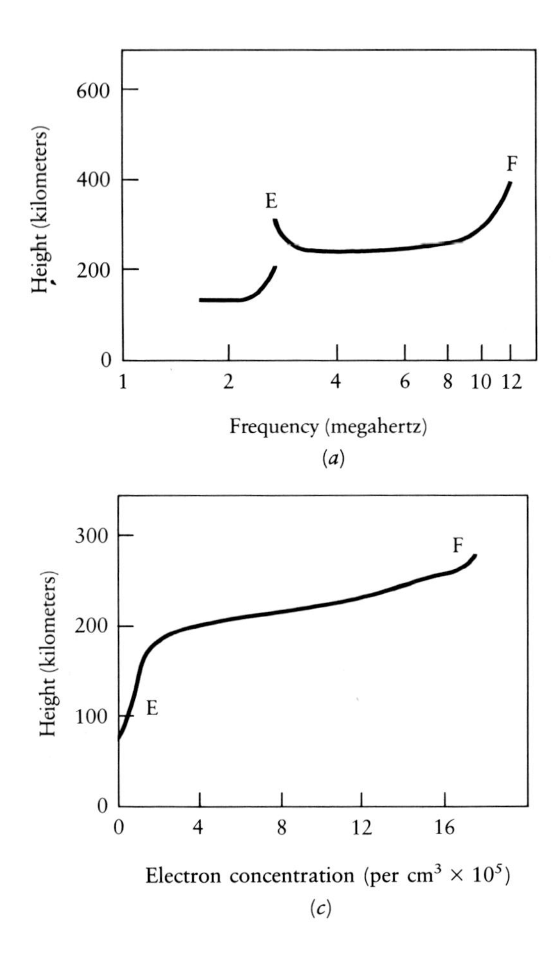

Ionosonde measurements of reflection heights versus frequency are shown in (*a*) and (*b*). The true distributions as observed from rockets at those heights are shown in (*c*) and (*d*). F-layer splitting into F1 and F2 occurs at times during the day, as shown in (*b*) and (*d*).

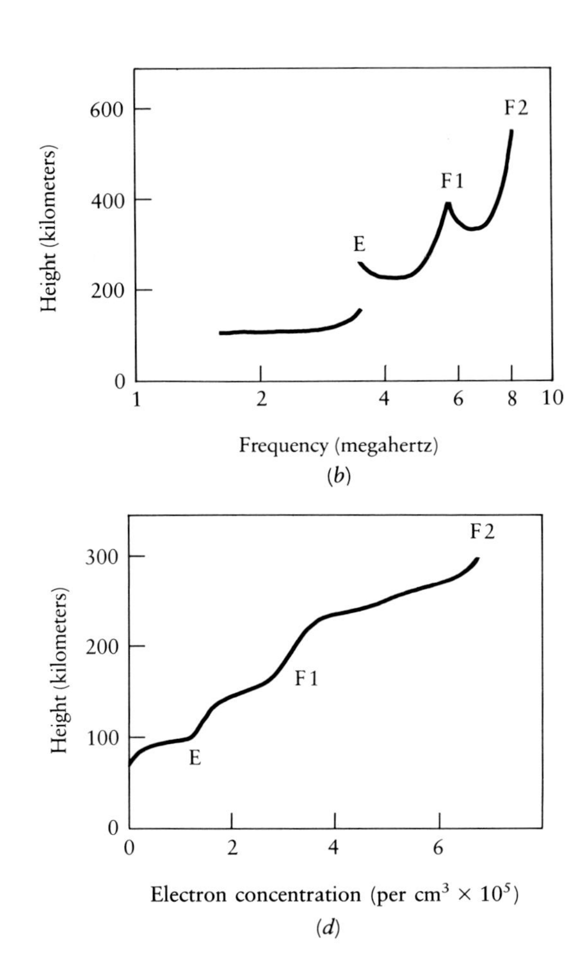

ionospheric D region. The maximum reflection frequencies are related to solar activity. At times during the day, especially in summer, an intermediate penetration frequency appears between the E-layer and F-layer peaks, as though F were split into two components, F1 and F2. Perhaps 160 ionosondes are now in continuous operation, worldwide, to record ionospheric behavior and provide a basis for communications predictions in the usable frequency bands.

light and every other electromagnetic wavelength from X rays to radio waves, by protons and electrons accelerated to more than half the speed of light and by clouds of ionized gas that sweep through space at hundreds to thousands of miles per second.

On rare occasions, a large flare can be seen as white light. Such a flare was probably the first to be observed and carefully recorded on September 1, 1859, by two Englishmen, Richard C. Carrington and R. Hodgson. A wealthy brewery heir, Carrington built his own observatory at Redhill in Surrey, where he and Hodgson studied sunspots. On the day of the flare, they were astonished to see, in the midst of a large sunspot group, "two patches of intensely bright and white light" suddenly break out. The flare lasted about five minutes and expanded some 35,000 miles across the sunspot group. Carrington thought he had witnessed the splash of a gigantic meteor as it slammed into the sun. At the moment he saw the flare, the magnetometers at Kew Observatory near London showed a marked disturbance. Eighteen hours later, one of the strongest magnetic storms ever recorded broke out, and the next night, an exceptionally brilliant aurora was seen overhead as far south as Puerto Rico.

Viewed in monochromatic light, for example the red light of hydrogen atoms, flares appear to develop with great suddenness and tremendous power. In a matter of minutes, an area on the order of hundreds of millions of square miles can increase tenfold in brightness. During moderately intense flares, shortwave radio communications are immediately blacked out over the entire sunlit hemisphere, and they do not return to normal until the flare disappears. Flare size ranges from barely detectable incidents, so-called microflares, to the largest events, which flood the interplanetary space with protons of energies comparable to those of cosmic rays in such abundance that they present a serious threat to the safety of astronauts. Clouds of slower-moving plasma surge out from the flare, taking a day or two to reach the earth's atmosphere, where they set off their displays of brilliant red and green auroral fireworks and set compass needles wiggling wildly.

Superflares

Once or twice in a solar cycle, a flare of extraordinary energy sets the entire sun-earth system reverberating with catastrophic impact. Such a superflare on November 12, 1960, provided a fascinating case history. By the time its particles bombarded the earth at 4000 miles per second, its collision front had spread 10 million miles across, and a trailing cloud of plasma stretched back halfway to the sun. This silent but violent intrusion into earth space dissipated more energy than the most powerful hurricane, and its impact was felt over the entire surface of the earth. Two precursor flarings on November 10 and November 11 had caused shortwave radio signals from New York to London and Paris to vanish abruptly. On November 12, at 2:37 in the afternoon, the McMath-Hulbert Observatory at the University of Michigan made the first report of the third and most brilliant explosion in the sequence. Six hours later, as the giant plasma cloud from the sun began to buffet the earth, a great radio storm in the ionosphere blacked out all long-distance communications. Transatlantic airline pilots lost contact with their control stations, and teletype systems printed out pages of garbled nonsense. Across Canada and other northern countries, coronas glowed on electric lines, lights flickered from surges in power transformers, and the sky was suffused with flaming auroral displays. Yet the violent overhead storm took place in eerie silence, undetectable by the human ear.

Superficially, flares appear to be a kind of lightning stroke in the solar atmosphere. On earth, lightning follows the accumulation of so much electric charge in clouds that the voltage difference leads to a discharge. But the association of flares with the strong magnetic fields of sunspots tells us that it is magnetic energy that is stored and discharged in solar activity.

X-Ray Flares and Sudden Ionospheric Disturbances

Solar-flare radiation has certain prompt impacts on the ionosphere, called sudden ionospheric disturbances (SIDs). Most dramatic is the onset of fadeout or blackout of shortwave radio communications. Radio scientists also observe reduced cosmic radio noise from the galaxy and an abrupt drop of as much as 16 kilometers in the height of reflection of long radio waves from the base of the ionosphere. All these phenomena require an increase in ionization at the lowest levels of the D region (the lowest layer of the ionosphere as designated by Appleton).

Because flares are most easily observed in hydrogen red light, which may brighten tenfold at the peak of a great flare, it used to be assumed that Lyman-alpha, the invisible ultraviolet resonance line of hydrogen, must also brighten enormously. But the evidence that a flare not only enhances ionization in the normal D region but also increases the depth of ionization by as much as 16 kilometers placed impossible astrophysical strains on a Lyman-alpha model of SID production. It occurred to me in 1954 that X-ray enhancement at the high-energy

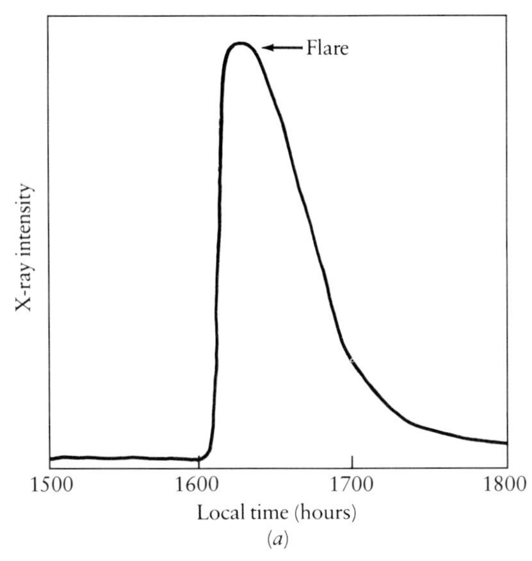

(*a*)

A large flare rises to peak intensity in minutes and decays in an hour or two.

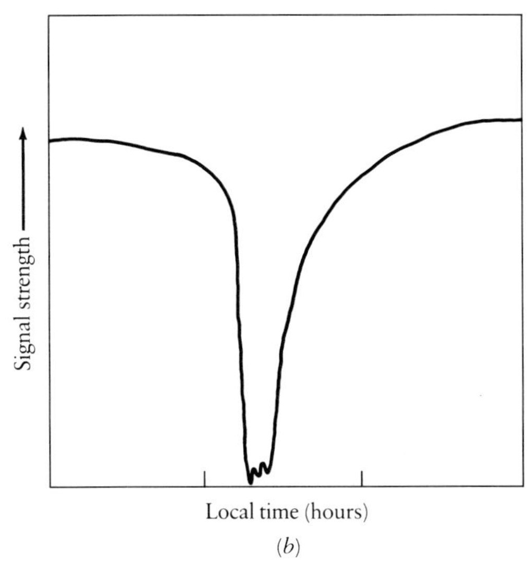

(*b*)

Simultaneously with the rise of the flare X-ray emission, radio signals fade out.

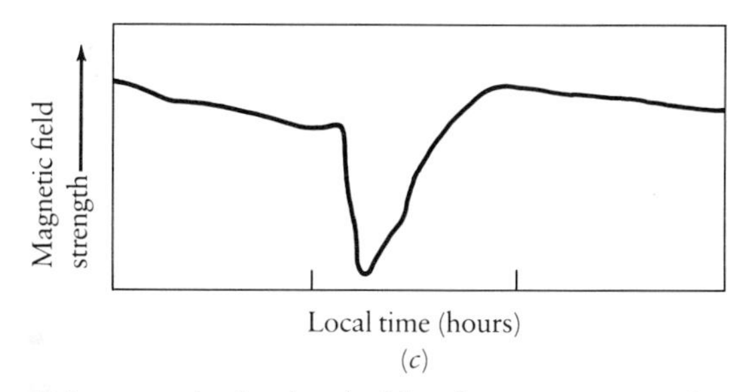

(*c*)

A decrease of a few hundredths of a percent occurs in the magnetic field strength during the flash of the flare.

limit of the spectrum was a much more plausible mechanism. To solve the puzzle of the solar-flare-SID connection, my colleagues and I set out in 1956 to measure solar X rays and Lyman-alpha at the time of a flare.

As the International Geophysical Year approached, rocket sondes were a revolutionary new tool for direct high-altitude probing of solar ionizing radiation and its photoelectric and photochemical interactions with the atomic and molecular constituents of the upper air. Talbot A. Chubb and I had concluded theoretically that only a flash of X rays at wavelengths shorter than 3 Å could produce the deep-seated ionization near 45 miles necessary for radio fadeout.

In 1956, as a rehearsal for the IGY, we planned a test series of rocket flights aimed at getting an X-ray and ultraviolet glimpse of the eruption of a solar flare. But flares are unpredictable. As many as a hundred microflares a day may produce imperceptible twitches in the ionosphere, but flares large enough to create a major disturbance occur only once or twice a week at sunspot maximum, and catching one in the act of producing radio fadeout is no simple task. An Aerobee rocket would need to stand in the launch tower, fully fueled with red fuming nitric acid and hydrazine and ready to be pressurized with helium 30 minutes before launch—clearly, not a minute's-notice system. Instead, we chose the Rockoon, a combination of solid-propellant Deacon rocket and Skyhook balloon conceived by James Van Allen.

Safety considerations required that we take this system to sea, away from inhabited land and ship traffic. In August of 1956, we set sail aboard the U.S.S. *Colonial,* an LSD that lumbered along at 15 knots, with a complement of 10 Rockoons instrumented for measuring X rays and hydrogen Lyman-alpha radiation at 1216 Å. Escorting us was the speedy destroyer U.S.S. *Perkins*. Our destination, 350 miles out of San Diego, was an atmospheric high-pressure region that gave the expedition its name, Project San Diego-Hi.

Each morning, a polyethylene plastic balloon was laid out on the helicopter deck and filled with 5000 cubic feet of helium until a 20-foot bulge appeared at its top. Standing off the deck, with its lower

NRL team for the IGY solar flare project, San Diego-Hi, in the ward room of the U.S.S. *Colonial.* On the table is the pen-recorder paper tape of four minutes of flight data. At left is a prototype instrument section. Examining the tape are Herbert Friedman, Talbot A. Chubb, and James E. Kupperian, Jr.

portion still draped in gathered folds, the balloon resembled a giant inverted onion. At an altitude of 80,000 feet, the helium would expand to fill out the entire balloon. The 12-foot Deacon rocket was suspended from the balloon by 100 feet of nylon line terminating in a steel ring that mated with a hook attached to the rocket motor. When the rocket was fired, the hook would slip out of the ring, and the rocket would tear through the balloon as though it were not there. Fired from 80,000 feet, the rocket could reach 60 or 70 miles in about 1½ minutes and collect about three minutes of useful data in the ionosphere.

Upon release, the balloon and its dangling rocket climbed at about 1000 feet per minute. When it reached floating altitude, winds as high as 30 knots would speed it westward while our radars held contact with it. If it got away from us aboard the *Colonial,* the *Perkins* would chase it like a hound after a fox to stay in radio range. Then it had to be cut down at the end of each day over the ocean. We expected to stand ready about six hours a day for firing at 80,000 feet.

Fourteen hundred miles away at the High Altitude Observatory at the University of Colorado, observers held a vigil on the sun to alert us by radio of the first hint of a flare brightening. As a backup, aboard the *Colonial* we kept four radios tuned to shortwave broadcasts from Tokyo, San Francisco, Mexico City, and New Mexico. If radio fadeout began on all four channels, we could be sure a flare was in progress.

The mission barely succeeded, but the result was convincing. On the fourth try, a small flare occurred, and the rocket detectors recorded a burst of X rays at a wavelength as short as 3 Å, sufficient to double the ionization near the bottom of the ionosphere. No increase in Lyman-alpha was detectable. Even this small flare significantly enhanced radio absorption as it delivered X rays that penetrated to the bottom of the D region. X-ray sunburn of the D region was clearly the cause of radio fadeout.

By 1957 and 1958, a two-stage, rail-launched rocket system that used a Nike booster instead of a balloon permitted ground-based launch on instant command. A very successful campaign of flare shoots was conducted from a pad high above a beach where hundreds of seals basked in the sun on San Nicholas Island off the coast of California. When large flares erupted, the increase in X-ray intensity outstripped all other radiation, and blackouts persisted as long as the X-ray bursts. Within a few years, it became possible to monitor the sun continuously from satellites, and the study of solar-flare X-ray emission became increasingly sophisticated.

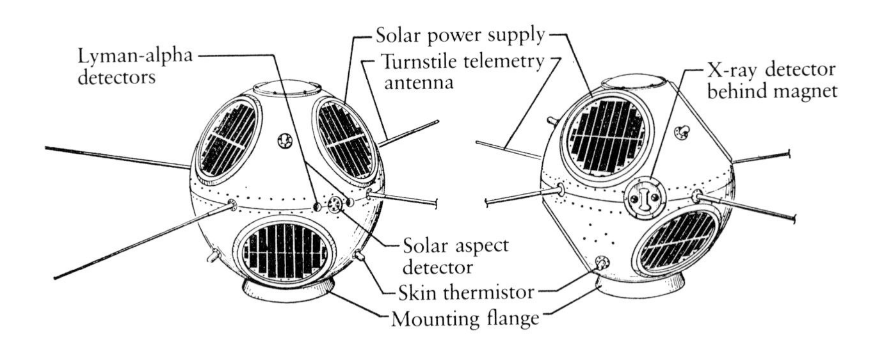

Solrad-1 (1960, Eta 2), the first astronomical satellite for measuring solar X-ray and Lyman-alpha emission.

Some flares, judged by their X-ray spectra, appear thermal in origin, characterized by source temperatures up to tens of millions of degrees. Others exhibit very impulsive flashes, implying rapid acceleration of electrons to high energies. Often it appears that X-ray flares originate in superheated kernels along the tops of magnetic arches rooted in sunspots. In spite of all the observational evidence, however, no theory of the flare process is as yet fully satisfactory.

Solar Radio Noise

As long ago as 1890, Thomas Edison anticipated solar radio astronomy. It occurred to him that solar disturbances appearing as visible light flashes might also produce radio waves. He decided to encircle a large deposit of iron ore in New Jersey with a loop of wires, expecting to detect an induced current on this antenna should radio waves from the sun hit it. He had the poles put in but never completed the loop. It was just as well. Although his intuition about solar radio emission was excellent, the apparatus was far too insensitive to have detected any radio waves that did get through the ionosphere.

In 1942, British radar detected intense static on a scale too large to be attributed to any local sources of interference. It was thought that the Nazis had come up with a new, superpowerful jamming device. When J. Stanley Hey analyzed reports from several sites, he concluded that the signals came not from Nazi jammers, but from the sun. The static rose and fell with the rising and setting sun, and disappeared when a large sunspot group passed to the backside. After the war, these observations led directly to the modern era of solar radio astronomy.

Whenever charged particles race by each other in near-collisions, radio noise is generated. Every plasma is a radio source, and the sun is a particularly bright one. Nature also seems to provide very powerful

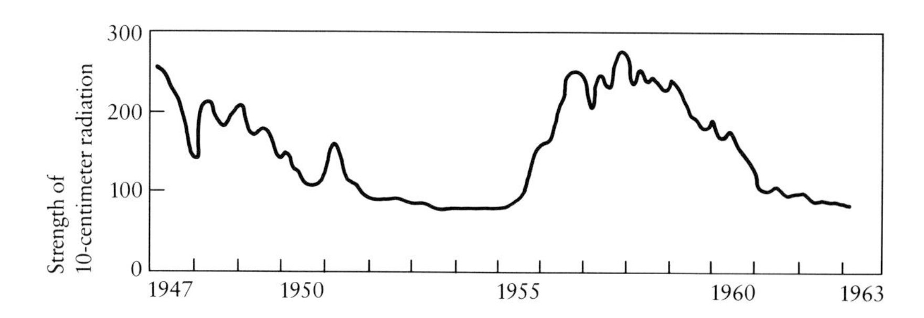

The intensity of solar radio noise on a wavelength of 10 centimeters follows in step with the sunspot cycle and provides a proxy measure from the ground of solar X-ray emission.

and efficient acceleration mechanisms for charged particles. When very fast electrons travel through a magnetic field, they are bent into circular paths around its lines of force and made to radiate all wavelengths, from radio to visible light. By analogy with man-made synchrotron accelerators, we call such magnetically induced radio waves synchrotron radiation.

In strong contrast to the image we see, the visible disk of the sun is almost dark at most radio wavelengths, and the corona is a very bright halo. Different regions of the solar atmosphere produce characteristically different radio waves because a critical plasma frequency is associated with any given electron density. Wavelengths from the chromosphere range from just millimeters to about 30 centimeters. As the wavelength increases, radio signals can escape only from higher and higher levels of the sun; 10- to 15-meter waves, for example, come from heights of 200,000 to 300,000 miles in the corona. By choosing a particular wavelength, it is possible to explore a specific layer of the sun's atmosphere.

The radio sun is highly variable, always flickering and pulsating. Its outbursts often amount to hundreds-of-times normal intensity and, on occasion, reach millions-of-times normal in only a few seconds. These sporadic radiations are related to other forms of solar activity—sunspots, plages, and flares. In general, however, the short-wavelength signals of the chromosphere are relatively calm and steady, reflecting only the mild chaos produced by bursting granules, erupting spicules, and shock waves. In contrast, the longer wavelengths may last for days, hours, minutes, or only seconds.

High-resolution-telescope mapping of slowly varying sources of radio microwaves shows that they arise from the same condensations above plages and spot groups that produce X-ray emissions. As the sun rotates, radio signals reveal that active regions pass around the limb in a characteristic 27-day solar period. Even if the sky were con-

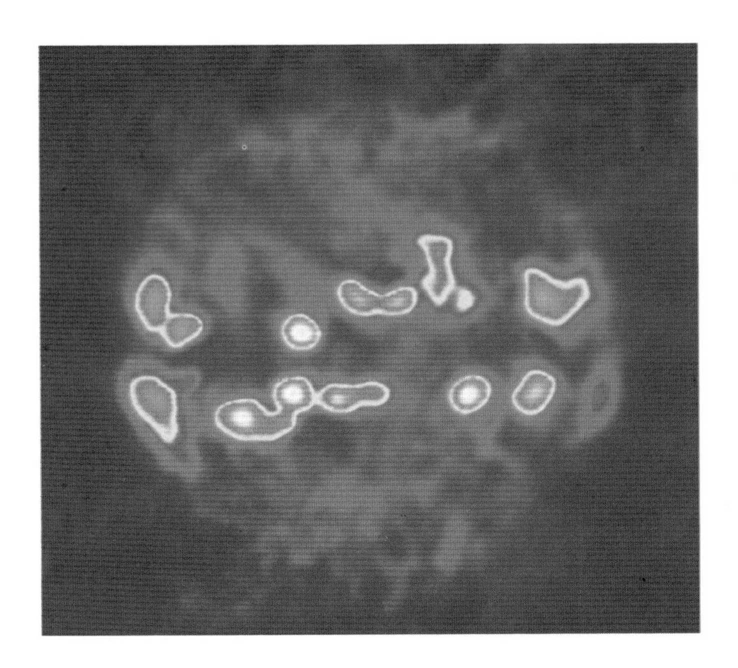

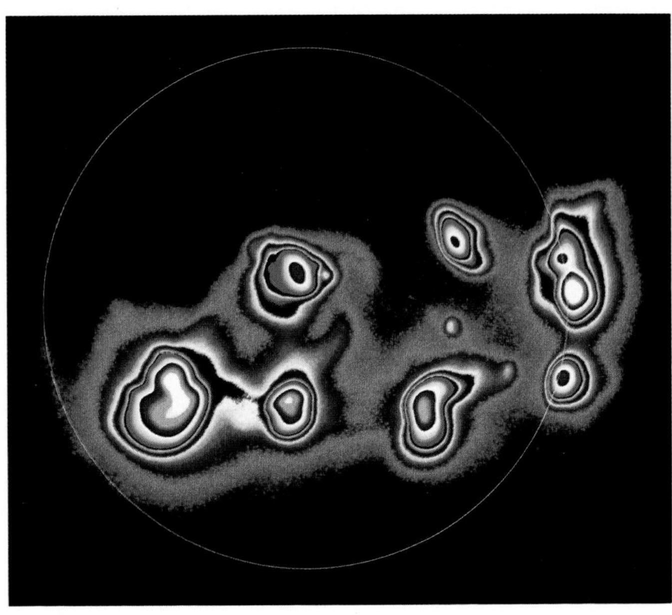

Left, radioheliograph from the Very Large Array at a wavelength of 20 centimeters and a resolution of 40 arc seconds (September 26, 1981). Right, X-ray image from Skylab (September 6, 1973). Both images show strong localized concentrations of emission. Intensity is color-coded by computer.

tinuously overcast, we could still infer sunspot numbers quite accurately from measurements of the microwave radio intensity at a 10-centimeter wavelength.

When a large flare develops, a complicated sequence of radio signals can be picked up, the most striking of which are those from one to 30 meters. Whereas random motions of electrons in plasmas at chromospheric and coronal temperatures produce normal, continuous radio noise, intense radio outbursts during flares imply temperatures millions of times greater than those normally due to random motion. The outbursts are more likely associated with directed streams of charged particles.

The radio outbursts from the corona often show a pattern of drift toward longer wavelengths. The emission may start at a wavelength of one meter, appear a couple of minutes later at three meters, and, in five minutes, drift to five meters. Because each wavelength characteristically appears from a particular height in the corona, we can compute that the radio source is actually moving outward at a speed of about 700 miles per second. This "slow-drift" radio emission is believed to be coupled to a shock front, which leads a blast of protons and electrons ejected at the chromospheric level. Electrons moving at nearly half the speed of light produce another group of fast-drift patterns.

Electrons of this energy directed downward into collision with denser chromospheric gas may produce the X rays that are observed at the same time.

In very large flares that eject particles of cosmic-ray energies, a third type of radio emission is almost always recorded. A very broad band, covering thousands of megahertz and with the character of waves emitted by very fast electrons spiraling in magnetic fields, is radiated for anywhere from half an hour to several hours. Radio telescopes have actually tracked such radiating clouds of gas as they carry their high-speed electrons, together with their frozen-in magnetic fields, to heights of several million miles.

By far the highest image resolution is obtained by an array of dishes, each one 85 feet in diameter, mounted in a Y-shaped configuration on the Plains of St. Augustine near Socorro, New Mexico. Each arm of the Y is a railroad track 21 kilometers long on which the 200-ton antennas can be moved at five miles per hour to any spacing. They combine to form the Very Large Array (VLA), the most powerful interferometer for solar research at centimeter wavelengths, with a resolution of about one second of arc, or a few hundred miles on the face of the sun, a performance that equals the best optical resolution obtained from the ground. The VLA's sensitivity is high enough to produce "snapshots" in only a 10-second exposure, revealing large-scale magnetic loops in detail that closely matches the quality of images in X-ray and ultraviolet wavelengths.

The Very Large Array (VLA) of radio telescopes consists of 27 antennas, each one about 28 meters in diameter, mounted on railroad tracks on the Plains of St. Augustine near Socorro, New Mexico. The antennas are arranged in a Y shape, each arm and the leg being 21 kilometers long. At centimeter wavelengths, the resolution is about one arc second.

What can high-resolution VLA radio images tell us about the flare mechanism? They contain information about the magnetic-field configuration at the site where energy is released by electrons accelerated during the development of a flare. These fields often undergo large changes beginning an hour or so before the flare explosion. In most events, the initiation of energy release seems to occur at the tops of isolated loops, which are clearly imaged at centimeter wavelengths. In some cases, the flare seems to be triggered in a current sheet that arises between loops arranged in the form of an arcade. Clearly, a number of configurations are capable of leading to the release of flare energy.

Exploring the Canopy of Air

> *. . . This most excellent canopy, the air, . . . this brave o'erhanging firmament, this majestical roof fretted with golden fire. . . .*
>
> William Shakespeare, *Hamlet*, II, ii

Aristotle postulated that air did not have weight and rose by means of "levity." Not until the fifteenth century did scientists begin to appreciate the mass of air and the phenomenon of barometric pressure. In 1643, Evangelista Torricelli invented the mercury barometer. The pressure of air supported a column of mercury 760 millimeters high, but from day to day the pressure varied slightly around this value. In 1646, Blaise Pascal, physically too weak to climb mountains himself, persuaded his young brother-in-law to perform an experiment by carrying a barometer up the side of a mountain in southern France. When, at about 1.5 kilometers above sea level, the mercury column dropped about eight centimeters, it became clear that atmospheric pressure decreases rapidly with increasing altitude.

Thermometers carried aloft on kites had somewhat successfully measured atmospheric temperature during the mid-eighteenth century, and in 1893, an Australian inventor built a box kite that lifted temperature and pressure instruments to 10,000 feet. But kites were soon superseded by balloons.

With the invention of balloons, scientists and instruments could at last rise above the mountaintops. The Montgolfier brothers, Jacques, Étienne, and Joseph, who were French paper manufacturers, launched their first balloon on June 4, 1783. It was 33 feet in diameter and made of linen lined with paper. Filled with smoke from burning straw and wool, it rose to 1500 feet. Very quickly, many flights followed in

The Montgolfier brothers, French paper manufacturers, constructed their balloon of linen lined with paper. It was filled with smoke from burning straw and wool. Their first launching took place on June 4, 1783, and the 33-foot balloon rose to 1500 feet.

manned balloons filled with hydrogen. Those early attempts often courted disaster, sometimes with comical results. In 1785, Jean Pierre Blanchard, inventor of the parachute, and John Jeffries, an American physician, took off across the English Channel with a collection of instruments to measure temperature, pressure, and humidity. Their balloon soon sprang a leak. Desperate to reduce weight, they first shed their outer garments and tossed them overboard, along with books and instruments. They managed to reach French soil in their underwear, somewhat embarrassed but undaunted by their narrow escape.

Among the early scientific balloonists was Joseph Louis Gay-Lussac, the French chemist best known for his law of the relationship between gas volume and temperature. In 1804, Gay-Lussac and the physicist Jean Baptiste Biot ascended to a height of 23,000 feet over Paris in a balloon originally intended for Napoleon's Egyptian campaign. Gay-Lussac collected air samples at various heights in evacuated glass vessels. His subsequent analysis showed that the composition of dry air remained unchanged and that the percentage of water vapor decreased with altitude. Investigations of the upper atmosphere with unmanned balloons began in 1893. Léon Philippe Teisserenc de Bort, a French meteorologist, perfected balloons that could reach 50,000 feet. His experiments revealed a region, which he named the "stratosphere," where the temperature no longer decreased with altitude but remained nearly constant. The early adventure of scientific ballooning has been exceeded in the space age by rocket and satellite exploration of the upper atmosphere. We can now describe the physics and chemistry of the atmosphere to heights that merge with interplanetary space.

The Neutral Atmosphere

The electrically neutral atmosphere is the ultimate sink for most of the solar energy that reaches the terrestrial environment. About one-third of the incident radiation is reflected or scattered back to space by the earth's atmosphere, its cloud cover, and its surface. The remainder is absorbed in the land surfaces and the oceans. In return, the heated surface sends infrared radiation back to the troposphere. This thermal radiation and the atmospheric winds it produces control the available energy and its distribution in the troposphere, and thus our weather and climate.

The mass of the atmosphere is less than a millionth of the mass of the solid earth, and its density at sea level is less than a thousandth that of the earth's crust. Still, the ocean of air above us weighs about 5000 million million tons (5×10^{15} tons), and it presses down with a force of 14 pounds on every square inch of surface. How high does the air reach and what lies beyond? Until relatively recent years, these questions were attacked in comparatively primitive experiments and crude theoretical modeling. Now instruments aboard balloons, rockets, and satellites have probed most of the physical and chemical characteristics of the terrestrial environment in great detail. Much of what we have learned is basically new, and the quantitative features observed often differ substantially from early estimates.

The atmosphere near the ground consists mainly of molecular oxygen (21 percent) and molecular nitrogen (78 percent). If the air were very cold, it would condense to a thin layer on the surface of the earth. But because the air is warm, the molecules are in continuous, rapid, random motion that counteracts the force of gravity. The competition between gravity and heat motion leads to an exponential decrease in density with height. If the composition of the air remained uniform, any increases in height would always correspond to a proportional decrease in concentration. For example, at a height of six kilometers, the density is half that on ground at sea level; at 12 kilometers, it is halved again to one-quarter sea-level density, and so on. It is more conventional to describe the density-versus-height profile on a logarithmic scale. The height interval in which the density falls to the reciprocal of the logarithmic base *e* ($1/2.72$) is known as the scale height *H*. At ground level, *H* is eight kilometers; two scale heights (16 kilometers) is about a factor of 10. At four scale heights, the residual air pressure is only 1 percent of its sea-level value.

The scale height is directly proportional to temperature and inversely proportional to molecular weight. If the temperature increases, the scale height becomes greater and the gas density at a given level increases; if the weight of a molecule is greater, its scale height is less and the gas does not extend as far. All the air above us would fill a shell only eight kilometers high, one scale height, if compressed to sea-level pressure.

If no turbulent mixing took place in an atmosphere containing a mixture of gases of different molecular weights, each constituent would have its own height distribution as if the other species were not present. The lightest gases with the greatest scale heights would dominate at the greatest heights. Of the gases at ground level, helium is only one part in 10^5, yet it becomes most abundant at 1000 kilometers. Such a distribution is referred to as diffusive equilibrium. If, however, the gases are mixed by winds, they do not separate diffusively. In the terrestrial atmosphere, turbulent mixing preserves the ground-level distribution of gases up to a "turbopause" at about 95 kilometers.

When rockets release fluorescent gases at sunset, the resulting vapor trails, glowing against the dark sky, reveal beautiful eddies at the boundary of the mixing region. Above the turbopause, smooth diffusion sets in, and the eddies disappear.

The various regimes of the neutral atmosphere in order of increasing height are the troposphere, which is the "changing" or "overturning" sphere; the stratosphere, or "settled" sphere; the mesosphere, or

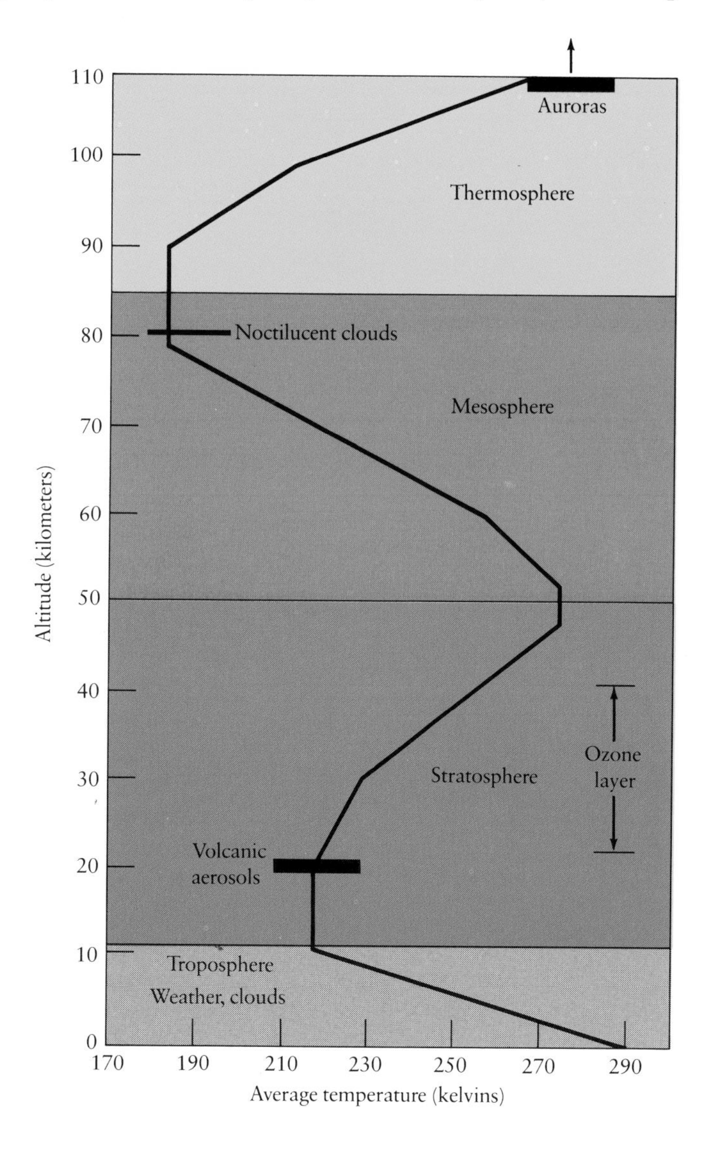

Temperature variation in the atmosphere from ground to 110 kilometers. Heights of various atmospheric phenomena are indicated.

"middle" sphere; the thermosphere, or "hot" sphere; and the exosphere, the "outside" sphere. The stratosphere and mesosphere are often referred to together as the middle atmosphere.

The Troposphere

All around us, water is rising
on invisible wings
to fall as dew, as rain, as sleet, as snow,
while overhead the nested giant domes
of atmospheric layers roll
and in their revolutions lift
humidity north and south
from the equator toward the frigid, arid poles,
where latitudes become mere circles
Molecular to global, the kinetic order rules
unseen and omnipresent, . . .

John Updike, "Ode to Evaporation"
The New Yorker, Dec. 31, 1984

The troposphere is the seat of all weather phenomena. Its thermal structure is governed by the heating of the earth's surface by solar radiation and the upward transfer of this heat by turbulent mixing and convection. Differential heating of the equatorial and polar regions drives atmospheric circulation. Evaporation of water from the land surfaces and oceans is accompanied by formation of clouds and rain. The resulting weather patterns are a complex of prevailing winds, shower clouds, thunderstorms, hurricanes, cyclones and anticyclones, depressions, fronts, and monsoons.

Most of us have experienced the cold temperatures of mountain altitudes and have heard the pilot of a jetliner report bitter subzero temperatures outside the plane. Temperature decreases steadily with height in the troposphere up to its boundary with the stratosphere—the cold tropopause. The average height of the tropopause varies from eight kilometers over the poles to 16 kilometers over the equator. In middle latitudes, day-to-day weather patterns produce large variations in the height of the tropopause.

The Stratosphere

Temperature in the overlying stratosphere rises again to a maximum at 50 kilometers because of the strong absorption of solar ultraviolet radiation by ozone. Early evidence of the temperature rise was deduced from the anomalous transmission of distant sounds. When guns from the fortress at Spithead saluted the cortege that carried the body of Queen Victoria from Carisbrook to the mainland in 1901, they were heard at Oxford, 70 miles away, though at closer distances silence prevailed. During World War I, a "zone of silence" was noted inside a radius of about 150 kilometers from booming artillery, which was audible closer in or farther out.

Two transmission modes of sound waves explain these effects: a ground wave traveling close to the surface, and a sky wave traveling upward from the source that is bent back by the atmosphere at a height of about 30 kilometers as a result of a temperature inversion. Sound in the near zone travels via the ground wave; in the far zone, it travels via the reflected wave out to a few hundred kilometers. In the silent zone, the ground wave has faded and the reflected wave has not yet returned to earth. Although the temperature falls about 70°C from the ground to the tropopause at 12 kilometers, the stratosphere becomes warmer above 30 kilometers and returns to ground level temperature again at about 50 kilometers.

The stratosphere is mainly in radiative equilibrium: it has no weather in the tropospheric sense and is comparatively stable. Tropospheric clouds constantly form between altitudes of two and 12 kilometers, but except for the occasional nacreous (mother-of-pearl) clouds in the 20- to 30-kilometer height range, the stratosphere is clear.

Nacreous clouds, incidentally, are a relatively rare scientific curiosity. Irridescent before sunrise or shortly after sunset, they appear near the 30-kilometer altitude over Scandinavia and other northern regions. Sometimes relatively thick, at other times gossamer thin, they

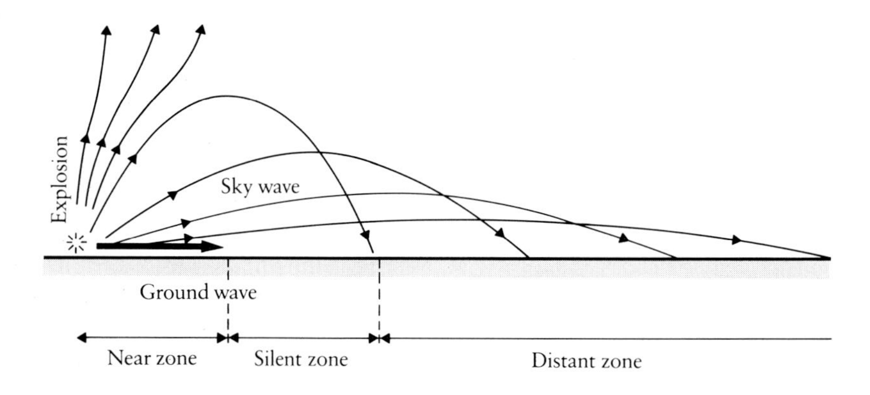

Sound from an explosion is heard in the near zone via the ground wave. In the distant zone sound is heard via the sky wave. Between the near and the far zone is a silent area.

consist of ice crystals or supercooled water droplets. The water must originate in the troposphere and somehow get carried into the stratosphere.

Although ozone constitutes only about one-millionth of the atmosphere, it absorbs the radiation that heats the stratosphere and provides the protective ultraviolet shield that is crucial to the preservation of life on earth. A form of oxygen composed of three atoms (O_3) instead of the normal two (O_2), ozone is continuously produced in the stratosphere with ordinary oxygen molecules. The oxygen molecule is first split into two atoms, and subsequently each atom attaches to a molecule of O_2 to form O_3. About one-thousandth of the solar flux is absorbed in its production.

Ozone is also continuously destroyed by collisions with reactive atoms and molecules called free radicals. Most free radicals are small, unstable molecules torn away from more complex parent molecules and left with an unpaired electron. A number of gases produced by human activities, such as using artificial fertilizers and burning coal, as well as the fluorocarbons lost from refrigerators and air-conditioners and released from cans of hair spray and whipped cream, also contribute free radicals to the stratosphere. The influence of sunlight produces the free radicals nitric oxide (NO), the hydroxyl molecule (OH), chlorine (Cl), and chlorine monoxide (ClO). Each of these can enter into tens of thousands of destructive reactions with ozone before finding a matching electron to stabilize it. Free radicals thus act as catalysts for ozone destruction.

Altogether, the total amount of ozone in the atmosphere is very small, sufficient to create a layer of gas only one-eighth of an inch thick if it were all concentrated at ground-level pressure. But that small amount of ozone is such an enormously effective barrier to the penetration of ultraviolet radiation below 2850 Å that essentially none of this radiation reaches ground level. Immediately longward of 2850 Å, ultraviolet starts to trickle through, and ozone filtering quickly disappears.

If the protective ozone blanket were depleted by 10 percent, the increased ultraviolet radiation at the earth's surface could cause serious crop damage, a rise in skin cancers and melanomas, and the death of plankton in the surface waters of the sea. Extreme depletion—by 50 percent, for example—could cause cataract blindness in nearly all animals. DNA and protein absorb far ultraviolet very readily, and so the formation of an ozone barrier must have been an essential condition for the emergence of life as we know it today.

Ozone is measured by various methods. From the ground, the

observed penetration of solar ultraviolet close to the wavelength edge of the ozone-absorption band near 2850 Å provides a sensitive measure of the total column of ozone. Dobson's photoelectric photometer, widely used during the IGY, employed a pair of wavelengths whose transmissions were very different for ozone but were insensitive to atmospheric extinction from other sources. The height distribution of ozone can also be obtained by observing not the direct transmission of sunlight but scattered light from high in the atmosphere when the sun is at low elevation (Dobson's *Umkehr* method). The early V-2 rocket flights gave precise height profiles of ozone by measuring the extension of the solar spectrum toward shorter wavelengths with increasing altitude.

Balloons that float in the stratosphere can make direct in situ observations of ozone, and they are especially suited to studies of reaction chemistry and temporal variability. Until recently, experimenters studying the chemistry of the stratosphere had relied on instrument packages parachuted from balloons floating near 140,000 feet. As the instruments plunged down at over 200 miles an hour, they radioed information on the concentration of ozone and free radicals. Now a promising new method of exploring the range of stratospheric ozone has been introduced by James Anderson and his colleagues at Harvard University. His 130-pound instrument payload can be lowered from a 26-million-cubic-foot helium-filled polyethelene balloon by means of a 2700-pound winch attached to the balloon. The payload, called "the monkey," can reel down a distance of 9.5 miles and ride back up again in about an hour and a quarter. Walter Sullivan of the *New York Times* has dubbed the monkey the "world's biggest yo-yo."

Technology for mapping global ozone distribution from space has developed very rapidly. Back-scattered ultraviolet (BUV) ozone measurements aboard Nimbus 4 in 1970 led to the total-ozone-mapping spectrometer (TOMS), now past its sixth year of operation aboard Nimbus 7. TOMS measures back-scattered solar ultraviolet radiation from the troposphere and the surface of the earth in a wavelength range that is mildly absorbed by ozone. Nimbus 7, in a noon polar orbit, enables TOMS to scan from horizon to horizon perpendicular to the orbital track, yielding nearly global coverage over the day hemisphere with a spatial resolution of 50 kilometers.

Modification of the stratosphere's ozone content by human activities has become one of the most profound environmental concerns of recent years. When atmospheric scientists realized that nitric oxide in the exhaust of supersonic aircraft could catalyze the destruction of

Under the direction of James Anderson, a novel payload is used as a spectrochemical laboratory in the stratosphere. While the balloon floats at 41 kilometers the payload is reeled down and up some 12 kilometers to study a wide height range. The 2700-pound winch platform hangs under the balloon and the 130-pound payload is reeled on a Kevlar (a synthetic fiber with 10 times the tensil strength of steel) line. The payload has been dubbed "the monkey."

An 800-foot balloon being inflated to carry the 2830-pound payload into the stratosphere. The balloon is filled with helium and rises to the stratosphere at some 900 feet per minute.

Distribution of total ozone column density over the United States and Canada. The color scale from dark blue to green, yellow, red, pink, and white indicates increasing amounts of ozone. In the 5.5 hours between the upper and middle curves the high-in-ozone column density moves from northern Minnesota to Lake Superior, and in the lower curves, 20.5 hours later, to eastern Canada. Measurements were obtained in October, 1981, from the NASA Dynamics Explorer I. Ozone concentrations are strongly dependent on latitude and season. The greatest total ozone content is found at high latitudes and the seasonal maximum is in the spring. The concentration peaks at about 15 miles and the variability occurs primarily below that level as a result of both horizontal and vertical mixing.

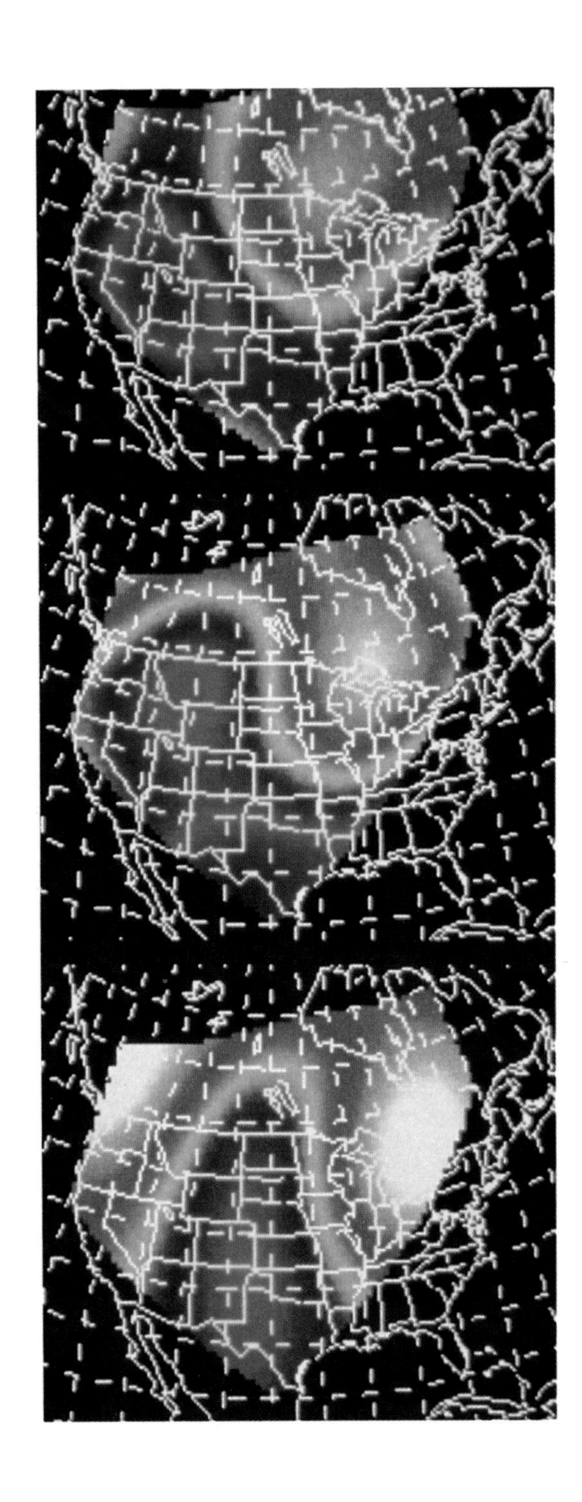

stratospheric ozone, they strongly urged that the United States abandon its development of the SST. Subsequent studies uncovered greater and greater complexity in the chain of photochemical and atomic reactions, and early predictions of ozone destruction have switched to expectations of a slight enhancement.

But hardly had the concern over nitric oxide pollution begun to calm down when a potentially more serious threat from the release of chlorofluoromethanes was predicted. These chemicals are used primarily as refrigerants and aerosol sprays. They are extremely inert and stable at ground level, but when they rise to the stratosphere by atmospheric mixing, they are decomposed by solar ultraviolet. The free chlorine that is released may start a chain of reactions that lead to destruction of ozone. The mixing process is so slow that gas now in the lower atmosphere may not have its full effect on ozone for another 50 years. Ozone concentration in the stratosphere must be monitored carefully from now on to detect any warning of trends with serious consequences.

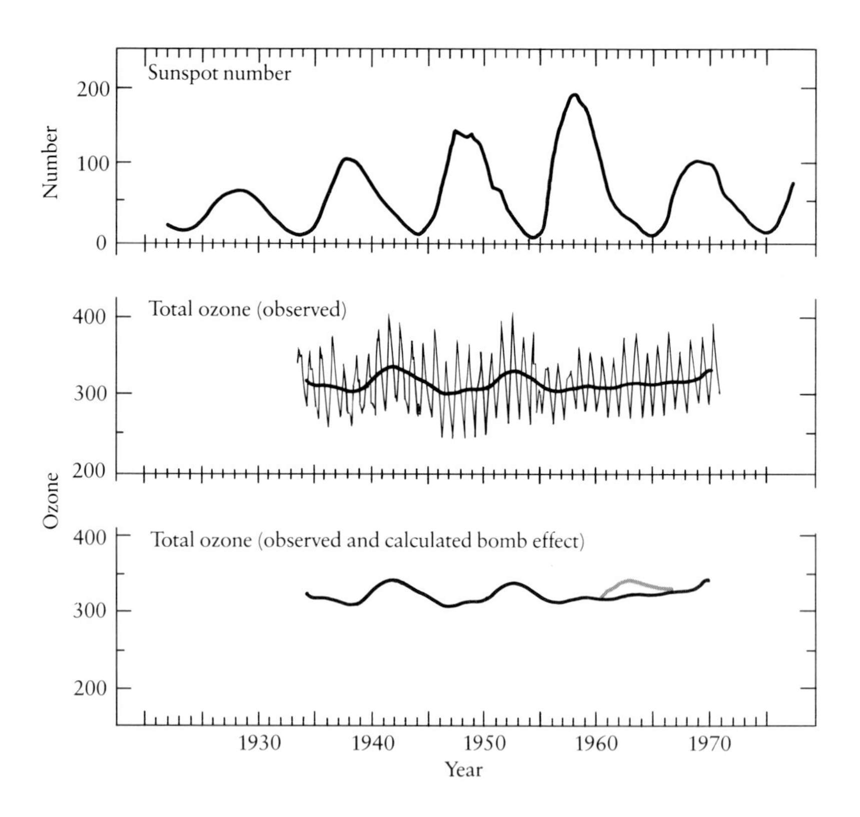

Ozone concentration varies over a sunspot cycle, but it lags behind sunspot maxima. In the bottom curve, the annual ozone variations are averaged out. A computed correction (colored curve) shows where the level would have been were it not for ozone destruction by nuclear testing in the atmosphere.

The Mesosphere

The term "mesosphere," coined by Sidney Chapman, comes from the Greek *meso,* which means "intermediate." Another often used term, "middle atmosphere," refers to both the stratosphere and the mesosphere. The mesosphere is a range of falling temperature, from about 271 K at 50 kilometers to 181 K or less at the 85-kilometer mark. Over northern Sweden, the temperature at times falls to 135 K, the lowest temperature ever measured in the earth's atmosphere. The main reason for the falling temperature with increasing height is that ozone decreases, and absorption of solar ultraviolet by ozone is the principal source of heat.

We know less about the mesosphere than any other range of the atmosphere because in situ measurements are so difficult. The region is too high for balloons and aircraft, including the U-2. Most rockets designed for meteorological studies do not fly high enough, and other rockets in the scientific stable fly through the region too fast. Efforts to parachute instruments from rockets have not had much success. Air drag would bring a spacecraft in the mesosphere down before it could complete one orbit. In a joint effort with the Italian Space Agency, NASA plans to test the concept of the tethered satellite, which can be "trolled" in the upper reaches of the mesosphere from a space shuttle at the end of a 60-mile-long line somewhat in the manner of the yo-yo balloon payload, in about 1987. The instrumented subsatellite, four or five feet in diameter, would be lifted out of the shuttle bay on a boom and let out on a tether made of Kevlar, a remarkably tough plastic. At the end of the mission, it would be reeled in on a pulley mechanism and returned to earth.

At altitudes of 85 kilometers in the mesosphere, noctilucent, or night-shining, clouds are the highest clouds known. Rarely seen except in summer at high latitudes, they are characteristically bluish white and very tenuous. Visible only in twilight when bathed in the light of the setting sun against the dark sky, they ripple with wavelike motions as they drift overhead. These clouds mark the boundary between the mesosphere and the thermosphere at the temperature minimum. When vertical winds bring water vapor into this region, the clouds begin to form, float to higher levels, and finally dissipate as they encounter warmer air. Scientists have collected particles of these clouds on specially coated plates exposed on board rockets to determine their composition. They appear to be tiny specks containing iron and nickel, surrounded by ice. Their origin is probably meteoritic dust, but how water can exist at such a high altitude and why the altitude regime is

Noctilucent clouds are occasionally seen after sunset at high latitudes in summer; they form at an altitude of about 80 kilometers. The clouds make visible the complicated wave patterns in the high atmosphere.

so narrow when meteorites appear at higher and lower altitudes in equal abundance remain a puzzle.

Did you stay up last night (the Magi did)
To see the star shower known as Leonid
That once a year by hand or apparatus
Is so mysteriously pelted at us?
It is but fiery puffs of dust and pebbles,
No doubt directed at our heads as rebels
In having taken artificial light
Against the sovereignty of night.
A fusillade of blanks and empty flashes,
It never reaches earth except as ashes
Of which you feel no least touch on your face
Nor find in Dew the slightest cloudy trace. . . .

Robert Frost, "A Loose Mountain"

Well before rockets were used to explore the upper atmosphere, simple observations of meteors gave good evidence of temperature and air density near the turbopause, the region where turbulent motion gives way to smooth diffusion. Shooting stars that flash across the horizon are produced by meteorites the size of a pinhead. They plunge into the earth's atmosphere at about 30 kilometers per second, and as they penetrate to the denser regions, air friction heats the particles to luminescence near 110 kilometers and to complete evaporation at about 80 kilometers. Theoretical modeling of this process required that air density near 110 kilometers must be much greater than had previously been supposed. The temperature had to be higher than it is at ground level, and the fall of temperature from ground to tropopause had to be reversed at higher altitudes.

The Thermosphere: The Region of Highest Temperatures

The thermosphere overlaps the E and F regions of the ionosphere. It begins at about 90 kilometers and extends upward to about 500 kilometers. Temperatures increase steadily to a constant value above 250 to 400 kilometers, depending on the level of solar activity. This rise in temperature occurs mainly because of the absorption of radiation up to a wavelength of about 2000 Å. Absorption causes dissociation and ionization, and both processes release heat.

In the early days of the space era, observations of the drag on satellites revealed a great deal about the structure of the thermosphere. Enough air remains at an altitude of several hundred kilometers to provide significant resistance to the passage of a satellite. As the satellite pushes aside as little as one gram of air for each full orbit, the drag is sufficient to cause a perceptible change in orbital velocity. The drag-induced energy loss makes the satellite fall closer to the earth, and the lower the orbit, the less time it takes to circle the earth. As a satellite moves across the star field of the night sky, an observer can note the time that it passes a stellar starting point on its racetrack and then mark the time of its return. Observations of drag, therefore, require no more than a watch and a pair of binoculars, but they must be made soon after the sun has set at ground level, when the satellite is still

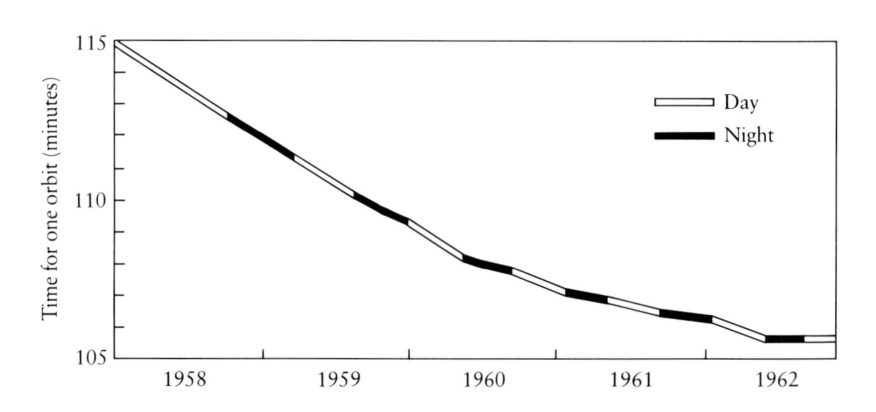

The orbital period of the Echo balloon satellite as air drag reduced its altitude from 1957 to 1963. Drag was higher during the day than at night and at sunspot maximum (1957) than at sunspot minimum (1962).

sunlit against a dark, starry sky. Because the observations are so simple, and because it is so pleasant to lie on one's back outdoors and contemplate the beauty of the night sky, much of what we have learned about drag has come from amateur observers all around the earth.

If a satellite moves in a circular orbit, drag acts uniformly throughout its path; when the orbit is elliptical, the most drag occurs at perigee, where the air is most dense. As the plane of the orbit rotates in space, the perigee point moves from night to day and back again. The perigee data over a period of time thus map out the diurnal variations in air density, which can be by as much as a factor of 10. Density also responds promptly to variations in solar X rays and ultraviolet radiation and to the arrival of plasma clouds signaled by geomagnetic storms.

The Exosphere: Escape from the Atmosphere

Above about 500 kilometers lies the highest region of the atmosphere, the exosphere. Here hydrogen atoms are traveling so fast that many of them have enough energy to enter into satellite orbits or to escape the earth's gravitational field entirely. The gas concentration in the exosphere is so low that an atom can make a complete orbit of the earth without bumping into another atom.

The Echo satellite that was launched in 1959 to test the possibility of radio communication via reflection from its balloon sent back a wealth of information about the structure and behavior of the exosphere. Passive reflectors were soon superseded by active communication satellites, but for three years a bright starlike object in the evening and early morning sky, Echo taught us much about the behavior of the atmosphere through which it moved. The drag on the balloon satellite in 1959–1962 indicated what seemed to be a surprisingly high density at 1500 kilometers if the major constituent of the air were atomic oxygen. To account for so much oxygen, the temperature had to be greater than 2000 K. Neither could an oxygen-hydrogen mixture explain such density; to do so, the concentration of hydrogen would have to exceed, by an order of magnitude, the best available evidence. However, when Belgian atmospheric scientist Marcel Nicolet introduced helium into the model, he showed that the observed satellite drag could be theoretically accounted for without exceeding a temperature of 1500 K. His calculations required a helium "bulge" in the composition of the atmosphere between a predominance first of oxygen and then of hydrogen.

The first direct measurements of neutral helium in the exosphere were made with a mass spectrometer carried on board the Explorer XVII satellite in 1963. Continuing measurements confirmed that helium is the major atmospheric constituent between about 500 and 1000 kilometers, but that the concentration at any time varies with solar activity. Near solar-maximum activity, helium dominates from 1100 kilometers to 5000 kilometers in response to increasing atmospheric temperature.

Helium enters the terrestrial atmosphere as a result of the radioactive decay of uranium and thorium in the ground. According to estimates of the abundance of these radioisotopes in the basalt and granite of the earth's crust, the outflow of helium is about 10^6 atoms per square-centimeter column per second, and only 10^6 years are needed to fill the atmosphere with its normal helium content. The rate of escape required to balance the ground-level release, then, led to the early estimate of an exospheric temperature of 1500 K.

The exospheric temperature is directly affected by energy delivered by solar ionizing radiations. It therefore varies with solar activity, such as flares, as well as with the more gradual changes over the sunspot cycle. Marked variations clearly follow the 27-day rotation period of the sun when the disk is highly spotted.

Calculations of the heat flow through the thermosphere and exosphere above 200 kilometers indicate a constant temperature nearly independent of height. Oxygen, helium, and hydrogen separate out as predicted by simple diffusion theory and in good agreement with the helium bulge predicted by Nicolet. Hydrogen, the lightest gas, diffuses upward most rapidly and escapes into interplanetary space. Solar ultraviolet dissociation of water vapor in the mesosphere is a continuous source of its replacement. As it leaves the earth, it creates a cloud of great extent and more than 50,000 miles high, known as the geocorona.

Knowledge of the drag effect on satellites is important in maintaining their lives. When solar activity is high, the atmosphere swells outward as short-wavelength radiation heats the absorbing regions. An extended atmosphere exerts greater drag on satellites and draws them down to lower levels. According to Kepler's law of orbital motion, the velocity of the satellite increases as its orbit shrinks, and it then loses energy yet faster to atmospheric drag. Skylab was such a fatality in 1978.

Still more recently, a large solar flare in April, 1981, occurred while the space shuttle Columbia, manned by astronauts John Young and Robert Crippen, was in orbit. The flare lasted more than three

Relative abundances of neutral atmosphere constituents at various temperatures and altitudes. Lines on the graph show where numbers of particles are equal—for example, where the number of oxygen atoms equals the number of nitrogen molecules per unit of volume.

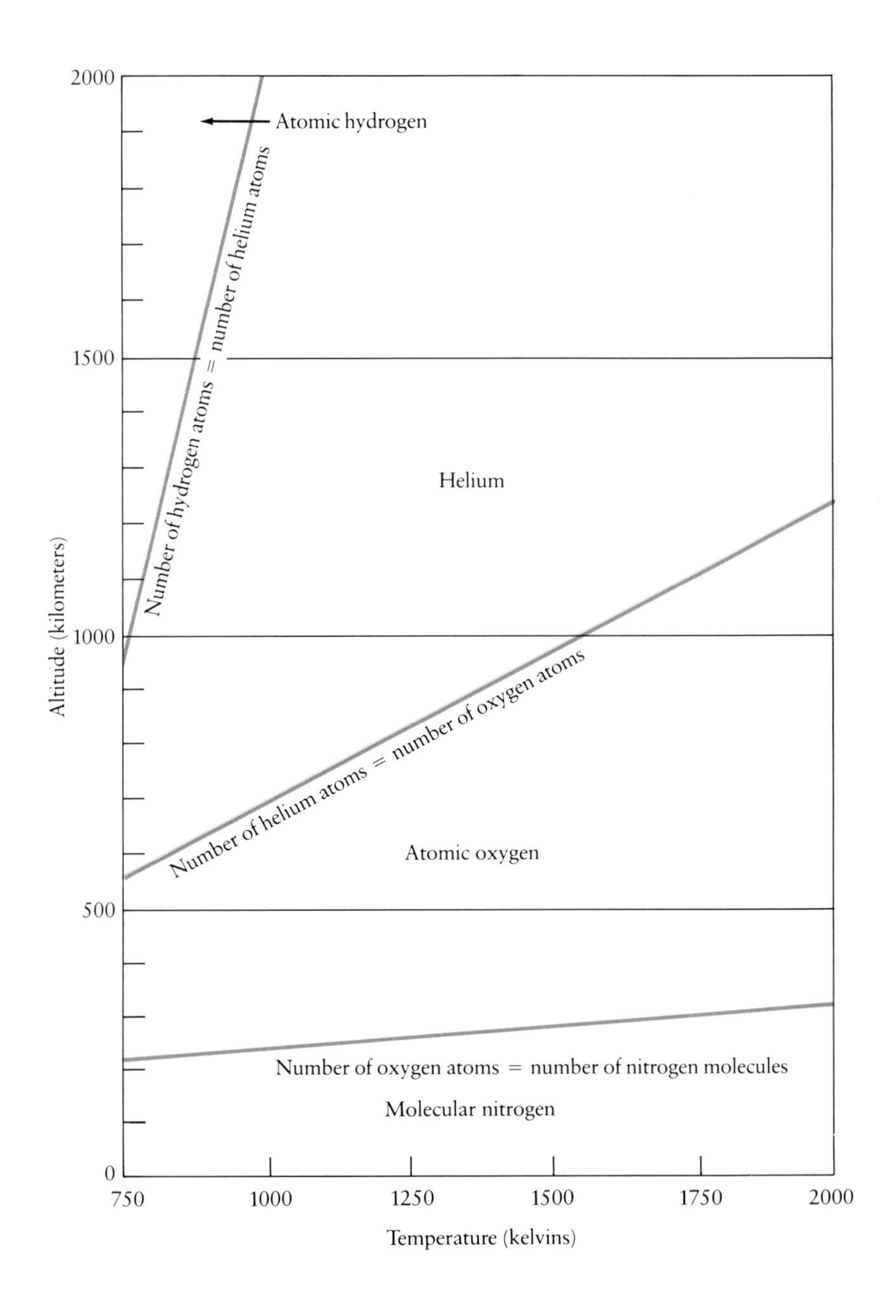

hours. The ensuing magnetic storm, which began only 15 hours after takeoff, was the largest of the current sunspot cycle. At 260 kilometers, the atmospheric temperature jumped from a normal 1200 K to 2200 K, the greatest increase recorded in 15 years of measurements. As the atmosphere heated up, the shuttle was dragged into a lower orbit 60 percent faster than expected. Unlike the derelict Skylab, however, the shuttle could be brought back to the required altitude by firing its retrorockets early enough to correct for the impending overshoot on reentry.

The stream of electromagnetic radiation from the sun—visible light, X rays, ultraviolet rays, and radio waves—travels in straight lines at the speed of light until it encounters the terrestrial atmosphere. The interplanetary medium between the sun and near-earth space offers almost no interference. But energy also flows from the sun in the form of particle radiation. Interplanetary magnetic fields guide the energetic particles, borne on the solar wind, in curving, spiral paths toward earth. At about 10 earth radii, the terrestrial magnetic field impedes their flow, forcing them to sweep around the barrier deep into the night side of earth. The following chapter treats this stream of particle radiation and its interaction with the earth's magnetic environment.

Light of the Night Sky

Visible airglow is a faint, diffuse light, barely visible to the human eye, even on moonless nights far away from the lights of cities. Produced by atoms and molecules that have been excited by solar ultraviolet radiation or chemical reactions, this internal energy can be quenched by collisions or radiated as airglow. Most of the glow originates at heights from 45 to 225 miles. It is about one-tenth as bright as the combined light of all the stars. In the near infrared, the glow is much more intense; if it were visible, it would be as bright as twilight.

Atomic oxygen is the source of green (5577 Å) and red (6300 Å, 6364 Å) airglow. Sodium produces a yellow glow (5893 Å), and various other colors result from excitation of molecular oxygen, atomic and molecular nitrogen, hydroxyl radicals, potassium, and lithium.

Airglow tells a complicated story of chemistry and dynamics in the high atmosphere. In Chapter 5, we discuss the most dramatic nighttime color spectacle of all—the aurora. The atomic and molecular radiation processes are much the same as in the airglow, but the stimulus for the aurora comes from energetic electrons rather than from ultraviolet excitation.

In 1955, my colleagues and I at the Naval Research Laboratory attempted an exploratory observation of extreme ultraviolet radiation in the night sky from an Aerobee rocket. Above 45 miles, we were startled to discover a flood of hydrogen Lyman-alpha radiation that quickly saturated the detector. In 1957 the observation was refined, and we found evidence that the Lyman-alpha glow was symmetrically distributed around the antisolar direction. It appeared that solar Lyman-alpha outside the cone of the earth's shadow was being returned to the night sky by scattering from atomic hydrogen in deep space. At first we thought the hydrogen must be interplanetary, but subsequent measurements and theory proved that it formed an earth-centered cloud extending as far as 50,000 miles—a geocorona.

For the Apollo-16 mission, George Carruthers of NRL built an ultraviolet camera that could photograph the geocorona, which astronaut John Young set up and used on the surface of the moon. In the Lyman-alpha photograph, the far-reaching

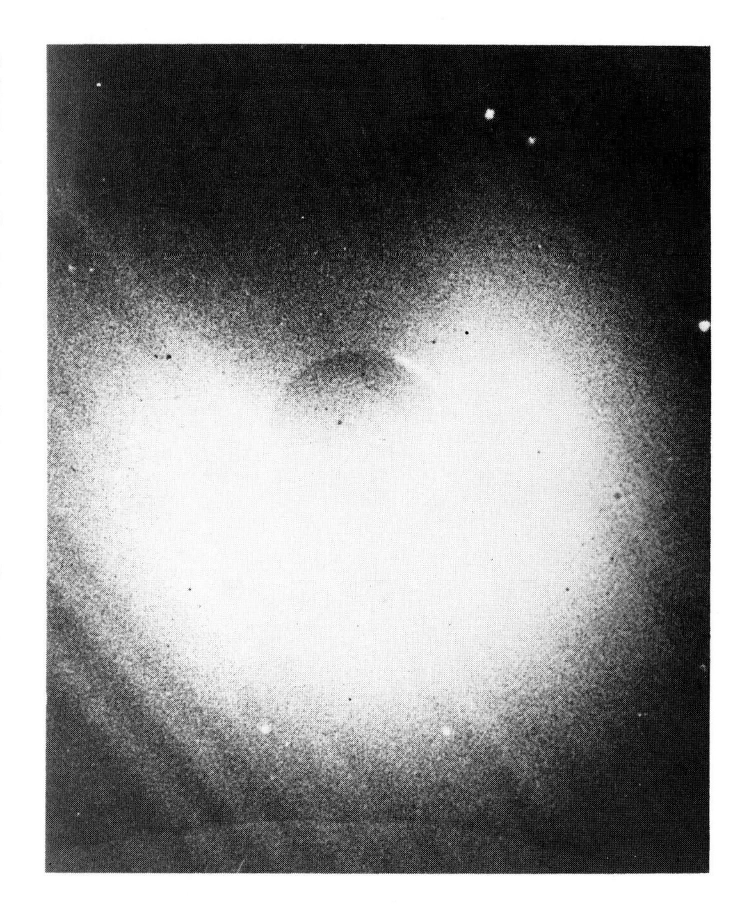

Lyman-alpha glow of the geocorona photographed from the moon. The earth-centered cloud reaches 50,000 miles, and its intensity exceeds the total of all visible airglow.

Apollo-16 camera-spectrograph on the moon. The electronographic camera had a three-inch aperture and was sensitive to far-ultraviolet light. It was designed by George Carruthers and operated on the moon by astronaut John Young. Photographs were obtained of the hydrogen geocorona and airglow emission from oxygen.

geocorona is fully revealed. The longer-wavelength image includes the principal resonance line of atomic oxygen at 1304 angstroms and emission from molecular nitrogen. Atomic oxygen above 60 miles scatters its resonance wavelength so strongly that the entire sunlit hemisphere glows with characteristic ultraviolet light. In the lower right part of the image, emissions from the polar cap on the night side are excited by energetic particles.

The spectacular arcs extending into the nighttime hemisphere shine in the ultraviolet airglow emission of oxygen ions excited in the process of recombining with ambient electrons in the high ionosphere. It almost appears as though the arcs are girdling the earth on the dark side, but, of course, the solid earth is in the way. The glow is aligned with the double-humped peak of the ionospheric F region that brackets the magnetic equator.

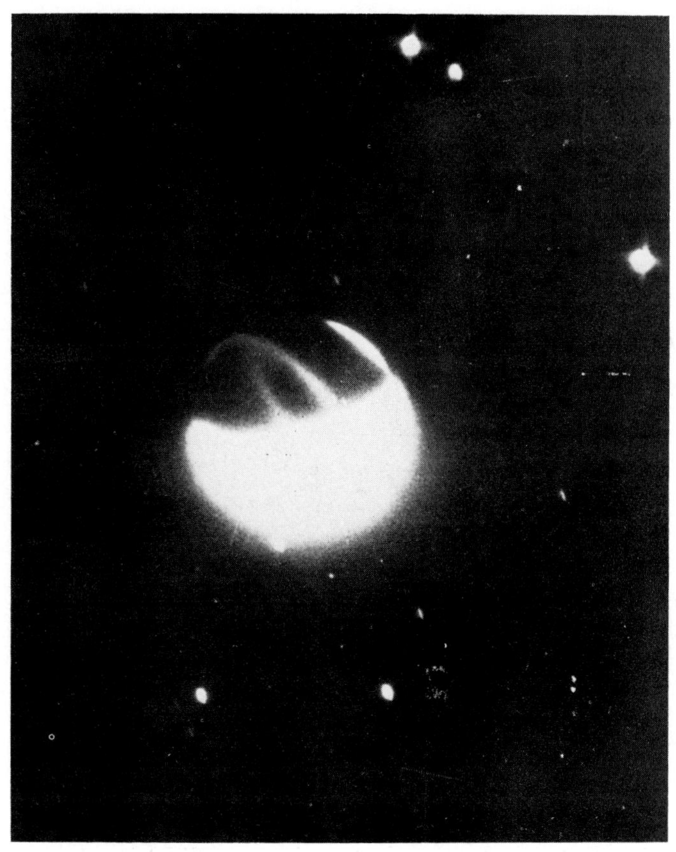

Extreme ultraviolet airglow photographed from the moon. Light is emitted above 60 miles. Arcs occur in the high ionospheric F region over the geomagnetic equator. Polar cap radiation visible on the night side is excited by energetic particles.

5

An Inconstant Sun: Solar Particle Radiation

A void was made in Nature; all her bonds
Crack'd; and I saw the flaring atom-streams
And torrents of her myriad universe
Ruining along the illimitable inane,
Fly on to clash together again, . . .

Alfred, Lord Tennyson, "Lucretius"

Painting of an aurora borealis by Frederic Edwin Church.

Although scientists had long known of a soft, radiant pillar of dust-scattered zodiacal light stretching from the earth toward the sun in the evening and early morning skies, until the middle of this century they believed that the earth and sun were isolated in a near-perfect vacuum. We have subsequently learned that interplanetary space carries a heavy traffic of particles and fields. The 93-million-mile gap between sun and earth is bridged by coronal streamers whose far-reaching fingers brush the fringes of the terrestrial environment and by winds of solar energetic particles that press and eddy against the edge of the earth's magnetic shield.

Solar Wind

As early as 1931, Sidney Chapman and Vincent Ferraro proposed that sudden changes in the geomagnetic field measured at the ground resulted when clouds of solar plasma collided with the earth's magnetic field. Twenty years later, observations of comet tails led to the concept of a continuous solar wind flowing across space from sun to earth. A thousand years ago, Chinese astronomers had noted that comet tails always point away from the sun, and, until relatively recently, it was thought that the pressure of solar radiation swept gas back from the comet head into its extended tail. Comet tails are so transparent that stars can be seen through them, and by the turn of this century, it had been concluded that gas atoms in the tail can absorb only a very minute portion of the radiation that falls on them. Finally, in 1951, the German astronomer Ludwig Biermann proposed that the sudden jumps observed in streaming comet tails could be attributed to interactions with directed, spasmodic streams of particles from the sun.

Comets are believed to have a solid nucleus that is about 10 to 20 kilometers in diameter and composed of dust entrapped in a block of frozen molecules—the "dirty ices" proposed by Fred Whipple. As a comet approaches the sun, it evaporates a cloud of gas molecules, known as a coma, that surrounds the nucleus to a distance of about half-a-million miles. Comets always trail one or more tails: a long straight tail of ionized gas that points almost radially away from the sun, and secondary tails of dust that are usually curved. The tails may be as wide as the coma and stretch as far as 10 million miles. Detailed studies of the acceleration of knots of gas and dust in comet tails suggest that gas tails are driven by a solar wind of a few hundred miles per second and dust tails by the pressure of sunlight. Ionized gas accelerates to speeds a hundred to a thousand times greater than that of the dust.

Comet Bennett seen in the evening over Gornergrat, Switzerland, on March 26, 1970.

Sidney Chapman was a leading contributor to the development of solar-terrestrial physics. He recognized that the coronal atmosphere of the sun reached as far as the earth and beyond, and advanced the concept of Chapman layers to explain the structure of the ionosphere.

In 1958, Eugene N. Parker calculated that the hot solar corona cannot be in static equilibrium but must expand as a wind into space. About two years later, the Soviet space probe Lunik 2, which impacted on the moon, reported the first tentative observations of a solar wind. Subsequent measurements aboard Lunik 3 and a Soviet Venus probe, together with evidence from NASA's Explorer 10, lent further support. But it was not until the results of Mariner 2 were reported in 1962 that the concept of a solar wind became widely accepted. We now know that wind flows from the lower corona at speeds increasing to about 400 kilometers per second at 20 solar radii and beyond. Although fluctuating in time and space, its speed is supersonic throughout the interplanetary medium. Solar-wind particles transport only ten-billionths the energy of that carried in light and all other forms of electromagnetic radiation, but their collective influence on the earth's magnetic environment is profound.

But just how could the sun dispense its matter into space? Until the mid-1950s, the corona was regarded as a great bag of hot gas held captive by the strong gravitational field of the sun. More refined observations, especially those from rockets, resolved a tight mesh of magnetic loops close to the base of the corona, and it became apparent that the corona is imprisoned in a magnetic cage rather than being held captive by gravity. Gene Parker's solar wind required an escape route from this magnetic trap.

Coronal Holes

Max Waldmeier, a Swiss astronomer who observed with a coronagraph atop Arosa in the Alps, had recognized persistent, extended gaps in the corona away from the poles where the light intensity was markedly reduced. He called these gaps *Löcher*, German for "holes." His observations drew no particular attention until the eclipse of 1970 revealed a spectacular coronal hole. Elongated east to west on the southwest solar limb, it was unmasked by streamers before or behind. The gap appeared so dark that it seemed to represent an almost complete absence of plasma. During the next few years, rocket observations in ultraviolet and X-ray wavelengths showed that coronal holes were commonplace, though not as prevalent as active regions. Regions 10 times less dense than the rest of the corona, they were not, however, complete voids. Most important, the magnetic field lines emerging from a hole appeared to be unipolar and open to space—they stretched radially outward and did not loop back.

Interest in coronal holes ran extremely high at the time of the Skylab launch in 1973. The mission covered the period from May 1973 through January 1974, the most opportune phase of the solar cycle for the production of magnetic storms, and it produced a wealth of information about coronal holes by means of X-ray photographs. Holes seem to be permanent features at the polar caps, and the charac-

The outer corona in white light was photographed from Kenya, Africa. Inside the eclipsed disk, a Skylab X-ray image of the inner corona is collaged. The outer coronal forms connect with the X-ray active regions near the limb. A brush of fine filaments stands straight up from the south pole. Covering most of the north pole is a large coronal hole that joins with a long, snaking coronal hole that reaches across the equator.

teristic polar plumes emerge above those holes. At lower latitudes, all the way to the equator, holes appear as dark islands between active regions that are bright in X-ray emission. Holes are so prevalent that at times they cover up to 20 percent of the solar surface. Some may last as long as a year, but four to six months is a more typical lifetime. Such persistence exceeds that of any other solar markings. The majority of low-latitude coronal holes seem to connect to the polar holes, and they can expand by as much as 20 million square miles per hour.

The growth of a coronal hole is always associated with the emergence of large, bipolar sunspot groups. Skylab observations dispelled many earlier impressions of the stability of bipolar groups. In particular, north and south polarity regions within a group do not remain coupled to each other. After a few days, spots of one polarity transfer their connections to other nearby spots of opposite polarity, so that unipolar groups are constantly coupling with different active regions. In the process, if a unipolar region finds no available opposite polarity adjacent to it, its lines of magnetic force may open outward, and a coronal hole is thus formed. Surprisingly, it appears that the holes develop most often in large active regions rather than in the magnetically quieter portions of the sun.

The flow of plasma in a magnetic field induces an electric current just as the rotation of the armature of a dynamo induces current in its wires. But every flow of current generates its own magnetic field, and if a current flow is displaced, it drags its magnetic field along with it. The magnetic fields drawn out of coronal holes are effectively "frozen" into the solar wind. By the time the solar wind reaches a few solar radii away from the sun, it has become so thin that its plasma particles are essentially free of collisions, and electric currents flow freely within it, with almost negligible resistance.

As the solar wind flows out of the sun, the lines of force of the solar magnetic field stretch outward like rubber bands. Up to about three solar radii, the magnetic fields attached to the corona are strong enough to dominate the flow pattern, and the streams of the solar wind corotate with the sun. At greater distances, the kinetic energy of the wind plasma becomes greater than the energy stored in the solar magnetic field, and the wind starts to slip out of the sun's magnetic grip. From there on out, the motion of the wind controls the shape of the stretched-out magnetic field. As the wind nears the earth, the energy in the terrestrial magnetic field gradually overbalances the energy of the moving particles in the wind, and this diverts the flow of plasma around the earth.

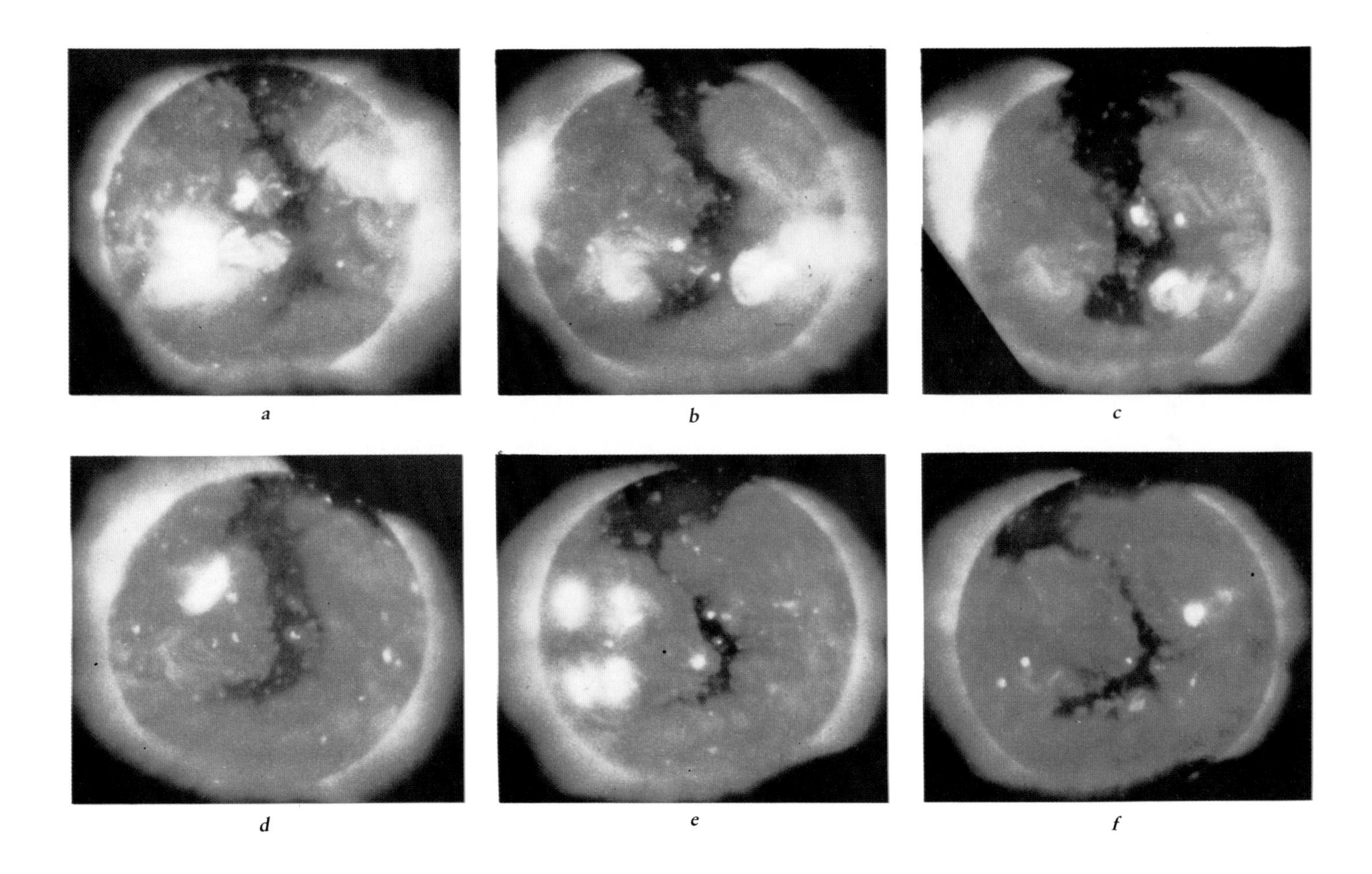

A persistent chain of coronal holes stretches across the solar equator. These photographs made with an X-ray telescope aboard Skylab show the sun at 27-day intervals over six solar rotations, beginning on June 1, 1973, in the upper left-hand corner and followed left to right by row.

The Flow Pattern of Solar Wind

X-ray and ultraviolet images show that magnetic field lines over the sun's polar caps are open to space. Plasma flows freely outward in the form of high-speed solar wind, and gradually turns back toward the ecliptic plane (the plane of the earth's orbit). At lower solar latitudes, the open field lines above coronal holes over much of the solar disk also offer an escape route. If we could look down on the solar pole, we would observe solar-wind streams rotating with the sun, the stretched magnetic field lines curved into graceful Archimedean spirals like jets from a rotating garden sprinkler. (The swirling pattern is evidence that the wind robs the sun of angular momentum and brakes its rotation.

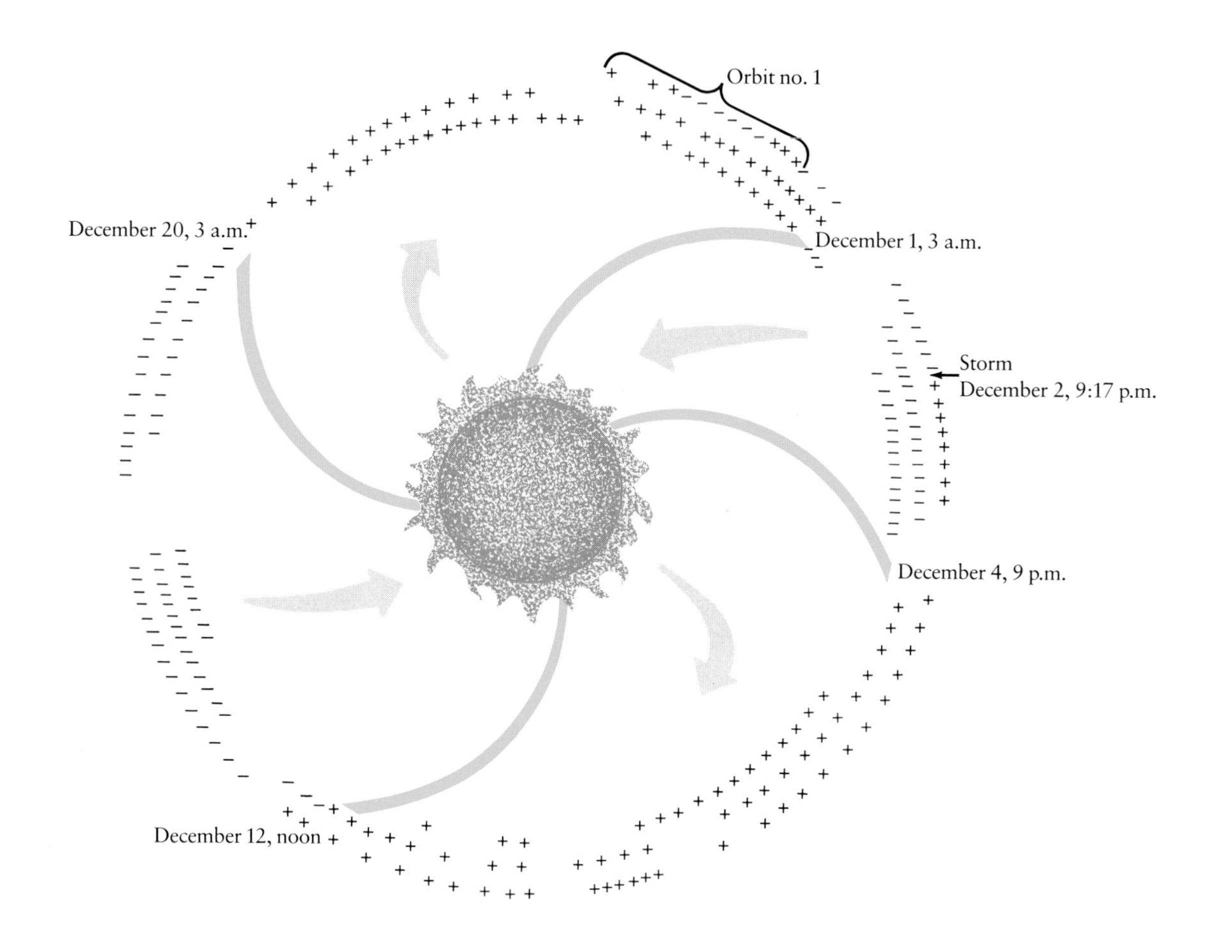
Orbit no. 1
December 20, 3 a.m.
December 1, 3 a.m.
Storm
December 2, 9:17 p.m.
December 4, 9 p.m.
December 12, noon

Solar magnetic sector pattern derived from the IMP-1 (Interplanetary Monitoring Platform 1) spacecraft. Plus signs (field directed away from sun) and minus signs (field directed toward sun) are three-hourly measurements of interplanetary magnetic field direction. The arrows in the four sectors indicate the predominant field direction. The pattern persisted over the three successive rotations shown here.

Early in its history, the sun must have rid itself of a great deal of rotational energy by this sort of slingshot release of a very much stronger solar wind.)

Over most of an 11-year solar sunspot cycle, the magnetic field assumes a spiral form in each hemisphere, but with opposite polarities that reverse from one cycle to the next. The oppositely directed magnetic fields are separated by a thin current sheet—a magnetically neutral layer along which current can freely flow—that lies close to the equatorial plane of the sun.

Magnetometers carried aboard spacecraft in the mid-1960s showed that the rotating flow had four magnetic divisions, or sectors, resembling the vanes of a pinwheel. The sectors alternate in polarity like oppositely aligned bar magnets: if the field of one sector is north-south, the adjacent one is south-north, and so on. A field directed away from the sun is termed positive; directed toward the sun, it is termed negative.

The simple sector pattern has a much more interesting three-dimensional shape. If the flow of wind were smooth and equalized from both hemispheres, the current sheet would lie in the ecliptic plane. But the current sheet is warped upward and downward as it extends into the interplanetary medium. As a result, the field at any point in the ecliptic plane is not a flat spiral but may be directed up or down at angles as large as 30 degrees to the ecliptic. This warped current sheet, then, cuts across the earth like the undulating skirt of a pirouetting ballerina. The sector pattern derives from four passages of the warped sheet past the earth every 27 days. At each crossing, the magnetic polarity switches from positive to negative or vice versa, depending on whether the earth is above or below the current sheet. From one rotation to the next, the size of each sector may vary, and the warp may flatten out near earth so that only two sector crossings occur in one rotation.

Why is the current sheet warped? A chart of the solar magnetic field over the photosphere from day to day reveals a basic pattern. Even though large areas of positive and negative polarity slowly vary in spatial position and extent, they tend to fit into four zones of alternately positive and negative polarity. The line of separation wanders up and down across the solar equator, and it often rises as far as halfway from the solar equator to the pole. As the solar wind carries these magnetic fields into space with the curving pattern of the neutral boundary line, the spread-out current sheet retains a characteristic undulation. This underlying pattern seems to be very long-lived, and the sectors that result from it may persist for decades.

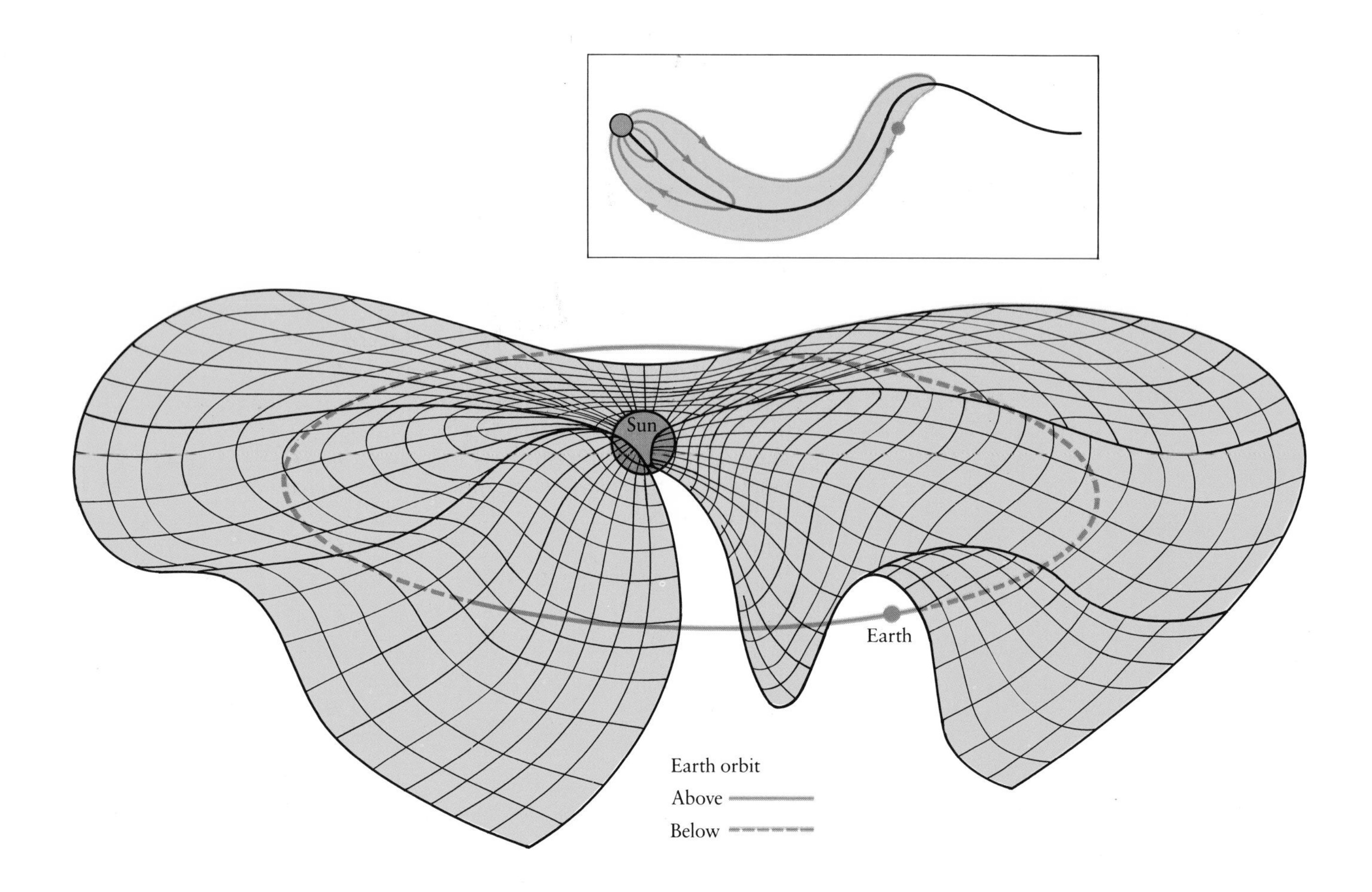

The so-called ballerina skirt model of the solar-interplanetary current sheet. The sun is at the center of an extensive and warped disklike sheet through which electric currents flow around the sun. The average plane of the disk is approximately the plane of the equator of the sun's average dipole magnetic field. The inset, which is a meridian cross section of the current sheet and magnetic field lines on each side, shows that the sheet separates solar-interplanetary magnetic-field regimes of nearly opposite, or at least greatly different, average direction.

Measurements from spacecraft show that the solar-wind speed is as low as 300 kilometers per second at a sector boundary and that it rises to about 750 kilometers per second at some distance away from the edge. If the warp in the current sheet is large, the solar-wind speed at earth will show a correspondingly large variation with each sector passage. Near the time of a sector-boundary passage, the speed and density of solar wind, the intensity of the "frozen in" magnetic field, and the geomagnetic storminess at earth all drop to a minimum. As the boundary sweeps by, all these characteristics start to increase in strength and reach a maximum in about two days.

Coronal Transients

Normally, the solar wind is quite smooth, but coronal activity can cause it to gust. Solar flares expel blobs of higher-speed plasma that plow into the slower flow of the wind and cause pileups that result in shock waves several million miles thick. The plasma becomes highly

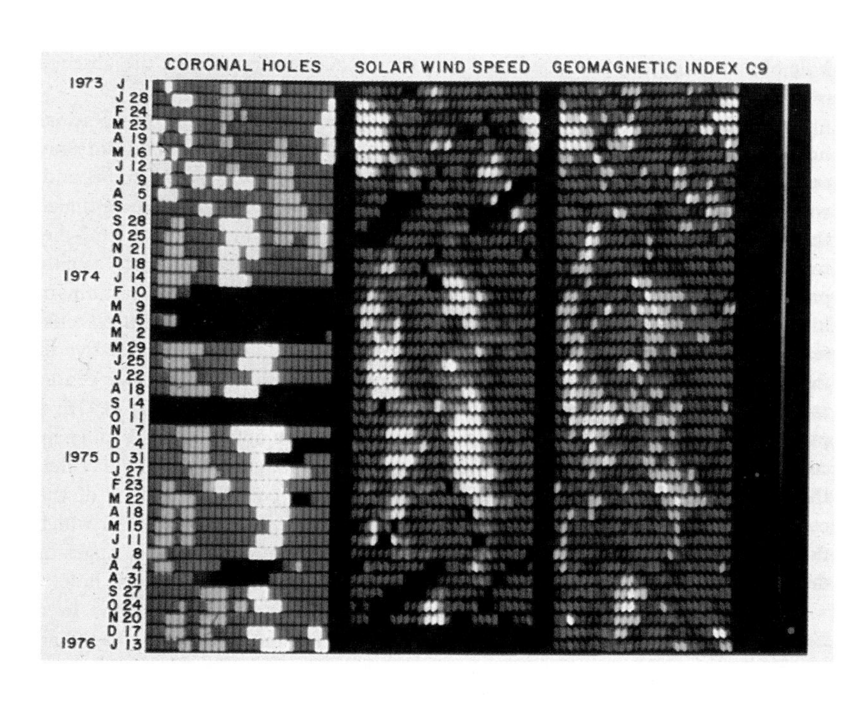

Comparison between coronal holes, central meridian passage dates, solar wind speed, and the geomagnetic disturbance index over three years from 1973 to 1976. Each 27-day row corresponds to one full solar rotation. The date of the first block in each row is indicated at the left. Orange blocks refer to coronal holes with outward-directed photospheric magnetic fields (positive polarity); white blocks represent inward-directed fields (negative polarity). Blue blocks indicate absence of a coronal hole. Black sections are missing data intervals. In the center and right columns the polarity of the interplanetary magnetic field is indicated by blocks slanted to the right for positive polarity, to the left for negative polarity, or not slanted for mixed polarity. The similarity of the large-scale patterns of holes, wind, and geomagnetic activity as well as the general correspondence between photospheric and interplanetary magnetic-field polarity confirms the strong connection between these features.

During the eclipse of 1860, German astronomer E. W. L. Tempel sketched a transient bubble in the corona.

turbulent, and frozen-in field lines get tangled. Such clumps of kinky field lines scatter galactic cosmic rays in random directions, away from the ecliptic and the neighborhood of earth.

Observations of the sun made with coronagraphs in space have shown a variety of transient forms of plasma expulsion—large loops, spikes, and great bubbles. A German astronomer, E. W. L. Tempel, sketched his impression of a gigantic whorl-like blister that appeared on the corona during the eclipse of 1860. Most of his contemporaries regarded his drawing as the product of artistic license, but more than a century later, coronagraph images from Skylab proved him right. The astronauts photographed 24 transient events, great gas bubbles blowing out of the corona at 150 to 500 kilometers per second. From that one-year sample, it appeared that such transients occur about once every 100 hours on the average.

Since 1979, an orbiting coronagraph mission prepared by the Naval Research Laboratory, Solwind, has been acquiring a large body of data on coronal transients. The largest transient expulsions have blown as much as 10 billion tons of gas out of the corona. Outward speeds have ranged from 150 kilometers per second to 900 kilometers per second. One event on November 27, 1979, completely surrounded the occulting disk of the coronagraph and appeared to be propagating

Coronal transients are billion-ton bubbles of gas ejected from the corona. The release often appears to be triggered by a surge prominence or flare. The sequence of ejection is shown from left to right and top to bottom, covering about two hours.

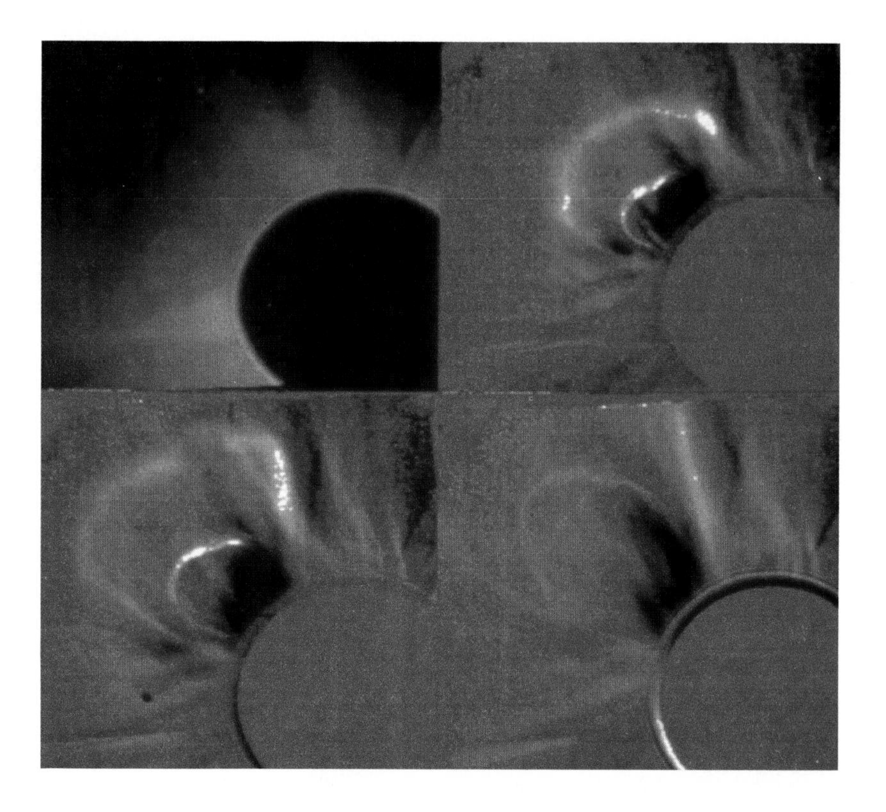

as a filled bubble rather than a shell. About half of the major mass ejections are accompanied by prolonged X-ray signatures, sometimes lasting as much as nine or 10 hours. Shocks in the interplanetary medium that travel with these blobs of plasma have been frequently observed by Helios solar satellites from distances of 60 to 200 solar radii above the sun's limb.

The Voyage of Pioneer 10

The entire solar system exists within a huge bubble swept out by the solar wind into the interstellar gas. Eventually, far beyond the orbits of the planets where it stalls against the pressure of interstellar gas, the solar wind must subside. No spacecraft has yet tracked the wind that far, but Pioneer 10 and Voyagers 1 and 2 will sense it as long as they continue to function. Pioneer 10 has now passed the orbits of Uranus and Neptune. On July 26, 1981, the spacecraft reached the 25-astronomical-unit mark (1 AU is the distance from sun to earth, 93 million miles). Launched on March 2, 1972, Pioneer 10 was expected

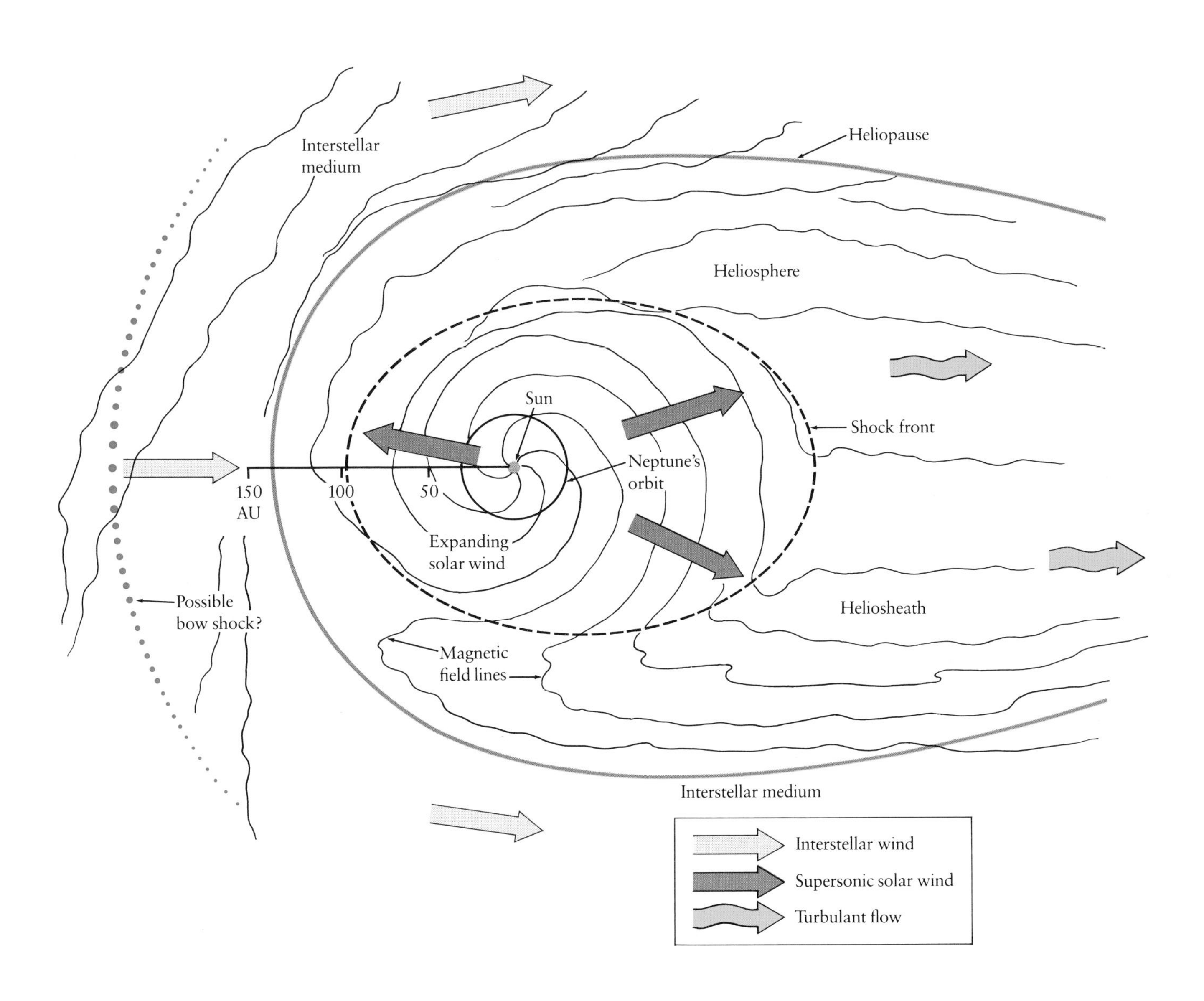

A tentative model of the heliosphere. The plane of the figure is the plane of the sun's equator. As the solar wind expands outward, its energy density diminishes until it is matched by the pressure of the interstellar gas and magnetic field. The heliopause boundary may occur at about 150 astronomical units. Pioneer 11, now well past Saturn, has not yet detected the heliopause.

to operate for about 30 months, just long enough to encounter Jupiter, but it is a remarkable survivor. It is now heading in the direction of Aldebaran at a speed of 2.8 AU per year but has not yet signaled the encounter with the stagnation boundary of the solar wind, the heliopause. Scientists now estimate that the heliopause may be as distant as 100 AU, far beyond the orbit of Pluto.

All of our spacecraft thus far have flown on trajectories close to the ecliptic plane, so that we have no direct measurements of the solar wind far above it or below it. What knowledge we do have of the solar wind blowing from high solar latitudes is gleaned from observations of comets and the scintillation of radio signals from cosmic sources, such as quasars. Irregularities in clumps of ionized gas in the interplanetary medium produce radio scintillation analogous to the optical twinkling of stars that is associated with turbulence in the atmosphere. Observations made in England, and more recently with the great 1000-foot radio dish at Arecibo, Puerto Rico, indicate that the solar wind flows radially from the sun at all heliographic latitudes and appears to increase in velocity toward the sun's poles.

The earth's magnetic shield is so perfect that only 0.1 percent of the mass of solar wind that hits it manages to penetrate it. But even that small fraction of total energetic-particle radiation generates a host of complex auroral phenomena and accounts for the storage of enormous numbers of particles in the Van Allen radiation belts.

The Earth's Magnetic Shield

The discovery of the lodestone—the ore of magnetite—led to the invention of the magnetic compass in the eleventh century. Its mysterious ability to attract iron particles was known to the Chinese perhaps as early as 2600 B.C. In searching for the explanation of its magical property, investigators discovered that when a lodestone was attached to a block of wood and floated on water, it assumed a north-south orientation. This observation led directly to the use of this apparatus as a compass.

That the compass did not faithfully point to true north from all locations was not generally known when Columbus set sail for America. While at sea, his crew noticed that the compass pointed 10°W of true north as identified by the pole star. There was considerable uneasiness and a threat of mutiny. To calm their fears, the next night Columbus surreptitiously shifted the compass card on its needle, a serious violation of maritime law that prohibited falsifying the compass. (One of the statutes of the time even charged mariners not to eat onions or garlic lest the odor "deprive the lodestone of its virtue by weakening it and prevent them from perceiving their correct course.") Penalty for violation was extreme. "If the culprit's life were spared he must be punished by having the hand which he most uses fastened by a dagger thrust through it to the mast."

Lodestones of magnetite encased in artistically craftcd keepers.

In 1600, William Gilbert, physician to Queen Elizabeth I, published the first generally correct description of the earth's magnetism in his book, *De Magnete*. After two decades of experimentation, he effectively disposed of lingering misconceptions:

> As to what some writers have related, that a lodestone will not attract iron if there be a diamond near and that onions and garlic will make it lose its vertue; these are contradicted by a thousand experiments that I have tried. For I have shown that this stone will attract iron through the very thickest diamonds and through a great many thick skins which an onion is made up of.

These and other experiments led Gilbert to the profound conclusion that "the terrestrial globe is itself a great magnet."

In its normal mounting, the compass needle is oriented solely by the horizontal component of the magnetic field. It swings aimlessly at the north and south magnetic poles, where the field is vertical, but locks on strongly at the equator, where the field is horizontal. Magnetic dip had been discovered in 1576 by Robert Norman, a seaman and instrument maker. He found that if he mounted the needle on a horizontal axle, enabling it to move in the vertical plane, it dipped below the horizon. This observation hinted that the source of the earth's magnetic field lay beneath the terrestrial surface.

Following Norman's observations, Gilbert asserted that the compass needle does not aim at a point in the heavens, as many still assumed. Because the dip needle pointed increasingly downward as it approached the north magnetic pole, he believed that the field was produced by permanently magnetized material within the earth. Modern scientists have had to reject any idea of a vast body of magnetic material beneath the crust, however. Almost all of the earth's mantle and all of its core are far too hot for magnetism to persist.

Today, the generally accepted theory is one proposed in 1946 by Walter M. Elsasser. Most of the core is molten and electrically conducting. In the outer core, columns of liquid metal rise and fall in a convection pattern generated by temperature differences. At the same time, they are twisted by the earth's spinning motion. Such motions of conducting fluid in a magnetic field generate electric currents that, in turn, produce an overall magnetic field lined up approximately with the spin axis. Although we can hardly claim a precise understanding of the process, in essence the earth is a rotating, self-exciting electrical generator.

This description bears a certain resemblance to that proposed for the generation of the sun's magnetism. Visible evidence of the operation of the solar dynamo appears in magnetic patterns of the sun's surface, while the terrestrial dynamo is concealed beneath the earth's crust. Every 11 years, the polarity of the solar magnetic field reverses, and the switchover can be seen in sunspot behavior on the face of the disk. It has been known for almost a century that the earth's field also reverses, but the time scale is more like a million years.

Worldwide surveys made during the nineteenth century revealed many puzzling variations in the geomagnetic field. While most permanent magnetism lies inside the earth, about 2 percent originates externally in the high atmosphere. Precise measurements showed that the direction of the compass needle fluctuates in a regular pattern over the course of the day, and these fluctuations differ at different locations. Occasionally, the regular swings of the compass became more rapid, more pronounced, and more irregular. These fluctuations, which were most intense in the auroral zones, were called magnetic storms.

Perhaps the earliest systematic studies in solar-terrestrial relationships were the searches for connections between sunspots, auroras, and geomagnetic storms. In 1747, Olav Peter Hiorter of Upsalla remarked,

> Who could have thought that the northern lights would have a connection and a sympathy with the magnet and that these northern lights, when they draw southwards across our zenith or descend unequally toward the eastern and western horizons could, within a few minutes, cause considerable oscillations of the magnetic needle through whole degrees?

Great Britain operated a network of magnetic observatories throughout her colonial empire during the early nineteenth century. In 1852, Colonel Sabine of the British Army, who had been systematically correlating the data, noted that, by "a most curious coincidence," the magnitude and frequency of magnetic disturbances were synchronized with variations in sunspot numbers. Globally averaged, both reached maximums in 1848 and minimums in 1843.

Before the discovery of the solar wind, geophysicists held a view of the remote geomagnetic field not essentially different from that sketched by Gilbert in 1600. The earth and sun were two isolated bodies with nothing but "empty space" between. Auroras and geomagnetic storms, they believed, were somehow connected with particle streams that arrived directly from the sun, uninfluenced by any interplanetary medium until captured by the earth's magnetic field. Locked on the lines of magnetic force converging toward the poles, the particles were guided, it was thought, to two ring-shaped regions around the north and south poles. The speed with which particles traveled was inferred from the time between a visual sighting of a flare and a subsequent bright aurora and magnetic storm. Typically, the delay time was about two days, implying an average speed of 500 miles per second over the 93 million miles from sun to earth.

A remarkably close correspondence exists between cycles of change in number of sunspots and in the intensity of magnetic disturbances.

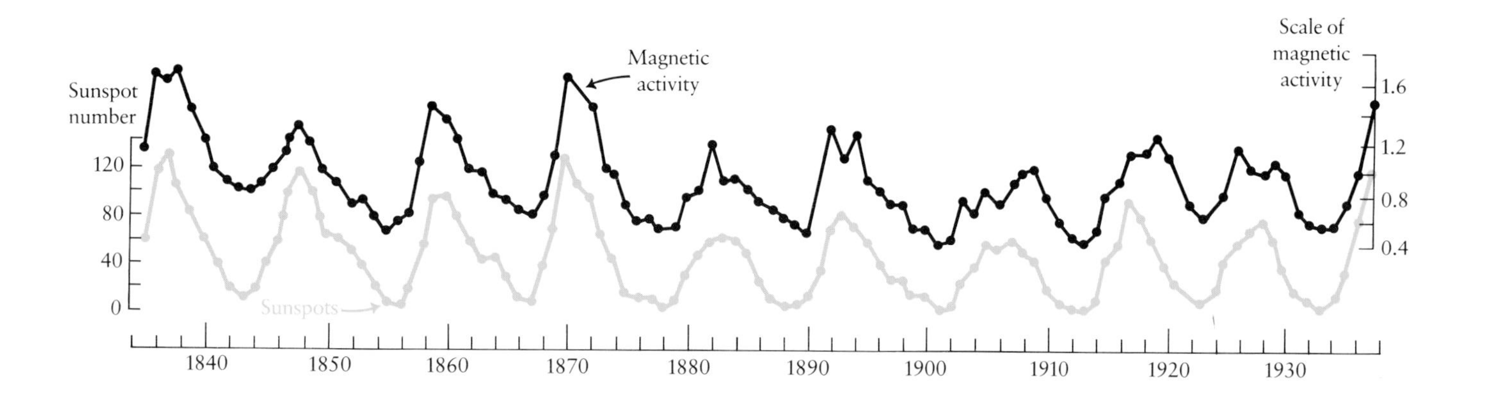

Our present understanding of the magnetic environment of the earth begins with auroral studies at the turn of the century. No substantial insights were gained, however, until the nature of the solar wind was recognized in the 1950s, followed soon after by the discovery of the Van Allen belts. As the concept of a terrestrial magnetic envelope, or magnetosphere, took shape, auroral theorists quickly proposed new models of solar-wind interaction with the earth's magnetic shield and a new mode of wind entry into the auroral zone by a back door rather than by direct flow to the polar caps. They soon saw that auroral current flows on many tracks, as though the magnetosphere were a gigantic switchyard. Indeed, it may be that the whole auroral mechanism is powered primarily by a trillion-watt natural generator created by solar-wind flow across the magnetosphere.

Northern Lights

> Gazing into the starry sky and across the sparkling bay, magnificent upright bars of light in bright prismatic colors suddenly appeared, marching swiftly in close succession along the northern horizon . . . . Though colorless and steadfast, the aurora's intense, solid white splendor, noble proportions, and fineness of finish excited boundless admiration. In form and proportion, it was like a rainbow, a bridge of one span five miles wide and so brilliant, fine and solid and homogeneous in every part. I fancy that if all the stars were raked together into one window, fused and welded and run through some celestial rolling-mill, all would be required to make this one flowing white colossal bridge.
>
> *Travels in Alaska*

With these words, John Muir described the most spectacular, the most beautiful, and until recently the most mysterious display in the night sky—the aurora. Credit for naming the northern lights after the Roman goddess of rosy-fingered dawn usually goes to Pierre Gassendi, the seventeenth-century philosopher, although the terms "aurora" and "aurora borealis" appear earlier in the writings of Galileo.

If we could look down from high above on the north polar region while it is cloaked in darkness, we would see a glowing ring around the geomagnetic pole. This auroral oval is constantly in motion, expanding toward the equator or contracting toward the pole, and constantly changing in brightness. When we look at the aurora from the ground, it most often looks like a curtain, luminous against the cold Arctic night. It spreads widely, stretching from the eastern to the western horizon. The lower edge reaches down to about 100 kilometers,

A computer-graphics image of an aurora over the earth's north magnetic pole as observed from the Dynamics Explorer I satellite at an altitude of 3.2 earth radii. The aurora's oval shape is projected on the green computer-generated map of the globe. The bright crescent on the left is the directly illuminated daylight side of the earth.

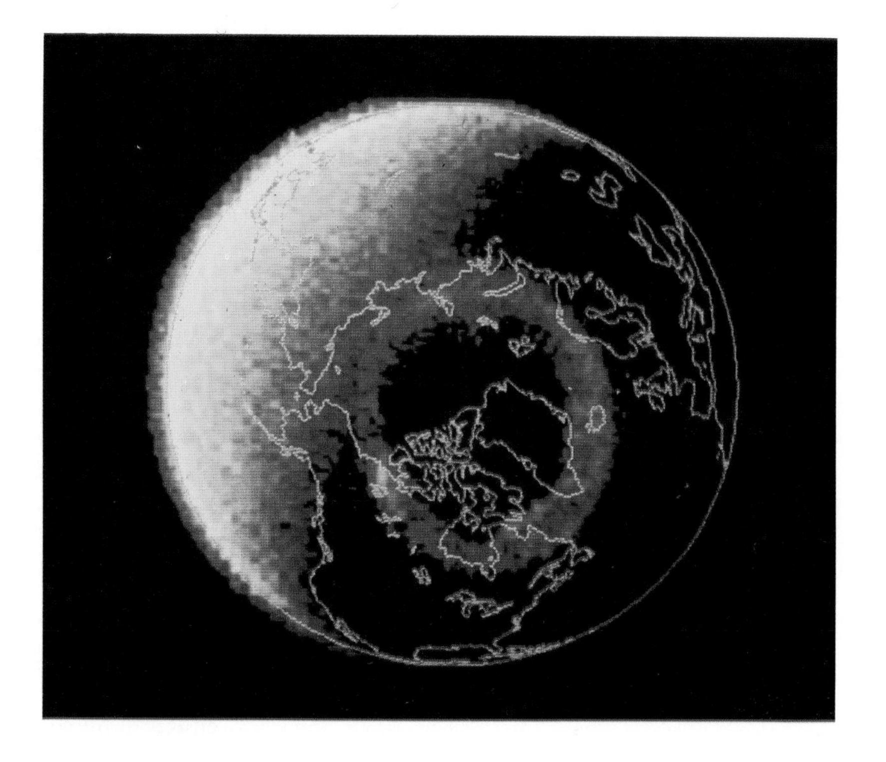

while the upper edge usually extends to about 400 kilometers and sometimes to 1000 kilometers. The aurora borealis, or northern lights, has its southern counterpart in the aurora australis. Images of the view from the moon obtained by Apollo 16 show simultaneous auroral ovals in both hemispheres.

Auroral forms are generally described in terms of five broad categories: *arcs* are gently curving streaks of light with smooth lower edges, *bands* are arcs that have evolved kinks or folds in their lower borders, *patches* resemble clouds of luminosity confined to small regions, *veils* are rather uniform sheets of luminosity spread over extensive regions, and *rays* are shafts of light oriented to the magnetic field at an angle to the vertical. These forms, along with *curtains, draperies,* and *rayed coronas,* may be homogeneous or striated. As events in time, auroras may be quiet or pulsating, flickering or flaming, and they may last only minutes or persist for hours.

No two auroras, like no two snowflakes, are exactly the same, but their brilliant displays tend to follow a characteristic sequence. A faint green glow creeps into the northern sky and gradually forms an arc that loops from horizon to horizon. As the arc moves upward, it brightens. New arcs rapidly develop across the zenith and down toward the northern horizon. When these bands start to thin out, a sequence of curtains and draperies of colored light begins to fold and unfold, while narrow rays dart here and there. As the very dazzling display draws to an end, the drama intensifies. Pulsating curtains collect in swelling clouds that brighten and fade in seconds. The wind-down is punctuated by sporadic bursts that prolong the fireworks until all light fades against the coming of dawn.

Most auroras are green with occasional fringes and patches of red. Excited oxygen atoms radiate both green, at a wavelength of 5577 Å, and red, at a wavelength of 6300 Å. Ionized nitrogen molecules emit ultraviolet and blue light in several spectral bands from 3914 Å to 4700 Å. Neutral nitrogen molecules glow deep red at 6500 Å and 6800 Å. Each color has a specific altitude range: oxygen/green and nitrogen/violet appear at about 110 kilometers and oxygen/red at 200 to 400 kilometers. The ionization and excitation that stimulates emission of most auroral light is produced by incoming electrons that have energies of under 10,000 electron volts, about half the energy of the electrons that hit the fluorescent screen of a TV tube.

Auroral studies have long fascinated scientists in Scandinavian countries ringing the Arctic Circle. In 1744, the venerable Proceedings of the Royal Swedish Academy of Science heralded the work of Sam-

Red aurora over Montana.

Green aurora over Nova Scotia.

uel Von Triewald, a poet and diplomat, for his insight into the mechanism of the aurora. Von Triewald's experiment reveals something of popular science at the time. In a dark room with a little hole in the wall to admit sunlight, he arranged a prism, a glass of cognac, and a screen. Rays of light passing through the aperture were refracted and dispersed as they entered the prism, much as in Newton's earlier experiments. When the refracted light skimmed the surface of the cognac, a pattern was projected on the screen. As Von Triewald put it:

> One was surprised to see a naturally occurring northern light on the screen that nothing could more resemble. As the cognac surface was warmed up by the colored sun rays, it began to evaporate and with that comes into existence a wonderful movement on the screen in which man sees all the phenomena like any natural northern light produces.

Von Triewald's experiment seemed to support a widely held notion that an aurora, rather like a rainbow, resulted from sunlight refracted by diffuse gases evaporated from the atmosphere. These gases set in motion by the wind, then, were supposed to create the dancing auroral patterns.

The great Norwegian physicist Kristian Birkeland led three expeditions to determine the height of the aurora in northern Norway between 1897 and 1903. His first expedition "to find out whether the northern light could . . . come right down to the tops of the mountains" nearly ended in death when he, his companions, and their reindeer were caught in a blizzard while ascending Berkades Mountain. Conditions on subsequent expeditions were equally fearful:

> In high winds it was impossible to get out, and more than once, on Sukkertop, it took three men with a great effort to close our little door . . . . Temperatures of −20°C accompanied by winds of from 20 to 30 m/sec were pretty frequent. No one who has not tried it can imagine what it is to be out in such weather.

On the second attempt, he succeeded in photographing an aurora from two sites 3.4 kilometers apart and, by triangulation, in establishing that its bottom edge lay at about 100 kilometers. During the last expedition, magnetograms and auroral observations from 25 stations encircling the auroral zone enabled him to conclude that large electric currents flowed along the earth's magnetic field lines into the aurora.

A keen observer of natural phenomena, Birkeland was also an ingenious experimenter in the laboratory. At the beginning of the cen-

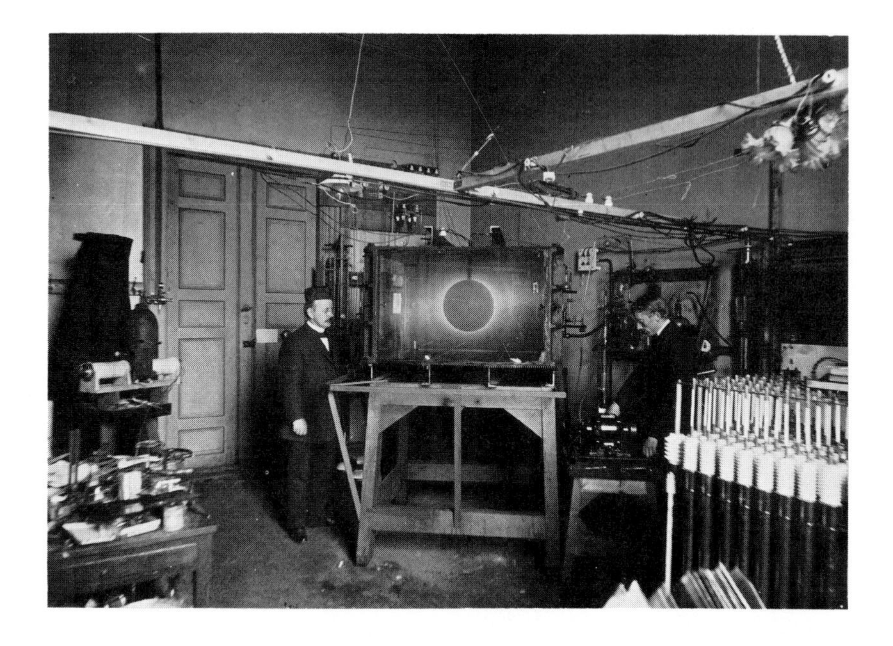

Kristian Birkeland in the laboratory with his terella apparatus, about 1910. An electron beam directed at the magnetized sphere in the vacuum chamber produced glowing "auroral" arcs. His assistant in the picture is Karl Devik.

tury, Sir William Crooke had demonstrated that cathode rays are deflected by magnetic fields. When Sir Joseph J. Thomson established that cathode rays are composed of electrons, Birkeland proposed that streams of fast electrons were ejected from sunspots. To model the phenomenon, he built a "terrella" apparatus, a sphere in which he embedded an electric coil to simulate the dipole magnetic field of the earth. He mounted the sphere in a vacuum chamber and projected a stream of cathode rays at it. When the strength of the electromagnet was varied, a change in the luminous glow pattern in the air of the vacuum chamber close to the terrella showed an astonishing similarity to the arcs of the aurora borealis. Birkeland believed that cathode rays came from the sun and were directed to the magnetic poles. Thus, he explained the daily occurrence of the aurora borealis and the already well-known coincidence between the 11-year period of solar activity and auroral occurrences.

Birkeland's ideas were not received favorably by his contemporaries, largely because Thomson argued "that the supposed connection between magnetic storms and sunspots is unreal, and the seeming agreement between the periods has been mere coincidence." Thomson's vigorous opposition squelched Birkeland's ideas for three decades. But Birkeland's terrella experiments excited Carl Stormer, a young theoretical physicist in Oslo, who undertook mathematical calculations of the motions of charged particles in a simple magnetic dipole field. Without the help of modern computers, he would spend almost 50 years on this problem, but as early as 1907, he published a paper describing how charged particles spiraled around the magnetic field lines of a simple dipole. He demonstrated that a charged particle moving in a magnetic field is subjected to a force at right angles to both the direction of particle motion and direction of the magnetic field. The strength of the force is proportional to particle velocity, its electric charge, and magnetic field intensity. Because the force acts perpendicular to the velocity, it bends the particle's line of flight but does not decrease its energy. As a result, charged particles move in helical paths around magnetic lines of force.

If the earth's field lines ran parallel to each other, any charged particle would continue to spiral along one line in loops of constant radius. But the earth's field lines are not parallel: they converge toward the magnetic poles and swell outward at the equator. Where field lines crowd together, charged particles spiral tighter and tighter until their forward motion finally stops completely. When all that remains is circular motion around the lines, the particle starts to unwind its spiral, moving along the field lines in reverse. As it travels back toward the equator, the loops of the spiral open up. At the equator, the process reverses, and these places where the reversal occurs are

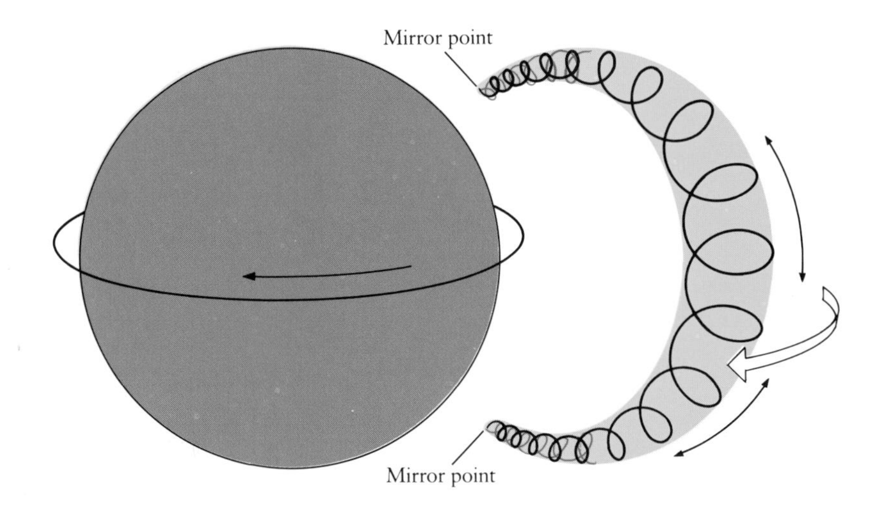

Charged particles in a magnetic field spiral about the lines of force. In the magnetic field of the earth the lines of force expand with height as they arch from north to south. As particles spiral up and down the field lines, they wind up more and more tightly closer to earth until the downward motion stops at the mirror points and the spiral starts to wind back outward.

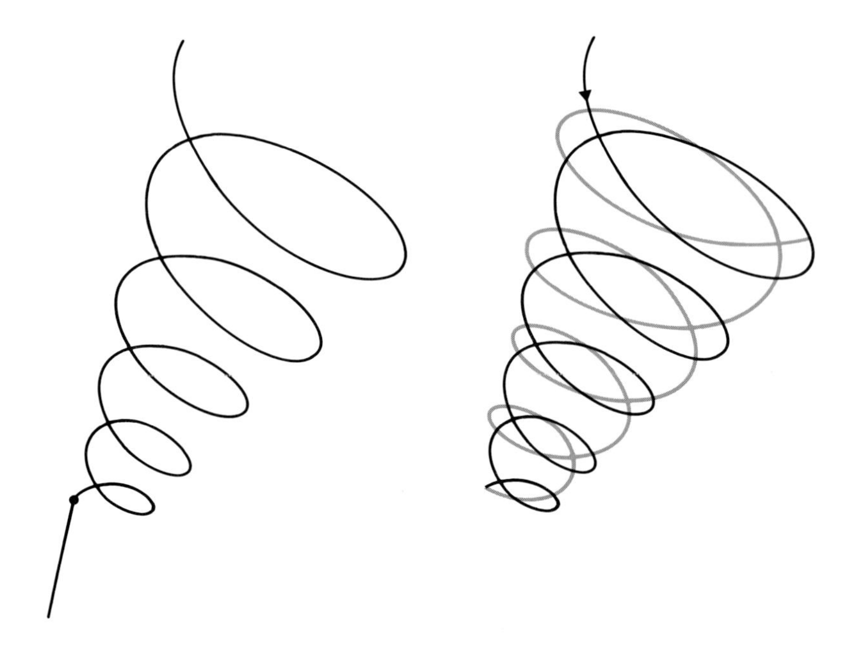

If a charged particle trapped on a line of force moves into the lower atmosphere, it may collide with an atmospheric molecule before mirroring backward. The collision disturbs the spiral motion and "dumps" the particle out of its trapped mode. When the mirror points are at relatively high altitudes, the particles may make a large number of back-and-forth trips without being jolted free.

called mirror points. If a mirror point is high above the ionosphere, the particle keeps bouncing back and forth without colliding with atmospheric molecules. But if a mirror point is nearer the earth, inside the ionosphere, the particle may quickly collide with an ionospheric particle and be "dumped" into the atmosphere.

Thus, 50 years before the discovery of the Van Allen belts, Stormer had theoretically modeled the behavior of then-newly-discovered elementary particles in regions so remote that they would not be physically accessible to measurement until the space age. Although the idea that auroras are directly excited by the entry into the atmosphere of particles shot from the sun was overly simple, the insights of Birkeland and Stormer were truly remarkable for the time.

Stormer's old models of the capture of charged particles in the geomagnetic field should have occurred, at least subconsciously, to space scientists while they prepared to conduct the first satellite mission. Indeed, in 1956 S. F. Singer had proposed the existence of trapped particles in the outer magnetic field, and plans were under way for a high-altitude nuclear burst that would test the hypothesis. Still, it came as a great surprise when the first American mission, carrying James Van Allen's Geiger counter, detected a huge flux of energetic particles in the high atmosphere.

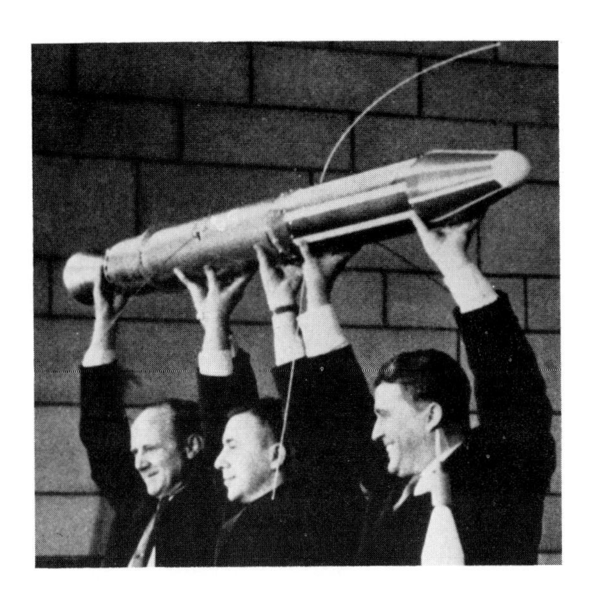

From left to right, William H. Pickering, James A. Van Allen, and Wernher Von Braun at the National Academy of Sciences, celebrating the successful launch of Explorer I, January 30, 1958.

Doughnut-Shaped Radiation Belts

The United States had planned to launch a series of earth-orbiting Vanguard satellites during 1957–1958, the International Geophysical Year. Among the IGY's programs, all of which were purely scientific, was Van Allen's global survey of cosmic-ray intensity. When public pressure following the successful Sputnik mission forced the preemption of the Vanguard schedule, Van Allen's cosmic-ray payload was transferred to the Army's Jupiter C missile for the earliest possible launch. The results from Explorer I, launched on January 31, 1958, were so puzzling that instrument malfunction was suspected. High levels of radiation intensity appeared interspersed with dead gaps, but the absence of tape recordings made the analysis very difficult.

On March 5, 1958, Explorer II failed. The final stage did not ignite, and the payload landed in the ocean. Three weeks later, Explorer III succeeded fully, and, most important, it carried a tape recorder. Simulation tests with intense X rays in the laboratory showed that the dead gaps represented periods when the Geiger counter in space had been choked by radiation of intensities a thousand times greater than the instrument was designed to detect. As Van Allen's colleague Ernie Ray exclaimed in disbelief: "All space must be radioactive!"*

Piecing together the available evidence, Van Allen concluded that the earth was encircled by a doughnut-shaped bottle of energetic particle radiation. When a reporter asked him if he meant "like a belt," the word caught on, and before long, everyone knew about Van Allen radiation belts. The term "magnetosphere" for this magnetic trap that holds energetic electrons and protons in the Van Allen belts was coined by Thomas Gold in 1959. Discovery of the radiation belts and the concept of a magnetosphere were undoubtedly the most important developments in the first decade of post-Sputnik space science.

The early measurements of trapped particles implied the existence of two belts, an inner and an outer one. Highly energetic protons (20 to 40 million electron volts, MeV) concentrate in an inner zone about 2000 miles above the earth. High-energy electrons (1 MeV) characterize an outer zone at 12,000 to 15,000 miles. The "slot" between

**Why did the Soviets fail to recognize the radiation belts before Van Allen? Sputnik II carried a Geiger counter, but over the Soviet Union, it was at low altitudes. Only over Australia was the satellite high enough to register the trapped radiation. Telemetry signals were recorded there, but the Soviets refused to reveal their codes, and the Australians withheld the data.*

these two zones contains roughly equal numbers of lower-energy electrons and protons. The most readily accessible region for safe orbital manned flight lies below 250 miles.

How do these belts fill with particles that have energies so much greater than those in the solar wind? Many scientists believe that high-energy protons in the inner belt derive from neutrons produced by cosmic-ray bombardment of oxygen and nitrogen nuclei in the upper atmosphere. Neutrons created at an altitude near 100 kilometers fly off in all directions, unimpeded by electric or magnetic fields. Some of them decay into protons and electrons while they pass through the magnetosphere. Immediately caught in the magnetic trap, protons with energies of tens of millions of electron volts may be stored for a decade, long enough to accumulate quite a dense high-energy proton population.

Rocket and satellite observations during the first few years after discovery of the radiation belts showed the inner belt to be comparatively stable, while the outer belt responded radically to solar-flare activity. A neutron source of the proton population would explain the stability of the inner belt. The variability of the outer zone, however, suggested that solar-flare particles somehow found entry to the belts through the earth's magnetic shield. Just how solar particles penetrate the magnetosphere is still the most puzzling problem in magnetospheric physics.

Confined by geomagnetic field lines, energetic particles spiral from pole to pole in about a minute if their mirror points are at altitudes of 500 to 1000 kilometers. But the back-and-forth journey of trapped particles is only part of the story. Trapped electrons not only spiral back and forth between north and south mirror points, they also drift eastward. Like any negatively charged particles, electrons spiral in such a fashion that at the top of each loop they are going east, whereas at the bottom they are traveling west. The strength of the earth's magnetic field decreases with altitude, and so, at the top of each loop, the eastward-moving electrons take off on a bigger loop than they make at the bottom of the loop, when they are moving westward. In effect, the loops are not circular; the radius of the curving motion is somewhat greater on the outside than on the inside. Therefore, their spiraling track drifts sideways, eastward around the earth, a passage that takes about half an hour to complete. For protons, the entire process is reversed so that the net drift is toward the west. These drift motions at three or four earth radii constitute a "ring" current, which, in 1962, S. F. Singer suggested was responsible for the main phase of magnetic storms triggered by the arrival of energetic particles from solar flares.

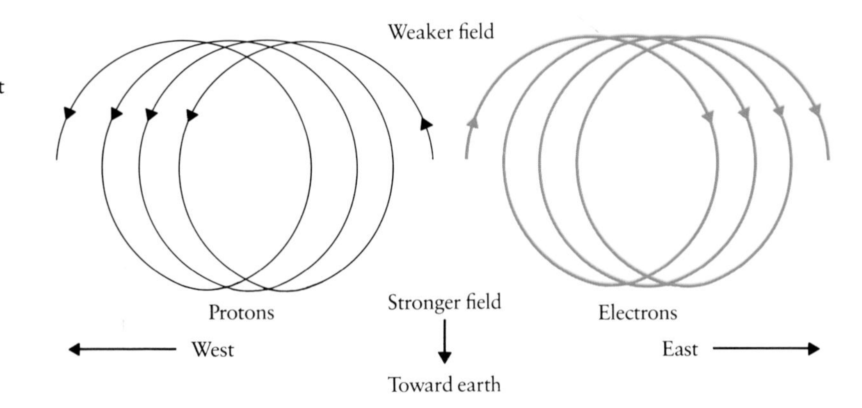

The motions of positively and negatively charged particles as seen along the magnetic lines of force from a point over the geomagnetic equator looking north. At greater heights the field is weaker and the trajectory is less curved, so that the spiraling paths move sideways. Electrons and protons drift east and west respectively.

Deep space probes by both the United States and the Soviet Union found that fluxes of particles in interplanetary space were insufficient to excite the aurora. It also became apparent that particles stored in the trapping regions and then dumped into the auroral zones by some storm process could not produce auroras. The numbers of particles observed at altitudes from which they could spiral down field lines to the auroral zones would be depleted in a matter of seconds, whereas auroras often last for hours. However, observations that trapped radiation, electron precipitation in the auroral zones, and auroral luminosity all increase simultaneously with solar flares imply that whatever disturbance sets off the auroral display also replenishes the radiation belt and makes the sustained aurora possible. How the solar wind, the magnetosphere, and the aurora interrelate is the essence of modern research on the magnetosphere.

Atmospheric Bomb Tests

The ability of human beings to disturb the universe has never been so dramatically demonstrated as in the test explosions of nuclear weapons at high altitude. Early tests in the atmosphere spread such damaging nuclear fallout around the world that an atmospheric nuclear-test ban became imperative.

When the Van Allen belts were discovered, civilian and military scientists immediately thought of ways to perturb them for scientific and military purposes. Edward P. Ney and Paul J. Kellogg suggested that exploding a hydrogen bomb within the belts and then observing the auroral effects, the shakeout of energetic trapped particles, and the injection of new particles into the belts would be revealing. (Earlier tests had created the first man-made auroras.) Military scientists had ideas of injecting particles into the belts at such high densities that they would seriously jeopardize any ballistic or orbiting object traversing the belts and would disrupt radio communications—a nuclear blackout. Ney and Kellogg's ideas never found support, but the military services did conduct a series of tests to satisfy their own objectives. Those tests were conducted before the horror of radioactive fallout was properly understood and atmospheric tests were banned.

On July 31, 1958, a group of New Zealand scientists at the Apia Observatory in Samoa looked on in amazement at a bright ultraviolet arc appearing in the western sky and stretching to the north. Over the next dozen minutes, the color changed to red and then green as it slowly faded away. Samoa is south of the equator, and the only aurora ever seen there is the aurora australis in the southern sky. Eleven nights later, the New Zealanders saw the sequence repeated. They had, of course, witnessed the aftermaths of two hydrogen bomb explosions above Johnston Island about 2100 miles to the north. The first shot, code-named Teak, was fired from a height of 48 miles, and its flash lit up the sky in Hawaii, 500 miles to the northeast. The magnetic arc that passed through the fireball came down in the ocean west of Samoa. The second shot, Orange, exploded at 27 miles. Both explosions produced radio blackouts for more than two hours on circuits within 2000 miles of Johnson Island, as well as magnetic storms and auroras.

Well before Teak and Orange, in October of 1957, a Greek emigré engineer, Boris Cristofilos, had proposed what later became known as the Argus experiment. Atomic bombs carried aloft on rockets would project electrons into the earth's magnetic field trap. Cristofilos had calculated that a one-megaton bomb could produce a serious radiation hazard in space and prolonged radio blackout. The military implications were obvious. To conduct the tests in secrecy, small-yield bombs of about one or two kilotons would be exploded at an altitude that would limit trapping lifetime to just a few days.

Preparations for Argus were undertaken immediately after the discovery of the Van Allen belts. On August 27, August 29, and September 6, 1958, rockets carrying bombs were launched from the deck of a missile-testing ship, the Norton Sound, to heights of about 300 miles. The first shot took place near the island of Tristan da Cunha in the South Atlantic. The brilliant flash of the explosion was accompanied by auroral streamers along arching magnetic field lines. For the next two shots, the Norton Sound moved about 800 miles south. Each shot produced a brilliant aurora.

The first Argus explosion emptied its energetic particles into the inner Van Allen belt. Explorer IV, which was already in orbit, clearly detected the artificially inserted shell of electrons. The following two explosions injected their particles into the "slot" between the two belts. A severe magnetic storm on September 4 so distorted the magnetic field that particles from the first two shots were rapidly shaken out. Explorer IV continued to measure the remnants of the third shot, fired two day later, until its data system failed on September 21. Explorer IV data indicated that the artificial radiation belt extended to an estimated 4000 miles above the earth's surface.

Many spot checks were also made with detectors carried to heights of about 600 miles aboard rockets. The second Argus shell was especially well situated for study by rockets fired from the Wallops Island Test Range in Chesapeake Bay. Five rockets from Wallops covered the shell from Argus II for 88 hours. It appeared to be almost stationary in space, and its thickness remained at a nearly constant 12 miles. During the observation period, injected electrons made more than a million trips between magnetic mirror points and circumnavigated the earth more than a hundred times.

On July 9, 1962, a 1.4-megaton thermonuclear bomb, code-named Starfish, was exploded by the United States at 400 kilometers over Johnston Island in the Pacific. As its high-energy electrons (1 MeV and greater) spread, a long-lasting artificial radiation belt formed between 300 and 5000 kilometers. The number of trapped high-energy electrons decreased by 50 percent within three months to

On July 9, 1962, a high-altitude nuclear burst over Johnston Island created a vast red aurora over the Pacific sky, as seen in this view from Honolulu.

a year in various parts of the belt, but the return to normal conditions took close to 10 years. Those involved regarded Starfish as a useful experiment, but it set off loud protest around the world from scientists who condemned it as reckless tampering with the upper atmosphere.

The Magnetosphere

Without the confining pressure of the solar wind, the earth's magnetic field would fade off indefinitely into interplanetary space. But the solar wind compresses the field toward earth, giving it a boundary on the sunward side. Some field lines that would have looped far out toward the sun are swept back downwind, and, together with the night-side field lines, form a teardrop-shaped tail stretching hundreds of thousands of kilometers in the direction away from the sun. The pressure of the solar wind shapes the magnetospheric cavity and gives it its size. As the pressure varies, the entire magnetosphere quivers like a bowl of jelly, sending magnetic storm signals to earth.

Many space probes have carried magnetometers across the magnetospheric boundary. Out to about 10 earth radii, magnetic field strength falls off inversely as the cube of the radius. At 10 radii, the field rather abruptly becomes smaller and highly irregular. At about 12 radii, it suffers another drop, but thereafter resumes a smooth, continuous decline. The region within 10 radii is the magnetosphere; the region beyond 12 radii is the interplanetary medium; in between is the turbulent magnetosheath.

When a stream of air encounters a solid obstacle, its flow changes direction. A reflected wave traveling back into the oncoming stream signals the barrier and tells the stream to avoid it by turning aside. This behavior occurs when the air stream is moving with the speed of sound. If, however, the air stream is traveling faster than sound, gas particles in the stream strike the obstacle before they receive any message that it is in the way. Flow lines come straight up to the surface and then suffer an abrupt change of direction. Pressure suddenly increases, as the gas molecules change direction and speed up. These changes take place in a thin layer called the shock wave. The most familiar shock waves that reach ground are thunderclaps and sonic booms when supersonic aircraft pass overhead.

The solar wind is so dilute that ordinary sound waves cannot propagate in it. Instead, energy is carried by a magnetohydrodynamic (MHD) wave that is typical of a plasma embedded in a magnetic field. Such waves are set up when plasma is displaced with respect to mag-

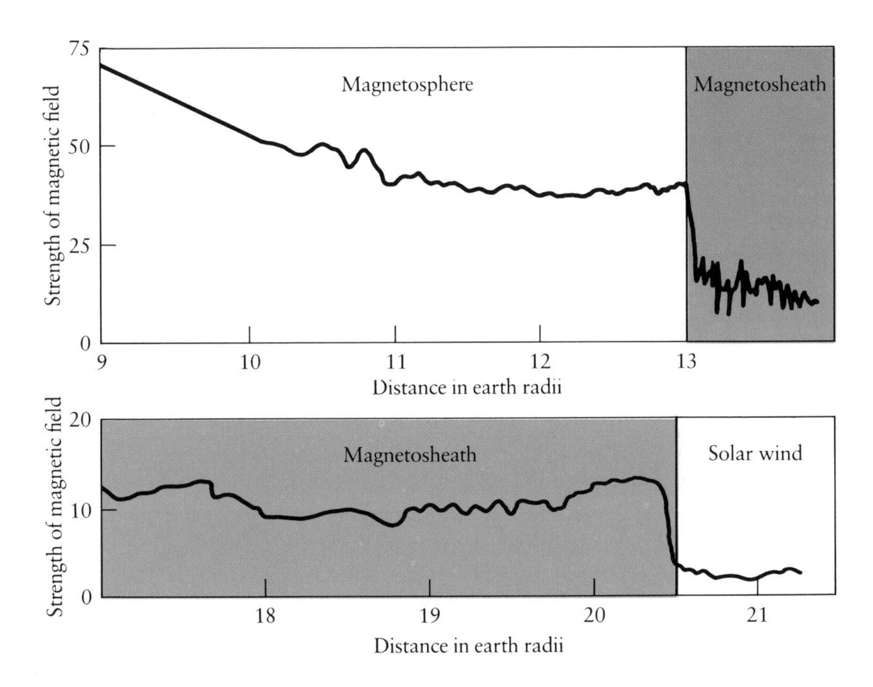

Space probe magnetometers traversing the magnetosphere show a steady drop in field strength with distance, marked by two abrupt decreases. One occurs at about 13 earth radii, the boundary of the magnetosphere, and the other at about 20 earth radii, the outer boundary of the magnetosheath. These distances correspond to a cut through the magnetosheath at a large slant angle to the sun-earth direction.

netic field lines. Tension in the field lines creates a restoring force that reverses the displacement, and the velocity of the plasma then carries the motion past the equilibrium position. In this way, a wave motion is generated. The MHD wave travels much more slowly than the solar wind, and so the arrival of the wind at the earth's magnetic barrier sets up a shock wave. The MHD shock that forms in front of the magnetosphere is "collisionless" because the energetic particles hardly ever bump into each other.

On the night side, magnetic field lines of opposite direction come together in the equatorial plane of the magnetotail to cancel each other and produce a "neutral sheet." Along this force-free channel, charged particles can leak from interplanetary space back toward earth and then follow field lines back into the polar cusps.

Closer to earth is the plasmasphere, where comparatively low-energy particles accumulate after escaping upward from the underlying ionosphere. The Van Allen radiation belts, which are filled with relatively high-energy particles, overlap and extend beyond the plasmasphere. While the tail maintains its orientation fixed in space along the sun-earth line, the plasmasphere corotates with the neutral atmosphere to which it is coupled by frictional drag.

A schematic view of the magnetosphere showing some of its major features (not to scale). The earth's magnetic field resists the entry of solar wind plasma, which flows around it into a magnetotail, perhaps a thousand earth radii in length. A collisionless bow shock upstream signals the impact of solar wind on the earth's magnetic shield. The boundary moves in and out between distances of 77,000 to 83,000 kilometers on the sunward side at speeds of 10 to 200 kilometers per second. The magnetosheath is a turbulent layer immediately behind the shock front in which electrons and protons reach energies several thousand times as great as in the incoming solar wind. The magnetopause is the outer boundary proper of the magnetosphere. On the sunward side, it is compressed to 10 earth radii or less. The sunward and tailward sectors of the magnetosphere are separated by cusps through which magnetic field lines funnel down to the earth's magnetic poles. Solar wind particles are believed to be transferred into the magnetosphere by various routes, including transport into the entry layer, cusp mantle, and tail boundary layer. As they find their way into the plasma sheet, they fill a reservoir for auroral precipitation during magnetic substorms. In addition to protons and electrons from the solar wind, the plasma sheet carries heavy ions, such as oxygen, which must be brought up from lower regions of the ionosphere. The plasmasphere is a doughnut-shaped region encircling the earth, populated by low-energy electrons and protons. Enclosing and overlapping with the plasmasphere are the Van Allen radiation belts in which particles of very much higher energy are trapped.

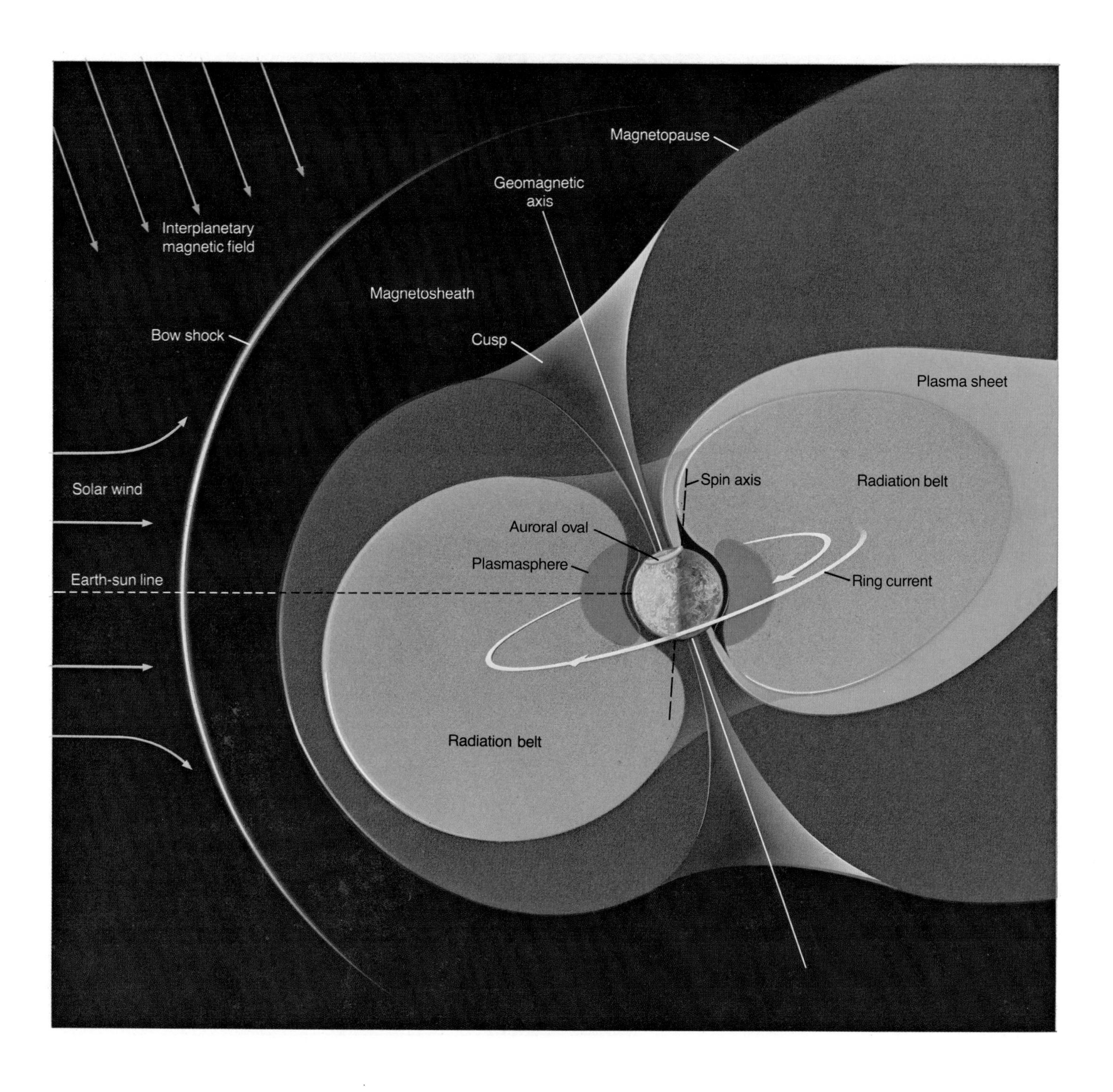

Magnetopause
Geomagnetic axis
Interplanetary magnetic field
Magnetosheath
Bow shock
Cusp
Plasma sheet
Spin axis
Radiation belt
Solar wind
Auroral oval
Plasmasphere
Ring current
Earth-sun line
Radiation belt

Most of the solar-wind plasma that enters the magnetosphere flows along the open field lines that lead to the magnetic tail and forms a shell known as the mantle. During this journey, a dynamo action generates an electric field that forces the flow into the central region of the tail and concentrates the plasma sheet. Some solar wind may also find access to the plasma sheet directly through the mantle. Still a third source of particles in the plasma sheet may be the ionospheric plasma that escapes along open magnetic field lines from the polar caps.

Under special conditions, the solar wind can wag the magnetotail and trigger a "magnetospheric substorm" that becomes a source of bright auroral displays. The earth's magnetic field direction is fixed, but the interplanetary field near earth changes direction from time to time. Some scientists believe that when it turns southward, it connects more readily with the earth's field. Solar wind then acquires easier access to the boundary layer and increases the compression of field lines and plasma in the tail. If this persists, the squeezing or merging of magnetic field lines in the tail may reach an explosive limit, either in the middle or at several points along the tail. Such intense compression

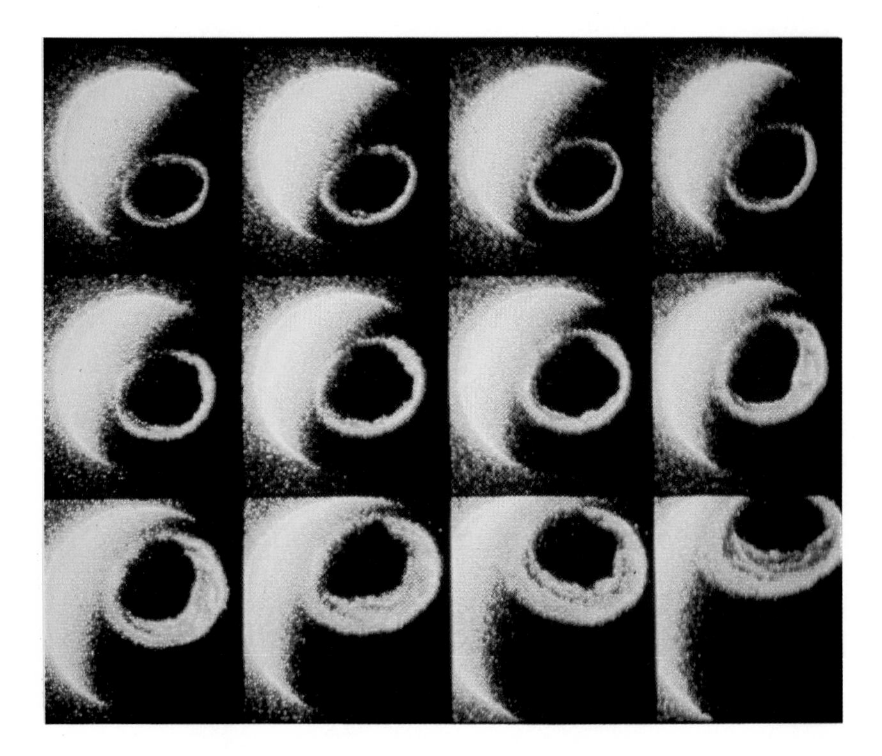

Prior to the onset of a magnetospheric substorm in the auroral zone, faint arcs spread around the auroral oval. Suddenly, an arc brightens in the midnight sector, marking the beginning of an auroral substorm. There follows a variety of auroral activity spreading in all directions. This sequence of images of the auroral oval high above the northern polar region, obtained from the NASA Dynamics Explorer, traces the development of an auroral substorm.

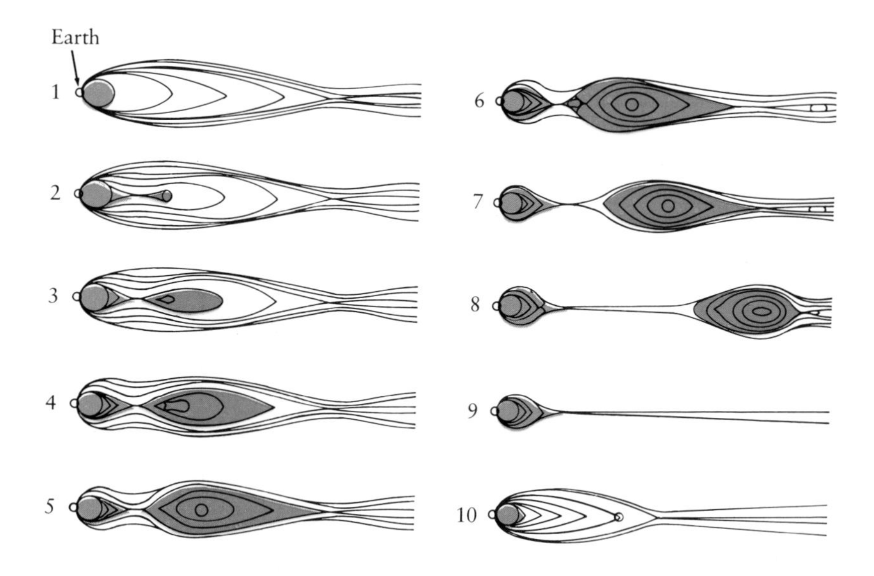

Computer-modeled stages in the reconfiguration of the magnetotail as a blob of plasma is pinched off near the nose of the magnetosphere and propelled downstream in the tail. The time scale is one hour.

may be released explosively in 10 minutes to half an hour, and plasma particles squirt out from the explosion point. Those particles directed toward earth may race down to the atmosphere at speeds of 1000 miles per second to create spectacular auroras or to be trapped in orbits within the Van Allen belts.

Of all the energy brought to the magnetosphere by the solar wind, only about 0.1 percent manages to cross the magnetic barrier. All of the energy expressed in the auroral ovals that circle the polar caps adds up to roughly 0.1 percent of the total energy in magnetospheric plasma. Auroral light itself amounts to only about 4 percent of the energy deposited in the atmosphere. The remaining 96 percent is dissipated in heating the atmosphere, producing ionization, and generating ultraviolet, infrared, and X radiation.

Throughout almost the entire 1970s, the idea prevailed that magnetic energy stored in the magnetotail could abruptly trigger magnetospheric substorms. It was puzzling, however, that so many different kinds of storms could originate from a common form of magnetospheric instability in the plasma sheet.

Syun-Ichi Akasofu prefers a model of direct, real-time control of the beginning and evolution of magnetospheric storms. According to Akasofu, when the solar-wind plasma blows across open field lines at the polar caps, the resulting electric dynamo action can generate a drop of 100,000 volts across the magnetosphere and drive currents of 10^{17} amperes. At times of solar flares, the wind may gust to 10 times

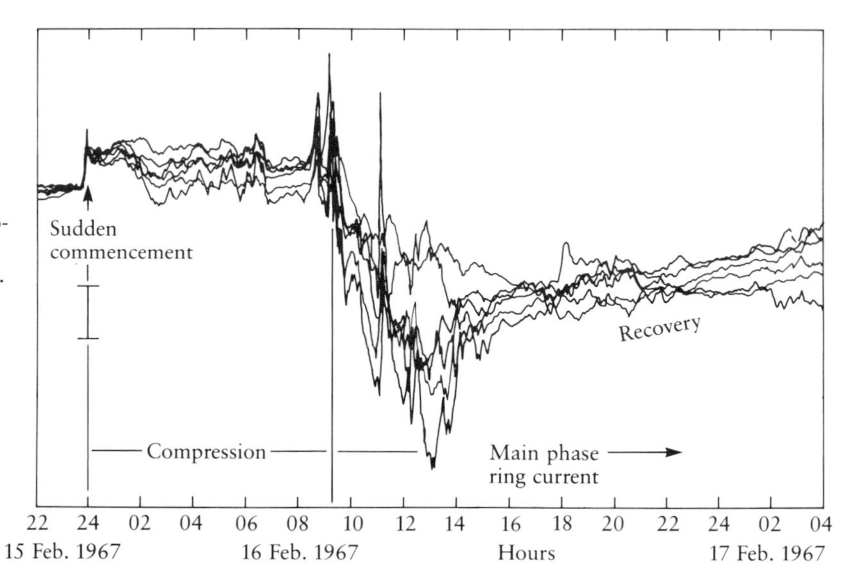

A magnetic storm detected by the variation of north-south components of the magnetic field at six different magnetic observatories around the world. The "sudden commencement" and following sustained high field intensity results from the compression of the magnetosphere by an enhanced solar wind flow; the pronounced decrease during the "main phase" is produced by the storm ring current that develops in the inner magnetosphere. Differences among stations are mainly due to longitudinal and local-time variations of the storm effects.

normal and generate as much as 10^{13} watts. Thus, magnetic substorms dissipate 10 times more energy than all the electrical energy generated by man-made power plants on earth.

Solar Proton Flares

Within one to two days after a burst of electromagnetic radiation from a moderately intense solar flare, particles of solar plasma begin to arrive at the earth. Because of their comparatively low energy, they enter the earth's atmosphere only along the magnetic lines of force that lead to the poles. At lower latitudes, their arrival is felt when they exert pressure on the more nearly horizontal lines of force, which they cannot cross. This pressure squeezes and ruffles the magnetic field, creating the magnetic storms that wiggle compass needles at the earth's surface and trigger auroral displays.

More energetic flares produce protons that reach the earth with energies from 10 million electron volts to half-a-billion electron volts. These energies overlap the range ordinarily associated with galactic cosmic rays. The very largest flares, of which there may be only a few per 11-year solar cycle, release protons with energies up to 100 billion electron volts. Meeting weak resistance from the earth's magnetic field, these protons crash into the atmosphere, exploding the molecules of oxygen and nitrogen that they hit into hosts of energetic

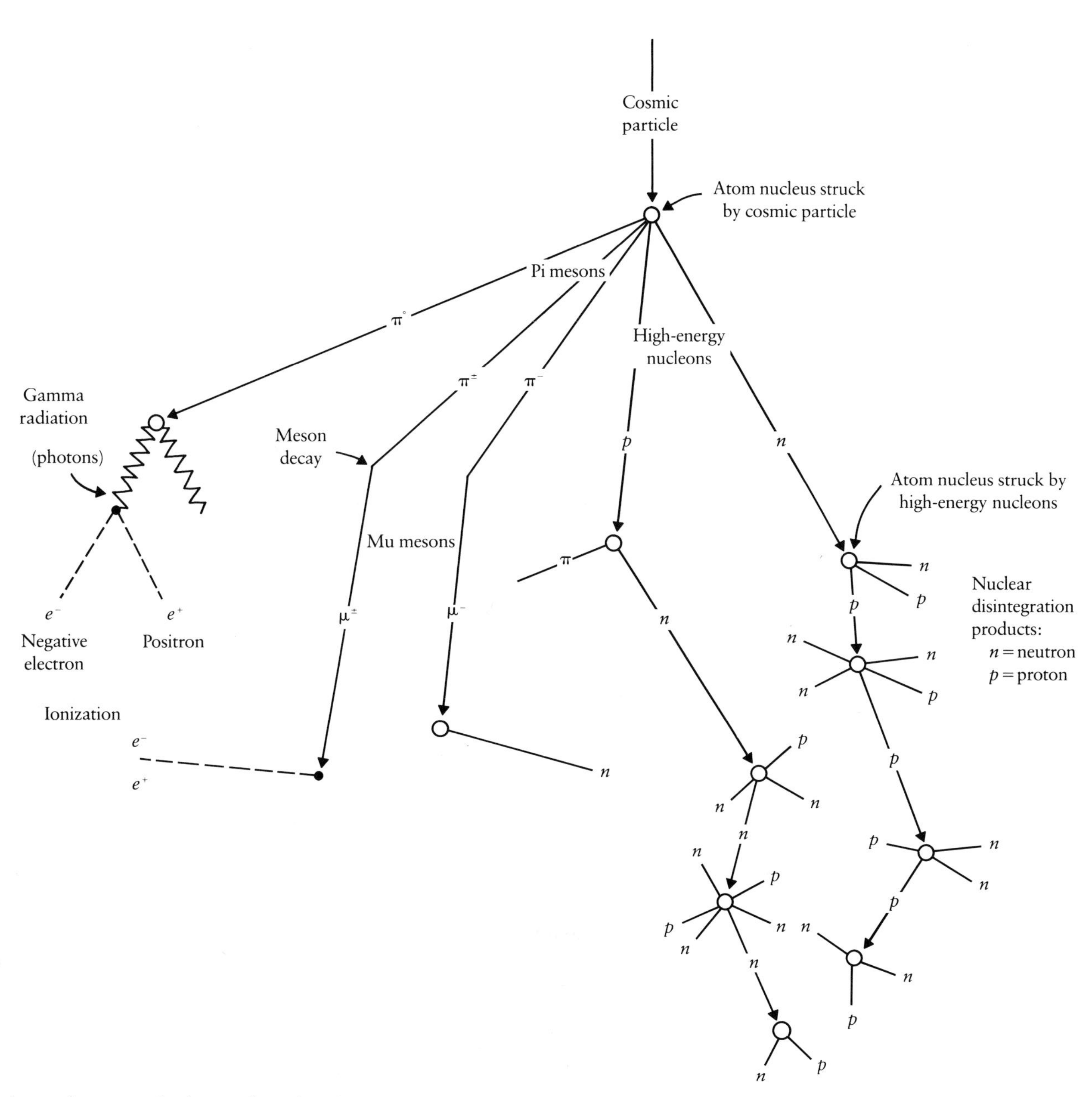

A cosmic ray shower results from a direct hit of an incoming energetic particle on an atmospheric molecule. The sketch shows a variety of disintegrations and the resulting particles that constitute the shower.

shower particles that reach the ground over a wide area. One of these superflares releases more energy than is stored in the heat of the entire corona and chromosphere, so obviously the flare's energy cannot be drawn from the heat of its surroundings. Only the energy stored in local magnetic fields appears to be an adequate source.

The origin of cosmic rays is still one of nature's great mysteries. The observation that the sun can produce billion-electron-volt particles has led to speculation that most cosmic rays may have stellar origins. But if all the stars in the Milky Way behaved on average like the sun, it would take a million times their number to produce the observed flow of galactic cosmic rays. Even then, it would be extremely difficult to explain individual cosmic-ray particles with energies billions of times as great as the most energetic particles ever produced by the sun.

Polar-Cap Absorption Events

During the IGY, scientists noted that some flares delivered protons of energy sufficient to penetrate deeply into the ionosphere over the polar caps. These events were marked by strong radio absorption throughout the interior of the auroral zones. Such polar-cap absorption (PCA) is attributed to protons of energy high enough to produce D-region ionization. The interval between the sighting of a flare and the start of a PCA is usually shorter than the flight-time of magnetic storm particles, and it implies energies consistent with penetration to 80 kilometers.

The flare particles that kick off PCA events do not always appear to take the fastest route from the sun. By timing the arrival of particles of different speeds, satellite measurements have shown that while all the particles have been produced simultaneously in the flare and have traveled the same distance, the distance covered was considerably greater than that from the sun to earth. These observations suggested a roundabout trajectory.

Possibly, particles are scattered in the direction of earth after numerous collisions with irregularities in the interplanetary magnetic field. Or perhaps high-speed energetic particles are trapped in some magnetic "bottle" that detaches from the sun at a speed lower than that of the individual particles. In effect, the particles would be like flies racing madly about in a jar that is carrying them toward earth at a speed corresponding to the time interval between the flare and the PCA. When it reaches the ionosphere, the jar breaks open, and the protons penetrate according to their individual speeds.

The study of proton flares is particularly challenging because, as yet, there are no adequate theoretical models for the production of such high energies in the solar atmosphere. PCA diagnostics is a useful tool for acquiring clues to the solar processes. More recently, the study of high-energy flare machanisms has been greatly advanced by spacecraft observations of gamma-ray spectra and neutron fluxes from flares.

Terrestrial Impacts of Geomagnetic Storms

Perhaps the first notice of an association between auroras, geomagnetic storms, and disruption of telegraph communications occurred during a sequence of large solar disturbances and visible auroras in August–September, 1859. Telegraph lines out of Boston stopped working for considerable periods. At other times, the system operated without connection to its battery supply—voltage drops associated with the appearance of auroras apparently provided sufficient power. When the aurora was visible, the operator noted increases and decreases in induced currents on the telegraph lines, with fluctuations lasting from about 30 seconds to several minutes.

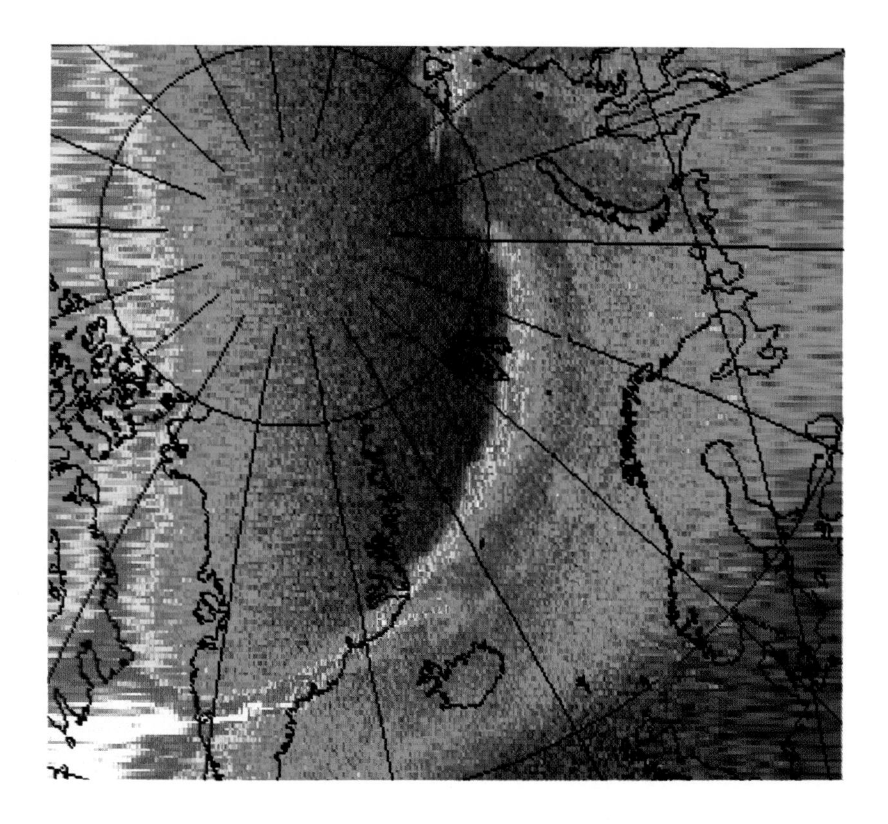

An aurora borealis imaged in full sunlight from the HILAT satellite on July 23, 1983. Shown here is a false-color image derived from a VUV (vacuum ultraviolet) camera at 1493 Å. The ultraviolet camera is very sensitive to this wavelength of atomic oxygen emission excited in the aurora but does not respond to visible light. From one extreme to the other, the auroral arc crosses the southern tip of Greenland, northern Iceland, Scandinavia, and Siberia.

We have many more recent examples of disruptions and damage caused by geomagnetic storms. A great storm on March 24, 1940, wiped out 78 percent of all long-distance calls from Minneapolis. Another storm on February 10, 1958, induced surges as high as 2650 volts on the Bell system's transatlantic cable from Clarenville, Newfoundland, to Oban, Scotland. Voices transmitted eastward alternately squawked and whispered, while those westbound were quite normal. A magnetic storm on February 9, 1958, tripped the circuit-breakers in an Ontario transformer station, blacking out Toronto. A great flare in 1972 burned out a 230-kilovolt power transformer in British Columbia. Because the northeastern United States is supplied with electricity by a complex grid that includes southern Canada as well, transformer failure in this net can lead to a massive power blackout like the one that occurred throughout New York State on November 9, 1965.

We have seen how human activities can disturb the natural system with, for example, high-altitude explosions. But even comparatively mild activities on the ground have an impact on the behavior of the ionosphere and magnetosphere. For example, high harmonics from power grids in North America and Europe radiate into the magnetosphere, where some unknown physical process appears to amplify them. As with natural waves, the man-made waves interact with electrons and so perturb their trajectories that they fall from the magnetosphere to the ionosphere. The process is accompanied by a "chorus"—a birdlike chirping of ascending frequency that occasionally appears on tape recordings. Although it has no significant impact on communications, the phenomenon does reveal a fundamental magnetosphere instability that amplifies waves and precipitates particles into the ionosphere.

One proposal to perturb the auroral system from the ground would use an intense beam of pulses from a radio transmitter to modulate the currents that flow continuously into and out of the auroral ionosphere. The radio source would trigger extremely low frequency (ELF) waves that could be used to communicate with submarines. Such a system would poach on the free power inherent in auroral currents, which carry millions of amperes.

The enormous energy generated in the auroral magnetosphere has led many people to dream of ways to tap in for ground-based power—a new channel of solar energy created by solar wind rather than sunlight. The power is tremendous but very diffuse, and very clever innovation would be required to draw on it. As Syun-Ichi Akasofu puts it, "If I took a copper coil and wrapped it around the seven-story building of the Geophysical Institute and rotated a magnet inside it to create a current, you might be able to light a flashlight."

Exploring the Magnetosphere with Spacecraft

During the first decade of space explorations, space physicists directed various probes through the magnetosphere or into highly elliptical orbits that repeatedly traversed the magnetospheric boundary. Such missions, however, could not separate the geographical differences from temporal variations during the time of traversal. Gathering of independent spatial and temporal information requires more than one spacecraft and careful placement of such craft in space.

The most recent effort involving a multiple array of satellites was the International Sun-Earth Explorer (ISEE) program, jointly conducted by NASA and the European Space Agency (ESA). In October of 1977, NASA's ISEE-1 and ESA's ISEE-2 went into nearly identical orbits. As the two satellites chased each other around the magnetosphere, they sensed the position and movement of the bow shock and magnetosheath about 130,000 kilometers above the earth. Where the magnetic field lines dragged from the sun by the solar wind merged with those of the earth's magnetic shield, the magnetosphere appeared to suffer a ripping of its surface. The solar wind's magnetic field merged with the earth's field on the sunward side of the magnetosphere and tore back the magnetospheric field, peeling it off toward the dark side of the earth, hundreds of thousands of kilometers into the magnetospheric tail. As the merger of the interacting fields progressed, the field lines became sharply bent and particles caught inside the bends were accelerated as though projected by a slingshot.

In August, 1978, NASA launched ISEE-3 to a vantage point 1.5 million kilometers above the earth, where it monitored the solar wind on its way to the magnetosphere. Instead of orbiting the earth, the satellite executed small circles in the gravitational well, known as the L_1 libration point, between the sun and earth. From this outpost, ISEE-3 sensed changes in the solar wind in time to give advanced warning of magnetic storms and auroras.

For the next round of magnetospheric research, plans are being developed for a joint effort by the United States, Japanese, and European space agencies known as the International Solar Terrestrial Physics (ISTP) program. It will explore all elements of the magnetosphere and its environment with a number of spacecraft simultaneously in orbits strategically intersecting the various magnetospheric domains.

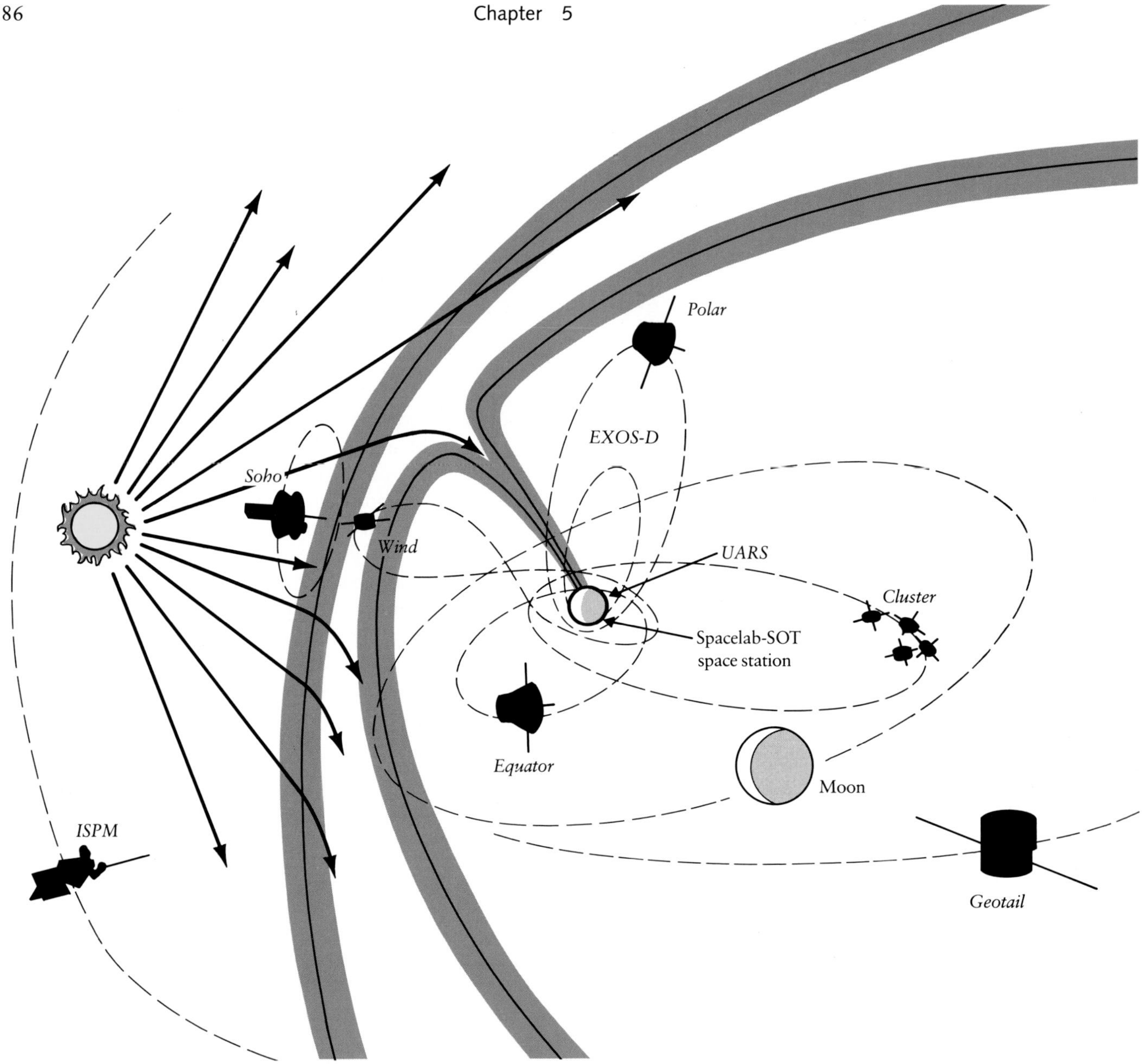

The plan for the International Solar Terrestrial Physics Program (ISTP) includes a minimum of six spacecraft provided by NASA, the European Space Agency (ESA), and the Japanese Space Agency (ISAS). The *Soho* spacecraft would be positioned ahead of the magnetosphere to monitor the incoming solar wind and record solar vibrations and coronal transients. Four spacecraft—*Wind, Equator, Geotail,* and *Polar*—would circulate through various regions of the magnetosphere. The *Cluster* mission would sample small-scale structure and plasma turbulence throughout the polar and middle magnetosphere as well as in the solar wind. Other satellites indicated here are planned or already approved programs that would complement ISTP. The entire configuration is contemplated for operations in the 1990s.

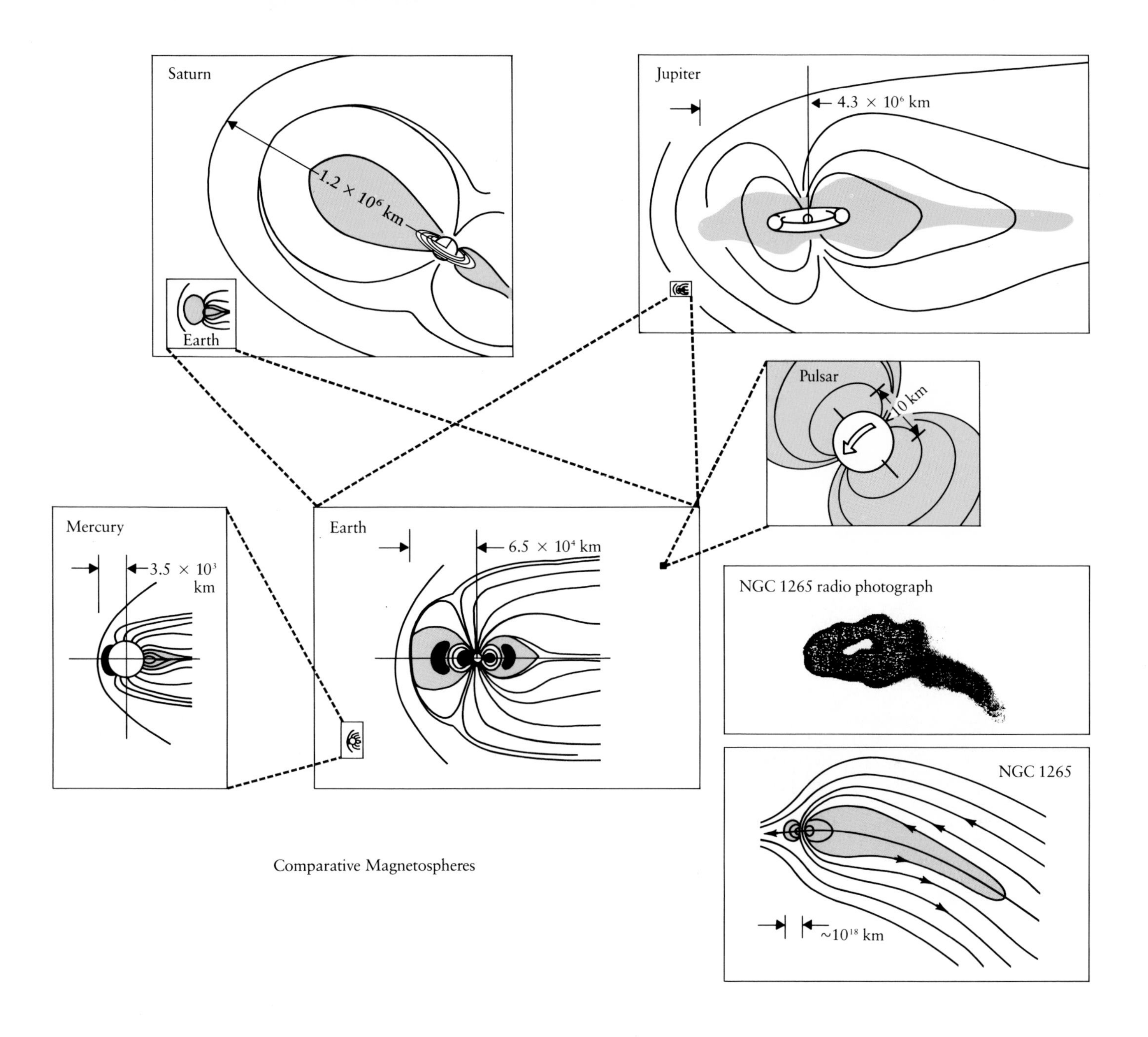

Comparative Magnetospheres

Fundamental similarities characterize the magnetospheric configurations of the planets in the solar system and various celestial objects of far greater and lesser scale in the universe at large. In each example, the shape and scale is intrinsic to the magnetic field strength, the plasma environment, and position in the wind stream. On the largest scale, NGC 1265 is a galaxy that exhibits a tail millions of light-years long, swept back by the intergalactic gas. On the smallest scale, a pulsar—the spinning collapsed remnant of a supernova—is only 10 to 30 kilometers in diameter, but its magnetic field is trillions of times as strong as the earth's.

6

In Search of a Climate Connection

All things belonging to the earth will never change—the leaf, the blade, the flower, the wind that cries and sleeps and wakes again, the trees whose stiff arms clash and tremble in the dark, and the dust of lovers long since buried in the earth—all things proceeding from the earth to seasons, all things that lapse and change and come again upon the earth—these things will always be the same, for they come up from the earth that never changes, they go back into the earth that lasts forever.

Thomas Wolfe, *The Web and the Rock*

Thomas Wolfe's images describe an eternal earth, resilient and resistant to any permanent change. But as we learn to read the geological record, the library of the earth, we gradually piece together a different story.

Most people think of the sun as a constant in the formula of life and of the earth as a "rock of ages," but nothing is permanent in heaven or earth. With modern planetary exploration has come a sense of the uniqueness and the vulnerability of the earth's ecological niche in the solar system. The tolerances of a living biosphere are so narrow that very small variations in the flow of life-sustaining energy from the sun spell the difference between a benign and a hostile environment for life on earth. The history of climate, for example, is locked to the tilt and wobble of the spin axis of the earth and the slow elongation and contraction of its elliptical orbit. Even minute variations in this sun-earth relationship drive a procession of glaciations every 100,000 years.

If solar light and heat appear deceptively constant, the notion of "terra firma" is just as insubstantial. The earth's crust is cracked into a dozen tectonic plates that float on a pliant mantle. Continents drift apart, mountains are uplifted and erode. This restlessness of the earth influences the partitioning of solar energy over the globe.

Necklaces of volcanoes ring the boundaries where the great plates that cover the earth converge. Mount St. Helens and the volcanoes of the Cascade Range in the northwestern United States lie on the "ring of fire"—a belt of volcanic activity that encircles the Pacific Ocean across the Aleutian Islands, the Kamchatka peninsula in the U.S.S.R., Japan, the Philippines, and the western coast of South America. Hidden deep under the sea, where the plates are separating huge mountain ranges form from basaltic magma welling up out of the mantle. At a spread-rate of only a few centimeters each year, a gross reorganization of the earth's land masses, inexorably driven by internal heat for hundreds of millions of years, is hard to imagine. Yet fossilized remains of tropical plants are found in icy Antarctica, and seashells are embedded in the rocks of the towering Alps.

The Sun-Climate-Weather Machine

Of all the mysterious links between sun and earth, the connection of sun to climate and its daily manifestation, the weather, has been the most sought after and the most elusive. The economic and social consequences of even small changes in climate can be profound. Historical

The glacier of Rosenlaui, painted by John Brette in 1856. It was only in the mid-nineteenth century that the evidence of the advance and retreat of glaciers in Switzerland was appreciated and related to an ancient ice age.

weather data and climate records derived from proxy indicators—tree rings, deep-sea sediments, ice cores, and other fossil remains—all show that climate has varied widely in past epochs. While the case for long-range sun-to-climate connections can be established in relation to ice ages and glaciations, no credible influence of the sun on day-to-day weather patterns has yet been demonstrated.

Over the past century, average annual temperature in the northern hemisphere increased by about 0.5°C. Peaking in the 1940s, it dropped slightly for the next 30 years. Now the temperature seems to indicate a new warming trend. The earth has not been as warm as it is now since the thirteenth century. According to atmospheric scientists, even a few tenths of a percent change in global temperature can produce a significant modification in climate.

Sun, air, oceans, and earth form a weather machine that produces clouds, winds, rain, and snow. Tropical regions of the earth receive the greatest amount of heat because the sun's almost vertical rays travel to ground through the least amount of intervening air. At higher latitudes, the angle of incidence increases until, almost parallel to the ground, the sun's rays merely graze the earth's surface. Near the poles, they must penetrate the greatest thickness of absorbing atmosphere. Those rays that gain admission bounce off the snow and ice and are reflected back to space. At the equator, the earth absorbs more radiation than it returns, while over the poles and over deserts, it sends back more radiation than reaches the surface.

As the sun heats the ground, the oceans, and the air, the earth's atmosphere adjusts to the latitudinal imbalance. Heat transported via the winds and ocean circulation from the equator toward the poles is countered by cold winds and water moving from the poles to the equator. Exactly how the global circulation patterns develop depends on the spin of the earth and the configuration of land masses, oceans, snow, and ice cover.

Short wavelengths of solar ultraviolet are absorbed high in the atmosphere. Closer to earth, clouds, dust, and liquid droplets intercept the sun's visible rays and reflect or scatter them back to space. Canopies of trees and other vegetation, as well as the oceans, are good absorbers; ice, snow, and sand are excellent reflectors. That portion of the solar radiation flux absorbed by the earth warms the overlying atmosphere and is thus partly reradiated as long-wavelength infrared waves. So finely adjusted is this balance that even a small disturbance can lead to significant regional and global climate changes. A shift of only 2°C up or down in global temperature can trigger the melting of ice caps or the onset of a new ice age.

Asiatic monsoons follow a northward shift of only a few degrees in warm, wet, stormy equatorial weather, which brings on heavy rains over India and southern Asia during the northern summer. This trend is followed by a southward shift that drives the monsoon down as far as northern Australia.

Climate may change in response to two different sets of forces: stochastic and deterministic. Stochastic processes drive internal interactions between the oceans, atmosphere, terrain, biomass, ice cover, and other factors within the terrestrial system. Asiatic monsoons, for example, are governed by a shift of only a few degrees northward of warm, wet, stormy equatorial weather that causes heavy rains over India and southern Asia during the northern summer. Following this, a southward shift brings the monsoon down as far as northern Australia. The lives of hundreds of millions of people are hostage to the vagaries of the wandering monsoon.

Deterministic factors that drive climate change include solar variability, orbital changes that control insolation, or exposure to the sun, and explosive volcanism, which fills the skies with choking ash, dust, and sulfuric acid aerosols. Particles that reach the stratosphere hang high above the convective weather currents of the turbulent troposphere. The stratosphere is so stable that acid droplets may persist there for years, blocking a substantial portion of solar radiation that would otherwise reach the earth and thereby significantly lowering global temperature.

Recently we have come to believe that the burning of fossil fuels and the wholesale destruction of forests is overburdening the atmosphere with carbon dioxide. Will the envelopment of the earth in a "greenhouse" of CO_2 raise global temperature enough to influence

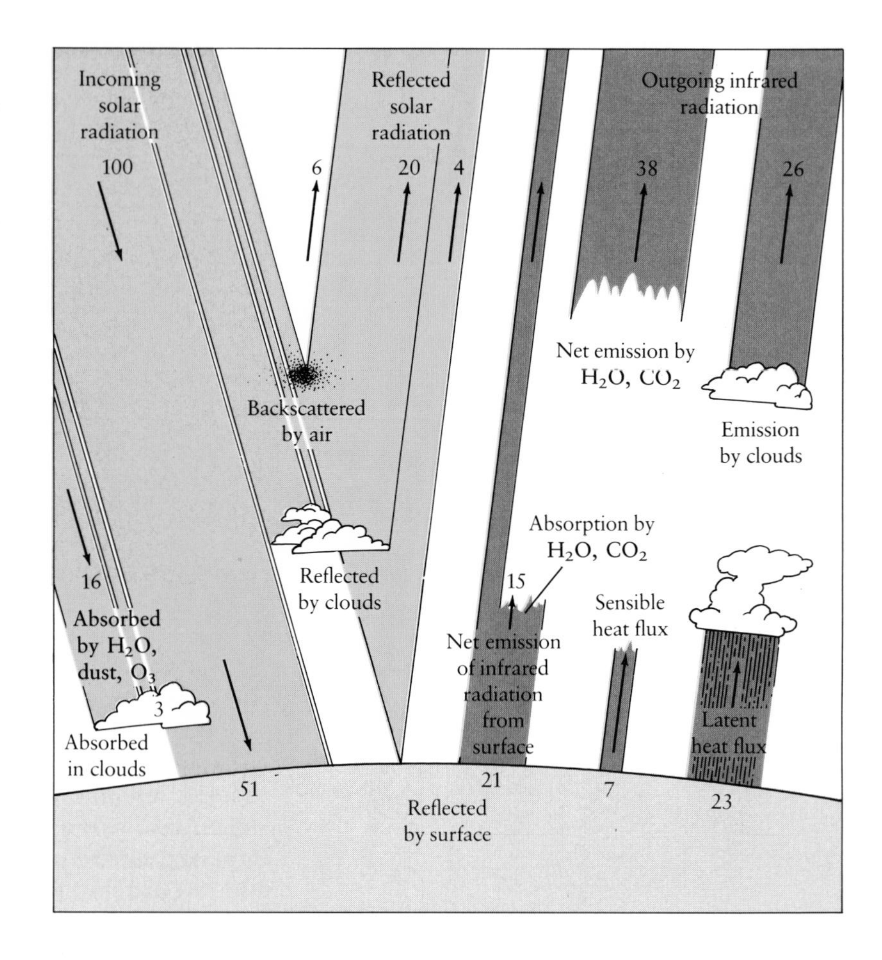

The global radiation balance. All the numbers in the diagram are percentages of the incoming solar energy flux. The incident radiation is attenuated by 16 percent in water vapor, dust, and ozone and by 3 percent in clouds. Thirty percent is returned to space from the surface, clouds, and atmospheric scattering, leaving 51 percent to be absorbed by the surface and reradiated in the infrared. Thermal equilibrium between the earth and the atmosphere requires a global mean surface temperature of about 13°C. To keep the earth in thermal equilibrium with space, the effective earth temperature must be about −18°C. This temperature is the average of that of the atmosphere at an altitude of about five kilometers.

climate dramatically within the next 100 years? What other anthropogenic alterations of the environment are or will become important factors in climate change? In asking how any of these deterministic factors might perturb the climate, we must recognize that an important property of the atmosphere is self-regulation, or homeostasis. The rest of this chapter examines the question of climatic forcing.

Sunspots, Tree Rings, and Climate

In 1801, William Herschel, the leading English astronomer of his time, believed he had detected a correlation between the absence of sunspots and the price of grain on the London market. Heinrich Schwabe's discovery of the sunspot cycle set off a wave of speculation about its connection with a host of indicators of weather, agriculture, health, and social and economic trends. By the end of the nineteenth century, numerous learned papers had drawn connections between measures of weather—rainfall, monsoons, the levels of lakes, and the flow of rivers—and the spotted condition of the sun's face. Sir Norman Lockyer's statement that "the riddle of the probable times of occurrence of Indian famines has now been read, and they can for the future be accurately predicted" is typical of popular science at the time.

That not a one of these connections has withstood analysis, statistical or otherwise, testifies to the self-deception that afflicts scientists, among other human beings, when they want very badly to find significant signals in very noisy data. This is not to say that some valid connection between solar variability and climate may not eventually emerge, but such a connection will take more precise and sophisticated observational efforts, carried out over long stretches of time, with unwavering dedication to objectivity.

In 1894, the Massachusetts textile heir and amateur astronomer Percival Lowell persuaded Andrew Ellicott Douglass to leave Harvard College Observatory and search for a southwestern site on which to place an observatory for viewing the planet Mars. Douglass established the Lowell Observatory near Flagstaff, Arizona, but his interest soon drifted from the study of Martian canals to the study of tree rings as a record of past climate. The native Ponderosa pine and Douglas fir were excellent material for this research. Tree trunks in cross section reveal a set of concentric rings, one ring for each year of growth. Favorable climate produces rapid growth and wide rings; relative drought inhibits growth and produces narrow rings. Leonardo da Vinci had commented on the connection between tree rings and climate history, but between da Vinci and Douglass the subject had escaped serious study.

Douglass eventually moved to the University of Arizona in Tucson where, at the age of 70, he established the Laboratory of Tree-Ring Research and founded the science of telling time by trees, or dendrochronology. From the beginning, he hoped to establish a relationship with the sunspot cycle, but his research was inconclusive: some trees in some locations showed an 11-year pattern, but no consistent, statistically useful sample could be found. His best evidence came from a forest of Scotch pines carefully grown and protected over an 80-year period in Germany. When he died at age 94, Douglass was convinced that the sunspot cycle was reflected in tree rings, but his evidence was viewed with considerable skepticism.

The content of the radioactive isotope carbon 14 (^{14}C) in tree rings provides a far more positive record of solar activity than their

A severe frost event occurred in September, 1884, and frost damage appears in that annual ring of the bristlecone pine. Frost damage zones in annual tree rings may indicate the climatic cooling effects of stratospheric aerosols resulting from volcanic eruptions. Damage has been noted at intervals of decades to hundreds of years over the past 4000 years in the rings of subalpine bristlecone pines.

width. Normal carbon has an atomic weight of 12; bombardment of the upper atmosphere by cosmic rays generates carbon 14. Most of the energetic particles borne by these rays come not from the sun but from deep in the galaxy, and they carry as much energy to the earth as all starlight, even though they enter the atmosphere relatively infrequently—just a few particles per square centimeter per second. Earth receives only a few particles at the highest energies, about 10^{20} electron volts, per square kilometer per century.

The production of ^{14}C is part of a chain of interactions in the atmosphere that culminate in an air shower of secondary particles. The core of a shower is a flood of protons, neutrons, mesons, and heavier fragments of air nuclei. Neutrons captured by ordinary nitrogen 14 knock out a proton, thus transforming it to carbon 14. Through photosynthesis, trees assimilate carbon dioxide containing some of the ^{14}C generated in the atmosphere, and this means that every individual tree ring contains a natural record of the ^{14}C-to-^{12}C ratio for the year it grew. It turns out that this ratio varies in rhythm with the sunspot cycle. With increasing sunspot numbers, the solar wind carries magnetic fields farther into space. These interplanetary magnetic fields deflect charged cosmic rays away from the sun, and the ^{14}C to ^{12}C ratio, therefore, decreases at sunspot maximum.

Carbon 14 is radioactive, with a half-life—that is, half of it decays—of 5000 years. It thus provides accurate dating of specimens as

The bristlecone pines of the White Mountains in California provide one of the longest tree-ring records. In live trees, the rings can be read for fully 5000 years. Dead trees and wood on the ground carry the record back about 9000 years.

old as 10,000 to 20,000 years. In the bristlecone pines of the White Mountains in California, the tree-ring record goes back fully 5000 years in live trees and perhaps 9000 years in dead trees and wood on the ground. From this forest, both standing and fallen, dendrochronologists can extract at least a dozen principal trends, each lasting 50 to 200 years. During this 5000-to-9000-year stretch of time, the record shows a predominance of quiet periods compared to periods of strong solar activity. Only a few brief bursts of intense solar activity follow the Late Bronze Age, and they are scattered at random over the course of time. No broader underlying periodicity can be discerned. A more nearly sunspotless condition seems to have been normal, with anomalous 11-year maxima during the last 21 cycles. As one wag remarked, if the close correspondence between fine vintages and sunspots noted by French wine connoisseurs were correct, life must have been especially dull during epochs of low solar activity, which would seemingly be marked by an unbroken record of indifferent burgundies and clarets.

Sunspots, Particle Radiation, Terrestrial Layering, and Climate

Nitrates frozen into the Antarctic ice, a by-product of the interaction of solar particles with the polar atmosphere, promise to reveal a great deal about past solar activity. Energetic particles arriving from the sun are funneled by way of the magnetic cusp into the polar atmosphere, where they manifest themselves in auroral activity. In the upper atmosphere, solar particle bombardment produces nitrogen oxides that react with moisture to form nitrates. Because the Antarctic ice sheet does not melt, snowfall, bearing nitrates with it, accumulates in layers year by year. Nitrate content, which directly follows the pattern of auroral activity, can be read in ice layers much as carbon content can be read in tree rings. Although nitrate dating is a new tool, it promises to provide a record of solar activity going back perhaps tens of thousands of years.

In the past five years, another important new technique for the measurement of extremely small quantities of naturally occurring isotopes has been developed. Called accelerator mass spectrometry (AMS), it makes use of a nuclear accelerator that focuses ion beams of the constituent isotopes in a sample and identifies the different nuclei by virtue of their rate of energy loss in absorbing materials. By mass spectrometry, we can now measure the naturally occurring radioisotopes carbon 14, beryllium 10, aluminum 26, and chlorine 36 with much greater sensitivity than was previously possible using radiation-counting techniques.

Beryllium 10 (^{10}Be) has a half-life of 1.5×10^6 years and, like ^{14}C, is produced in the atmosphere when cosmic rays bombard nitrogen and oxygen. Most of this production occurs in the stratosphere, where ^{10}Be quickly attaches to aerosols.* From the stratosphere, it moves to the troposphere, and thence is precipitated to the ground rather rapidly. Its mean residence time in the atmosphere is only about one or two years. Once it reaches the ground, ^{10}Be is preserved in snow and ice layers, in surface soil and biomass, and in the ocean and lake bottoms to which it flows in sediments. Its rapid descent from the upper atmosphere and its long half-life allow scientists to extend back the record of solar activity at least a thousand times past carbon-dating records and thus to cover possibly thousands of sunspot cycles.

The earth's crust, its deep-sea sediments, and its ice sheets make up a library of global change. By using several isotopic clocks, the geological record can be traced back through prehistoric times with gradually diminishing precision to almost 3.8 billion years ago, when the oldest rocks known on the surface of the earth were formed.

**Aerosols are tiny particles of solids or liquids that are suspended in air and that range in size from clusters of a few molecules to particles with radii of 20 micrometers or more. They include such diverse substances as dust, fog, smoke, bacteria, pollen grains, sea spray, and volcanic ash.*

Ice and the Climate Machine

In studying historical sunspot records, John A. Eddy uncovered an interesting coincidence between two-century-long minima in the numbers of sunspots and the two coldest stretches of European temperature that occurred during the "Little Ice Age" (1450–1850). During the earlier dip, known as the Sporer minimum, the Arctic ice pack expanded so far that a colony of Norsemen were isolated on southwest Greenland and finally became extinct. Grain that had grown for centuries disappeared. Even in North Africa, snow cover persisted on the high mountains of Ethiopia, where it has not appeared since. The second chill lasted throughout most of the seventeenth century in Europe, roughly coinciding with the reign of Louis XIV and rendering his appellation as the Sun King somewhat inappropriate. The Thames and the English Channel, which rarely freeze, became the sites of winter skating carnivals. By the end of the seventeenth century, the sunspot cycle returned to full strength, but cold lingered on through the eighteenth century in North America. Most Americans know of the great hardships suffered by Washington's Continental Army at Valley Forge in the winter of 1776. Two years later, the severe winter was described

During the Little Ice Age of the eighteenth century, the Thames River froze over every winter. Frost fairs were held in tent cities, complete with carnival entertainment and food stalls.

by David Ludlum as "the most hard difficult winter . . . that ever was known by any living person." A sheet of ice entirely covered upper New York Bay for many days. According to Ludlum, people walked the five miles from Staten Island to Manhattan over the ice. "Heavy loads and even large cannons were dragged across the iceways to fortify the British positions on Staten Island which had been subject to cross-the-ice forays from Washington's outposts in New Jersey."

The seventeenth century is known as the Maunder minimum, after an Englishman who reported to the Royal Astronomical Society in 1890 that for a period of about 70 years, ending in 1716, there seemed to have been a remarkable interruption in the ordinary course of the sunspot cycle. During several years, no spots were seen at all, and when, in 1705, two spots were seen on the sun at the same time, the event was recorded as most remarkable—nothing like it had been seen during the previous 60 years. Noting a spot in 1684, John Flamsteed, the Astronomer Royal, wrote, "These appearances, however frequent in the days of Scheiner and Galileo, have been so rare of late that this is the only one I have seen in his face since December 1676." Had Galileo lived 50 years earlier or later, he might not have seen spots when he examined the sun with his new telescope.

In the course of his tree-ring research Douglass had noted a curious anomaly in seventeenth-century tree rings. Neither any sign of cyclical variation nor any clear pattern matching one tree to another could be discerned. Baffled, Douglass could only surmise that the seventeenth-century environment was in some way more uniform and that the uniformity of tree-ring growth reflected that condition. When in 1922 he read Maunder's paper on the absence of sunspots over exactly the same period of time, he immediately wrote to Maunder, who reported the tree-ring–sunspot connection to the British Astronomical Association.

The "Little Ice Age" is difficult to describe in global terms. In many parts of the world, the extent of snow and ice appears to have been as great as it was at any time since the last great ice age, but temperatures seem to have bottomed out at different times in different regions. Lowest temperatures may have arrived earliest in North America, China, Japan, and the Arctic; Chilean glaciers reached their most advanced point during the eighteenth century; Alpine glaciers were most extensive around 1850; the Antarctic ice pack reached its maximum around 1900. Surface temperatures in Europe and North America may have been a full degree Celsius lower than the present mean temperature, but global variation could have averaged much less.

Cyclic lamination in the Elatina Formation at Pichi Richi Pass, Flinders Ranges, South Australia. These rocks were deposited in a glacial lake during an ice age in late Precambrian times about 680 million years ago. Darker bands of clayey siltstone spaced seven to 13 millimeters apart separate groups of 10 to 12 graded (coarsening-upward) laminae of very fine sandstone (paler) and clayey siltstone (darker). The graded laminae are up to two millimeters thick, and each is interpreted as an annual increment or "varve" deposited in a glacial lake. The varves vary systematically in thickness, being thickest near the center of each group or "cycle." These varve cycles reveal strong climatic periods of about 11 to 12 and 22 to 24 years, as well as longer periods, suggesting an influence of the solar sunspot cycle. The bar scale equals one centimeter.

The need for caution has been further emphasized by Minze Stuiver, who found that the coldest period in Switzerland came precisely midway between the Sporer and Maunder minima. Stuiver also concluded that if these two minima were sunspot-related, there should have been as many as three or four little ice ages every millennium, more than climate records show. The carbon-14 record does indicate that Maunder-type sunspot absences are not rare. Perhaps research on the scale of centuries is simply insufficient to solve the problem of a sun-climate connection.

The Oldest Sunspot Record in the Rocks

In Precambrian times, 680 million years ago, sediments were deposited on bottoms of lakes fed by glacial meltwater. As these glacial sediments hardened into rock, they preserved a striking banded pattern of 11 regular, light-colored layers repeated every 11 years, much in the manner of tree rings. Each of these siltstone layers, which are called varves, corresponds to one year of deposition. In summer, when glacial meltwater was plentiful, the sediments were coarse grained. In winter, when the meltwaters ceased, the finer grains in suspension slowly settled to the bottom. As a result each has a graded grain-size pattern that distinguishes it from its neighbor on either side. George Williams, a prospecting geologist with the Broken Hill Mining Company in Australia, has found evidence for a sun-climate connection in these glacial sediments.

The varves in Australian rocks, Williams found, also vary in width, presumably in synchrony with changes in meltwater volume, which, in turn, reflect changes in climate. Warmer summers brought more meltwater and thicker varves; cooler summers produced thinner varves. These variations show cyclic changes, with 11, 145, and 290 laminations per period, remarkably close to other evidence of solar sunspot periods. It appears that even when the sun was hundreds of millions of years younger, it may have gone through essentially the same cycles it goes through today. What remarkable constancy!

If we accept these cyclic patterns in varves as evidence of solar periodicities, the climate must have been extremely sensitive to solar variations during the late Precambrian era. Williams suggests that the earth's magnetic field may have been near minimum at the time, allowing the solar wind and cosmic rays much greater penetration and influence than they have today.

Time scale (Millions of years ago): Present, 50, 100, 150, 200, 250, 300, 350, 400, 450

Era	Period	
Cenozoic	Quaternary	Early man
	Tertiary	First primates
Mesozoic	Cretaceous	Flowering plants; Subdivision of continents
	Jurassic	Dinosaurs
	Triassic	First mammals
Paleozoic	Permian	
	Late Carboniferous (Pennsylvanian)	Fern forests; Coal deposits
	Early Carboniferous (Mississippian)	Early reptiles
	Devonian	Amphibians; Age of fish
	Silurian	First vertebrates
	Ordovician	

The Advance and Retreat of Glaciers

Surface weather-observing networks were first established in the mid-nineteenth century. In the 1930s, observations were extended above ground by networks of balloons that measured upper-air winds. The modern approach is greatly augmented by global views from weather satellites. Still, forecasting the weather seems more like a guessing game than scientific analysis. Understanding climate change, however, is a different challenge. Although the ingredients for its scientific analysis seem to be at hand, its numerous processes compete with each other in complex ways, and subtle feedback mechanisms need to be sorted out.

Sources of climate variability clearly exist within the atmosphere, the oceans, the biosphere, and the cryosphere (regions of ice and snow cover). Volcanism and human perturbations—the burning of fossil fuels, deforestation, and the widespread use of fertilizers—are strong forcing factors. Even the slow movement of continents—plate tectonics—over hundreds of millions of years modifies the heights and relationships of land masses, the depths and shapes of ocean basins, and, consequently, the distribution of ice cover.

From comments in historical diaries referring to the breakup of river ice and the first signs of spring, we can derive gross climate features as far back as two millennia. The evidence of past climate in ocean-bottom cores, polar ice cores, and coral beds has led scientists to deductions about temperature records millions of years back and to theories about the events that shaped the climate. For most of its 4.5 billion years, the earth has been free of ice, even over the polar caps. Ice first began to appear about a billion years ago; the earliest major glaciation appears to have begun more than 500 million years ago, in the Precambrian era. For the last 50 million years, ice has dominated global climate processes. Evidence points to at least four great ice ages in which polar ice caps stretched almost to the equator. We are now living in the fourth ice age, which began about 2 million years ago. During this period, a partial thaw has halted the advance of glaciers about every hundred thousand years.

Eighteen thousand years ago, glaciers crept out of the frigid north, relentlessly burying forests and fields and crushing mountains in their paths. The weight of ice depressed the crust of the earth, and so much water drained from the oceans and became locked up in glacial ice that sea level fell about 350 feet, exposing large reaches of the continental shelf as dry land. Over the Great Plains of the United States, glaciers a mile thick ground their way southward. Globally, ice covered about

Louis Agassiz (1807–1873), Swiss-born U.S. naturalist and geologist, carried out a comprehensive study of fish fossils in which he identified nearly a thousand species. Later he studied the movements of Swiss glaciers and reached the conclusion that "great sheets of ice resembling those now existing in Greenland once covered all the countries in which unstratified gravel (boulder drift) is found."

11 million square miles of land that today is free of ice, and about 5 percent of all the oceans' water turned to ice over land.

The retreat of the ice sheets began about 14,000 years ago, eventually leaving only the Greenland ice sheet and scattered fingers of glacial ice in arctic Canada. By the eighteenth century, reminders of the ice age had faded, and geologists speculated instead about boulders swept before the great flood of water from the biblical deluge of Noah's time. Recognition of the glacial epoch returned late in the nineteenth century, largely through the perceptiveness of Louis Agassiz, the Swiss-American geologist, who described his vision of the ice age in vivid prose:

> The development of these huge ice sheets must have led to the destruction of all organic life at the Earth's surface. The ground of Europe, previously covered with tropical vegetation and inhabited by herds of great elephants, enormous hippopotamuses, and gigantic carnivores became suddenly buried. The silence of death followed . . . springs dried up, streams ceased to flow, and sunrays rising over that frozen shore . . . were met only by the whistling of northern winds and the rumbling of the crevasses as they opened across the surface of that huge ocean of ice.

The beginning of modern civilization corresponds to a warm spell, the climatic optimum, that began about 8000 years ago. This long warming has been interrupted from time to time by little ice ages. Which way is climate changing now? Are we heading toward a little ice age again? The evidence of recent years has tempted some weather observers to hazard an affirmative guess. In 1972, frigid temperatures and icy winds ravaged the Soviet Union's winter-wheat crops. Severe droughts occurred in India and South America, while the American Midwest was inundated by floods. In June of 1975, a blanket of snow covered London for the first time in this century, and the most severe blizzards in history hit Canada and the northern parts of the United States. January of 1982 saw 75 national weather records broken in one week from California to New England. But do these vagaries signal an accelerating trend toward greater cold?

Significant variations in the cryosphere are reliably observed from year to year. It is postulated that fluctuations in ice-sheet dimensions have some effect on long-range climate trends through an albedo feedback mechanism. (Albedo is the fraction of radiation reflected back from a surface.) And what about cloud cover? About half the planet is covered by clouds at any time and clouds have a high albedo. How important are clouds compared with snow and ice cover?

Albedos of Various Surfaces

Surface	Albedo (percent)
Water	
Rays vertical	2
Rays 40° from zenith	2½
Rays grazing surface at low angle	Very high
Snow or ice	46–86
Grass	14–37
Fields	3–25
Forests	3–10
Bare ground	7–20
Clouds	40–80

Satellite pictures of the earth have begun to provide a record of annual variations in snow and ice cover. For the past 15 years, the average trend does appear to have been toward increasing snow cover. A sharp increase in 1971 may have produced the weather extremes, such as the drought in the U.S.S.R., of 1972. A larger increase in late 1975 may have been connected to the severe weather in North America.

The natural patterns of earlier times are now being modified by anthropogenic influences as well. Agriculture, overgrazing, and the clearing of forests have had profound influences over the past several thousand years. How much do these man-made changes affect the radiation budget and influence climate? Their cumulative effect on global albedo is estimated to be about half a percent, enough to produce a global temperature decrease of as much as 1°C. How much might carbon dioxide from the burning of fossil fuels tend to increase global temperature through the greenhouse effect and thus compensate for the effects of an increasing albedo? Present projections anticipate a rise in global temperature of perhaps 2°C by the year A.D. 2050.

This painting of Taos Valley (1935) by Victor Higgins gives a vivid impression of the variety of landscape and sky features that contribute to albedo in any local region. Modern remote sensing from satellites is making it possible to assess global albedo, including that of oceans, ice, and snow, as well as dry land and biomass.

Milutin Milankovitch (1878–1958), in a portrait by Paja Jovanovic painted in 1943. From about 1910 until his death, the Yugoslavian scientist dedicated himself to developing a mathematical theory that would explain the long history of the earth's climate in terms of the variations in shape of the earth's orbit and the precession and tilt of its spin axis.

Toward an Astronomical Theory of Ice Ages

With all these complex factors operating on climate, can any identifiable control be clearly associated with the variability of solar insolation? The persistent efforts of a Yugoslavian mathematician, Milutin Milankovitch, have brought one convincing connection to light. Milankovitch dedicated his entire scientific career from 1921 to 1941 to describing the connection between the changing shape of the earth's orbit, the tilt of the spin axis and its slow wobble, and the variations in global climate over the ages. His goal was to produce an astronomical theory of ice ages, and a growing body of evidence in recent years tends to support his basic ideas. But Milankovitch figures only later in the story.

In the seventeenth century, Johannes Kepler showed that although the earth's orbit is nearly circular, it is not perfectly so. This slight ellipticity produces a significant variation in the sun-earth distance over the course of the year. On about January 3 each year, the earth arrives at perihelion, when it is closest to the sun. Six months later, it reaches aphelion, the most distant point from the sun. It is then 3 million miles farther away than at perihelion.

At the start of winter, December 21, called the winter solstice in the northern hemisphere, the day is shortest because the north pole tips farthest away from the sun. In the southern hemisphere, it is the longest day of the year. June 21, the longest day for the northern hemisphere, is the summer solstice, when the north pole points closest to the sun. The seasons are reversed in the southern hemisphere.

Twice a year, on March 20 and September 22, the two poles of the earth are equidistant from the sun, and the days are equally long everywhere on earth. These two points are the equinoxes, the vernal equinox for the beginning of spring and the autumnal equinox for the beginning of fall in the northern hemisphere. Again, in the southern hemisphere the situation is reversed.

In the northern hemisphere, spring and summer have seven more days than fall and winter. Total daylight hours exceed those of night by 168 hours each year. In the southern hemisphere, darkness hours exceed daylight hours, and cold seasons are seven days longer than warm seasons.

Joseph Alphonse Adhémar, a French mathematician, seems to have been the first to search for a connection between orbital characteristics and climate change. In a paper published in 1842, Adhémar reasoned simply that more hours of darkness than light each year would cause the southern hemisphere to become progressively colder.

He interpreted the Antarctic ice sheet as evidence that the southern hemisphere is now in an ice age. Next, he had to explain how ice ages would have occurred in the northern hemisphere in past time. To answer this, he proposed a coupling of oscillations in ice ages between the northern and southern hemispheres with the nodding of the earth's spin axis. (As far back as 120 B.C., Hippocrates had identified this variation in tilt of the spin axis by comparing his own observations with those of Timocharis, 150 years earlier.)

Adhémar's theory, however, had a basic flaw: the idea that one hemisphere heats up while the other cools down is incorrect. As Baron von Humboldt showed in 1852, the factor controlling the amount of Antarctic ice is the total amount of sunlight received in a year. The decrease in solar heating during the season that the sun is farthest from the earth is exactly balanced by an increase during the opposite season when the sun is closest. The reason that the Antarctic is cold enough to support a permanent ice sheet, Humboldt explained, is that the continent centers on the south pole, far removed from the moderating influence of warm ocean currents. Furthermore, its high albedo reflects back much of the incoming radiation. But even though Adhémar's hypothesis was wrong, his ideas did encourage scientists to think further about orbital effects on climate, including precession of the spin axis and changing eccentricity of the elliptical path traveled by the earth around the sun.

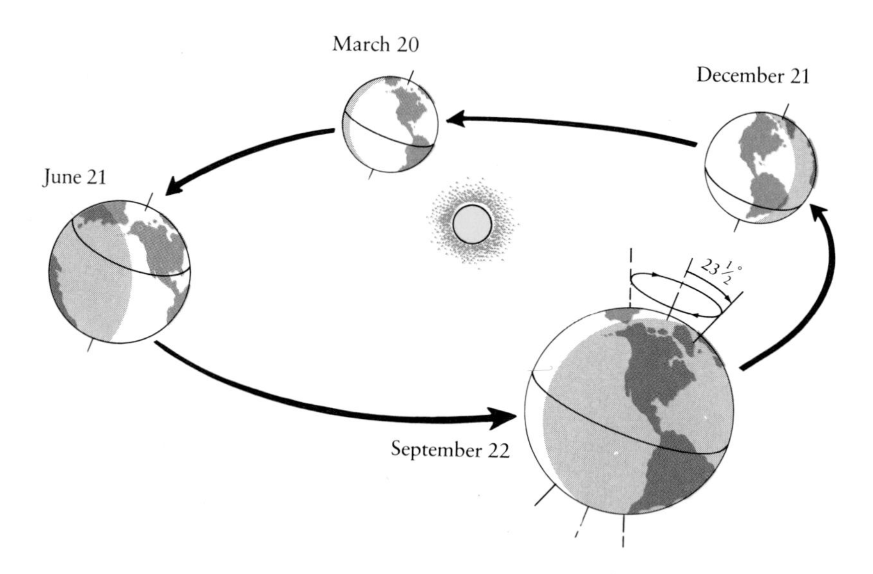

The spin axis of the earth tilts 23.5 degrees from perpendicular to the plane of the orbit. Changes in the amount of sunlight received at different latitudes produce the succession of the seasons. As a result of the gravitational pull of the sun and moon on the earth's equatorial bulge, the axis of rotation precesses with a 26,000-year cycle. Independently, the tilt varies by about 1.5 degrees from its average value of 23.5 with a 41,000-year cycle.

The Wobbling Top

As the earth spins, it precesses like a wobbling top about to fall, but, unlike the top, the earth is not spinning down appreciably, and it will not lie over on its side. Because it spins, the earth has developed an equatorial bulge of 21 kilometers (1⁄298 of the earth's radius). Furthermore, the spin axis is tilted 23.5 degrees from perpendicular to the plane of the earth's orbit, the ecliptic. The sun, moon, and planets tug on the earth's equatorial bulge to pull it toward the ecliptic plane. But the gyroscope effect prevents the spinning earth from tilting into alignment with the plane of the solar system. So it drags its tilted axis around in a perpetual precessional motion that describes a cone (actually two cones, tip to tip) around the direction perpendicular to the ecliptic. This precession is very slow—25,800 years for one cycle, in which the earth's axis traces out a circle about 47 degrees in diameter.

The celestial pole does not now point to Polaris, the North Star, but about 50 minutes of arc away. Every year it draws closer by 17 arc seconds, and it will reach its closest alignment in A.D. 2105. Stone Age pole stars lay in the constellation Hercules. In the near future, the pole will point to Cepheus, and in A.D. 14980, it will aim directly at the bright star Vega, which now appears almost directly overhead in summer.

Axial precession also causes the equinoxes to move slowly clockwise as seen from above the ecliptic plane, while, at the same time, the elliptical orbit itself turns in the opposite direction with respect to the stars. The net effect is a residual cycle of equinoxes lasting about 22,000 years. Winter in the northern hemisphere now comes when the earth is close to the sun near one end of the elliptical orbit. Eleven thousand years ago, winter began when the earth was near the farther point on the ellipse. And 22,000 years ago, the earth's position relative to the sun was the same as it is today.

A Glaciation Every Hundred Thousand Times Around the Sun

A French astronomer, Urbain Leverrier, detected the next clue to a possible role of orbital mechanics in climate. In 1843, he showed that the shape of the orbital ellipse itself changes. Using orbital eccentricity as a measure of the orbit's elongation, Leverrier demonstrated that eccentricity varies with a 100,000-year cycle. As the planets revolve about the sun, the gravity of each tugs on every other. Although it took him 10 years, when Leverrier succeeded in calculating the combined effects, he found that they produce a constantly changing ellipticity. The magnitude of variation ranged from near zero to as much as 6 percent over 100,000 years. Right now, the eccentricity of the earth's orbit is 1.7 percent and heading toward zero—almost perfectly circular. As Leverrier showed, however, the total amount of heat received by the earth during an entire year is affected almost insignificantly by variations in orbital eccentricity. Global insolation has changed by less than 0.3 percent over the past million years. The corresponding global average temperature could only have changed a few tenths of a degree Celsius.

After Leverrier came James Croll, a self-taught Scotsman from an impoverished background. Forced to leave school at age 13, Croll went from millwright to shopkeeper to hotelkeeper to insurance salesman, drifting from one disappointment to another and pursuing his obsession with scholarly studies as best he could. In 1859, he took a job as curator in the Andersonean College and Museum in Glasgow. The job paid little, but it gave him access to a fine scientific library, where he could indulge his powerful urge to study. From 1867 to 1880, Croll was in charge of the Edinburgh office of the geological survey of Scotland.

Learning of Leverrier's work, in 1864 Croll published a paper in *Philosophical Magazine* in which he proposed that changes in orbital eccentricity were the link to ice ages. For the next 20 years, he worked diligently on the problem. Using Leverrier's formulas, Croll calculated orbital eccentricity back over 3 million years. His solutions showed that orbital eccentricity varies cyclically, with intervals of high eccentricity lasting many tens of thousands of years, alternating with long intervals of low eccentricity. One hundred thousand years ago, eccentricity was near maximum, but for the past 10,000 years, it was very small. Leverrier had shown that the total heat received by the earth in any year was independent of the shape of the orbit. Croll set out to prove that eccentricity does have a significant effect on the seasonal variation.

Croll hypothesized that ice ages developed when winters were colder than average—when, for example, winter occured in conjunction with a greater-than-average distance from the sun. Such situations are governed by the precession of the equinoxes and the eccentricity of the orbit. If the orbit were circular, the amount of heat received during any given season would not vary. If it were markedly elongated, however, seasonal insolation would vary according to whether the earth was nearer or farther from the sun in that season. Croll's calculations of the change in orbital eccentricity showed peaks at about 100,000, 200,000, and 300,000 years ago. The most recent maximum elongation had been followed by a steady decline to a nearly perfectly circular orbit about 10,000 years ago, a time that also marked the beginning of the present interglacial period.

Great skepticism that orbital effects were strong enough to promote global glaciation still prevailed, however. By the end of the nineteenth century, most geologists had put aside Croll's theory, and it was soon forgotten. It remained for Milutin Milankovitch to revive and complete the astronomical theory of ice ages half a century later.

The Milankovitch Cycles

Croll had had to start with calculations of precession and eccentricity that went back only 100,000 years. Milankovitch immediately had access to calculations made in 1904 by a German mathematician, Ludwig Pilgrim, which included variations in axial tilt as well as eccentricity and precession and went back a million years. Milankovitch set out to calculate how all the orbital variables combined to affect the amount of sunlight reaching the earth during each season of the year and at every latitude. By 1914, he had convinced himself that eccentricity and precession were potentially sufficient to influence the advance or retreat of ice ages and furthermore, that a 41,000-year variation in axial tilt was equally important.

Although his work was briefly interrupted by World War I, Milankovitch persisted in sharpening his calculations. By 1930, he published radiation input curves for all latitudes from 5°N to 75°N. His results confirmed that the various astronomical factors had their maximum effects at different latitudes. For example, the 22,000-year precession was most effective at the equator and least significant at the poles; the 41,000-year tilt cycle worked in an opposite fashion, with maximum effect at the poles and minimal influence at the equator. In 1941, Milankovitch wrapped up his work with the publication of a book in which he also explained how the effects of each factor could be amplified by albedo—the increased reflection from growing ice sheets.

Scientific Breakthroughs

Too many uncertainties afflicted the sedimentary record of the advance and retreat of glaciers on land to provide a truly reliable basis for testing Milankovitch's astronomical theory. A far more detailed temperature record of the past is locked up in the sea bed upon which the skeletal remains of microscopic organisms rain down, becoming layer upon layer of relatively undisturbed sediment. Away from the

The Milankovitch cycles of orbital eccentricity, precession, and obliquity. Changes in these geometric characteristics of the earth's orbit induce cycles of 100,000 years, 22,000 years, and 41,000 years in global temperature. Below, the orbital eccentricity is compared with global ice volume over the past 730,000 years.

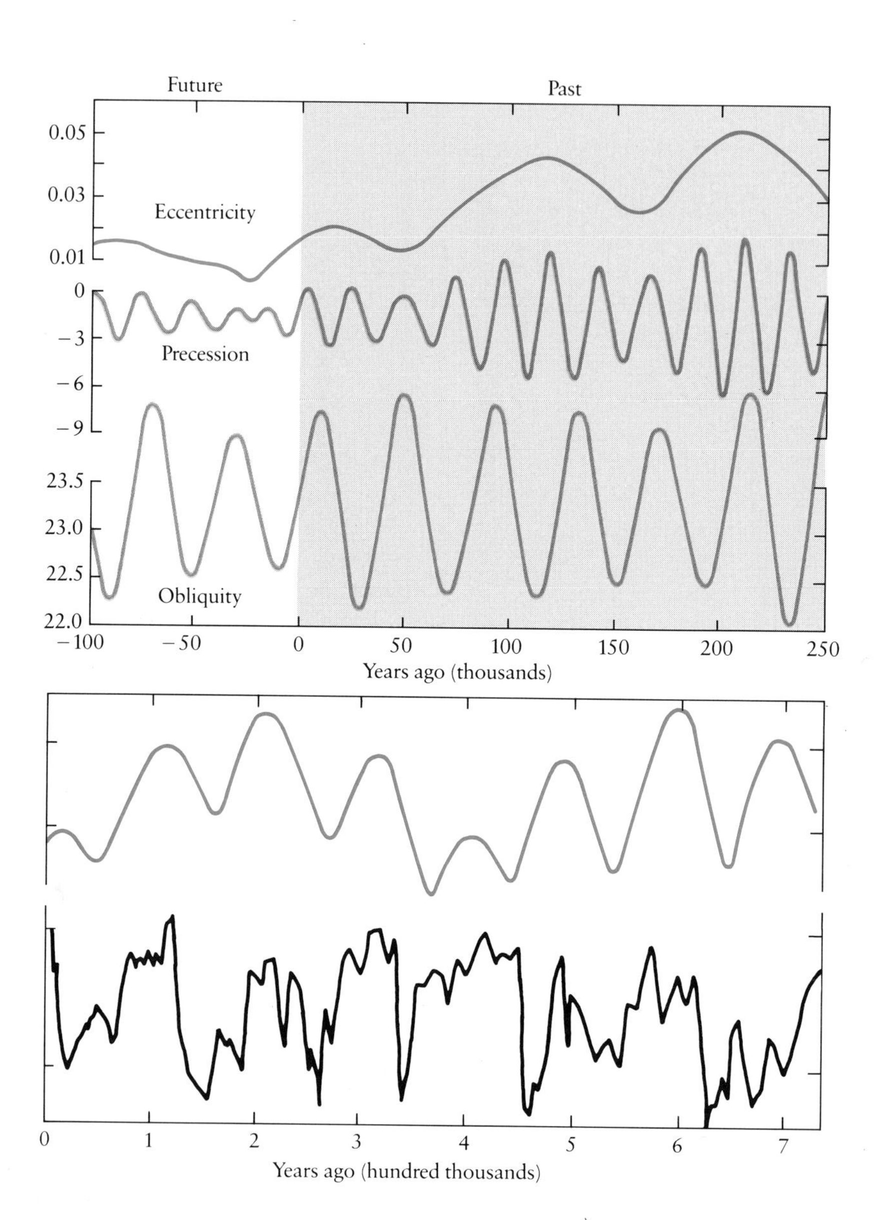

continental margins, finely divided oozes, composed almost completely of fossilized plant remains and minute sea creatures, cover the ocean floor.

Live floating organisms, known as plankton, abound in the oceans, and their mineral content is identical to that of the skeletal remains on the ocean bottom. Early investigations found the remains of one type of plankton, known as foraminifera, to be prevalent in temperate and tropical seas; another type, known as radiolaria, preferred colder waters of the Arctic and Antarctic. Subsequent studies showed that some species of forams were found only in cold waters, while others existed only in warm waters. The geographic extent of this spread or retreat of temperature-sensitive species with climate change held the key to developing a record of climate history. The layering of skeletal remains at any given spot should reveal the temperature sequence.

The earliest method of deep-sea coring was simply to drop a weighted, hollow, steel pipe onto the sea floor. Cores about a meter in length were extracted, but these were not long enough to reveal the complete cycle of the ice ages. Using dynamite to drive the pipe deeper badly distorted the retrieved cores. A Swedish oceanographer, Bjore Kullenberg, solved the coring problem in 1947 with the hydraulic piston corer, a device that sucked up sediments while the tube was driven into the ooze. With this equipment, the drilling ship *Glomar Challenger* brought up relatively undisturbed samples of deep-sea sediments in cores 10 to 15 meters long. In the Atlantic Ocean, sediments have accumulated at the rate of two to three millimeters per century, and the long cores were expected to contain a rather complete record of ice-age climate.

While sea-bottom coring was being developed, a study of the coral-reef terracing in Barbados, Hawaii, and New Guinea was offering another source of climate dating. Coral grows near sea level. If the melting of polar ice in a warm interglacial period causes the sea to rise, a coral reef builds up. With a colder climate, glaciers lock up more water and the sea subsides, leaving the coral terrace high and dry. A sequence of advance and retreat of glaciers leaves a legacy of coral terraces, which can be dated by their radioactive thorium content.

Progress in the reading of deep-sea cores came from studies of the ratios of oxygen 16 to oxygen 18. Most of the oxygen in the atmosphere and oceans is the lighter isotope ^{16}O, but a few molecules out of every thousand incorporate the heavier ^{18}O, which has two extra neutrons in its nucleus. Molecules containing the ^{18}O evaporate more slowly and tend to be left behind. When the extent of the ice sheets is

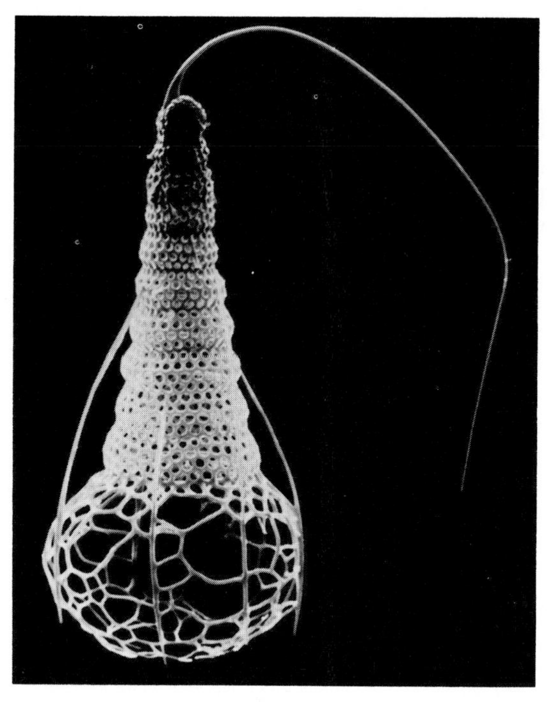

a

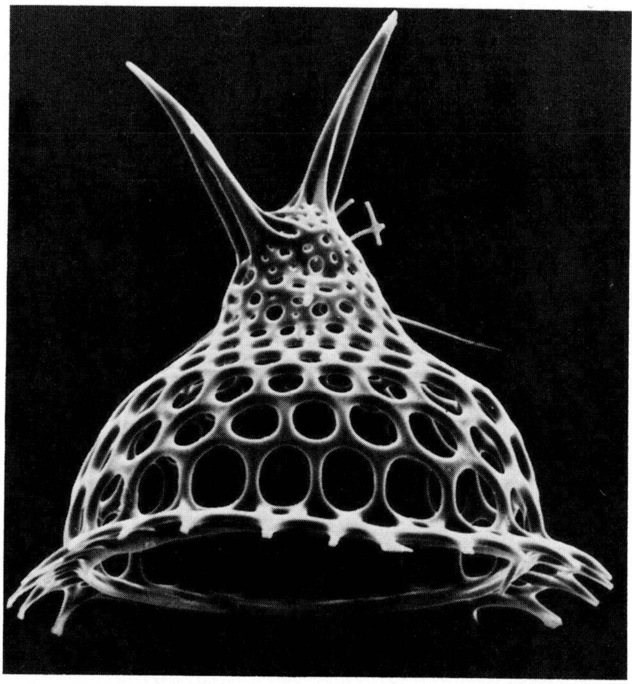

b

Common skeletons of radiolaria from the Gulf of Alaska (*a, b*) and of foraminifera from the Mediterranean (*c*) found in deep-sea sediments. Adult specimens of these organisms are about the size of small grains of sand (0.1 to 0.3 mm).

c

The *Glomar Challenger,* a deep-sea drilling ship that explored the floors of all the world's oceanic basins from 1968 to 1973. Its 122-meter hull resembles that of a tanker, with the bridge and living quarters at the stern. Amidships is a well open to the sea, surmounted by a derrick, 59 meters from waterline to top. The drill pipe is constructed of 29-meter sections, making a string long enough to reach the ocean floor several kilometers below the sea surface. Core barrels are lowered through the pipe and bring back 10-meter lengths of core, six centimeters in diameter. A single core may hold more than a million years of geological history. By the end of 1973, the *Glomar Challenger* had brought up cores totaling 35 miles in length.

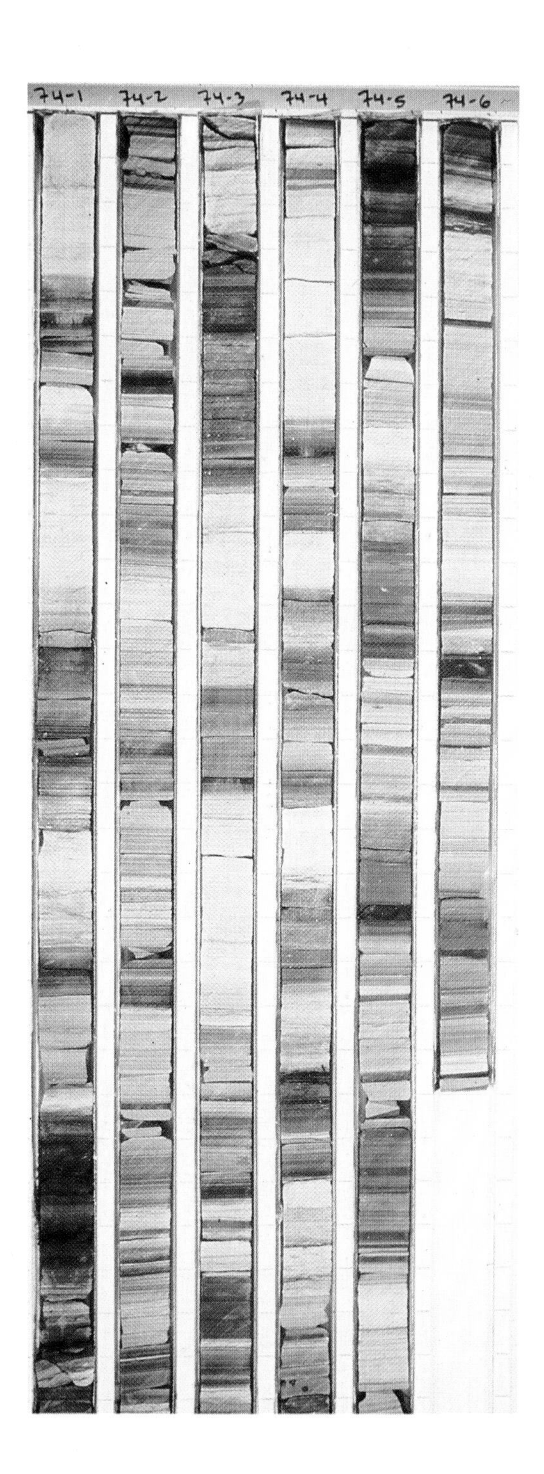

A core obtained from the Deep Sea Drilling Project. Hole 534-A was drilled about 300 nautical miles east of Florida to reach the oldest sediments of the North American Basin. Core 74, shown here, is from the early Cretaceous epoch. It shows alternating light-gray or beige limestone and dark-gray to black marls. The cyclic nature of these deposits probably results from fluctuations of the deep-sea circulation connected with climatic oscillations.

constant, the ^{16}O that evaporates from the oceans is returned to the oceans in rain and snow, and the isotopic composition of the sea remains unchanged. However, when the ice sheets are expanding in a glacial period, some of the ^{16}O gets locked up in ice and fails to return to the oceans. The ratio of ^{18}O to ^{16}O in the sea water then increases. Accordingly, a relatively high content of ^{18}O in the ocean indicates an ice age; a low concentration of ^{18}O signifies an epoch of minimal global ice volume. Following a lead suggested by Harold Urey, his professor at the University of Chicago, Cesare Emiliani began a study of the oxygen isotope ratio in the calcium carbonate ($CaCO_3$) shells of forams contained in deep-sea cores. Urey, a Nobel laureate for his work on heavy water, had suspected early on that the skeletal remains of the microscopic creatures would reveal isotope ratios characteristic of seawater temperature at the time they were alive. Emiliani did indeed find that isotope ratios in the sediment layers showed cyclic variations. In cores taken from the equatorial Atlantic, he found evidence for seven complete cold cycles in the last 700,000 years. This record provided a much more finely resolved history of past climate than anything that had been available before.

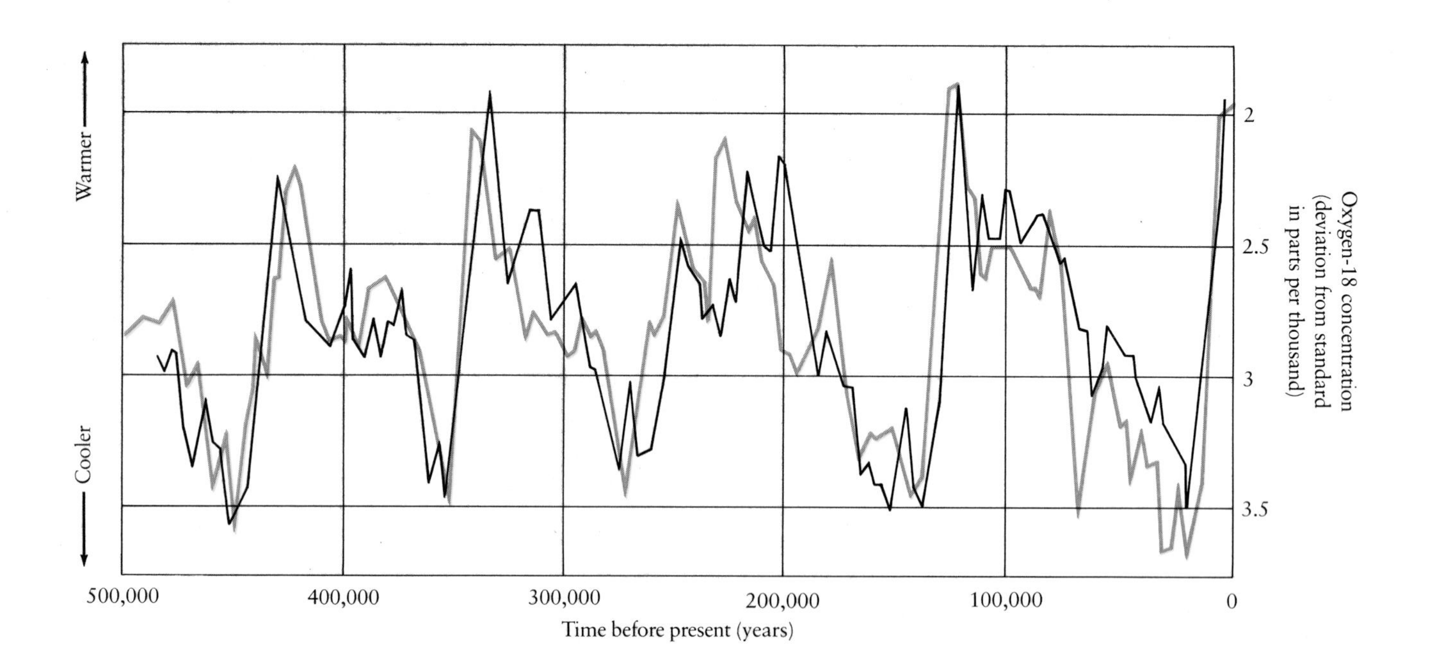

Climate variation over the past 500,000 years. The ratio of ^{18}O to ^{16}O is a measure of past global ice volume. Two sets of isotope data are shown, derived from cores extracted from widely separated places on the ocean bottom. These indications of cyclic variation in global climate fit well with the Milankovitch astronomical theory of the ice ages.

By the 1970s, it had become clear to other scientists that Emiliani was detecting qualitative evidence of variations in global ice volume. The skeletal remains of microscopic plankton were providing a reading of the advance and retreat of glaciers, but an accurate timing marker for the oscillations revealed in the fossil layers was urgently needed. The solution to this problem lay in the reversals of polarity of the earth's magnetic field.

Few geophysical discoveries of this century have aroused as much scientific excitement as the recognition that the polarity of the earth's magnetic field has, for millions of years, repeatedly reversed itself. Astonishing as it may seem, at various times the north magnetic pole has appeared near the south geographic pole.

The original discovery was made in a brickyard by the French physicist Bernard Brunhes in 1906. Brunhes found that as the brick cooled, iron-rich mineral particles aligned themselves parallel to the earth's magnetic field, so that the brick acquired a slight overall magnetization. When Brunhes studied volcanic lavas from Pontfarein, Cantel, in France, he found that some of those rocks were magnetized in a direction opposite to the earth's field. When lava cools and solidifies, by virtue of its hematite (Fe_2O_3) and magnetite (Fe_3O_4) content it retains a magnetization parallel to the earth's field and of the same polarity. No matter what later changes occur in the earth's polarity, this polarity remains fixed in the solid lava. Brunhes correctly concluded that the earth's field must have been reversed at some time in the past, but his contemporaries largely ignored his ideas.

By 1907, many examples of variations in remnant magnetism had been found. Eighth-century-B.C. Etruscan vases, Greek pottery of the seventh century B.C., and neolithic pots from 1500 B.C. as well as volcanic lavas showed evidence of magnetic-field variation. Because the pots had been baked upright, their remnant magnetism revealed the local magnetic dip at the time when they were fired.

Twenty years after Brunhes' discovery, a Japanese, Montonauri Matuyama, studied a succession of lava flows in Japan and Korea. He found at least one reversal during the Pleistocene epoch, and detected several more that must have occurred during much earlier geological epochs. In the late 1950s and 1960s, worldwide correlations positively confirmed the work of Brunhes and Matuyama, and the Pleistocene epoch of "normal" polarity was named the Brunhes epoch, while the earlier epoch of reversed polarity was called the Matuyama epoch. Why the earth's magnetic field reverses is still very mysterious, but attempts to model disk dynamos in the liquid core of the earth seem to show the possibility of intermittent reversals.

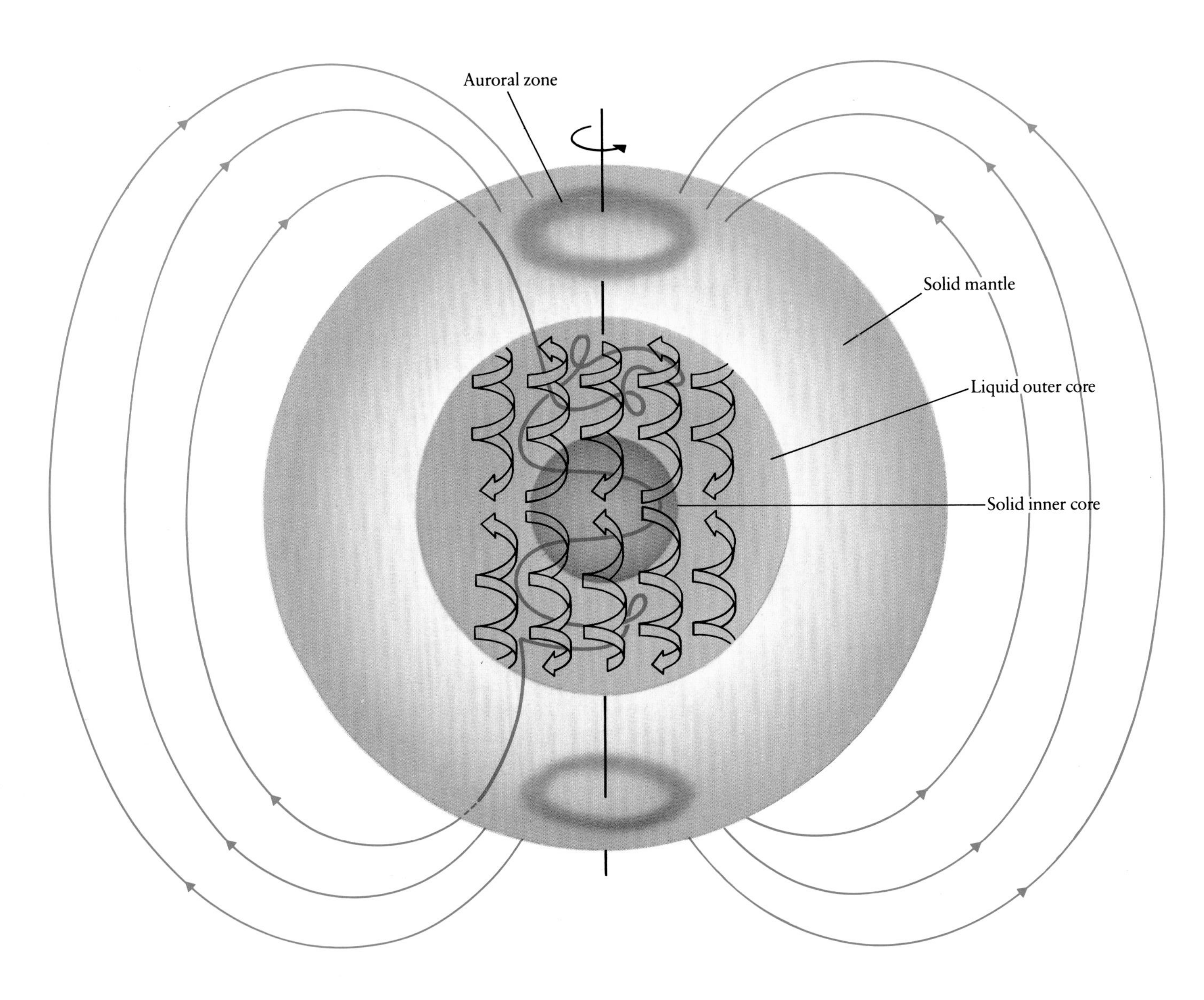

Models of the generation of a magnetic field by dynamo action in the sun and in the earth bear many similarities (see the figure on page 65). Electrically conducting metallic fluid in the earth's core may circulate in screwlike rollers. Lines of force are threaded through the rollers as shown here for a single line of force entering from the north and emerging from the south (heavy line).

Each reversal of the earth's field can be clocked by a potassium-argon dating technique. Potassium 40 constitutes only 0.012 percent of natural potassium, but it is radioactive. It is present in all rocks, and we even carry a slight potassium radioactivity in our bodies. At a very slow rate, potassium 40 decays into argon 40, an inert gas that does not participate in any chemical activity but simply accumulates in the interstices of the rock. If the rock is heated to a high temperature, the argon quickly boils off. When the rock cools, the buildup of argon starts all over again. By measuring the amount of potassium in a rock specimen, the amount of potassium 40 that was present when the rock last cooled can be fixed. A determination of the argon-40 content then indicates how long the decay of potassium 40 into argon gas has been going on. That interval is the measure of time since the rock cooled and captured the magnetic field that then existed.

The last flip of the earth's field occurred 700,000 years ago, and the preceding one about 1.1 million years earlier. This 700,000-year reversal established the time marker on which a calendar for that portion of the Pleistocene epoch covered by Milankovitch's calculations could be based. Ocean cores were also found that contained both the climatic and magnetic evidence that permitted the precise dating of glaciations.

Orbital Variations Clock the Ice Ages

Beginning about 1970, determined paleoclimatologists began to attack the interpretation of deep-sea cores with all the isotopic, magnetic, and statistical techniques available to them. Statistical studies of radiolaria populations were combined with measurements of oxygen isotopes in foraminifera to give evidence of water temperature and global ice over the past 300,000 to 450,000 years. A large national effort, the Climate: Long-Range Investigation, Mapping, and Prediction, or CLIMAP, program, was formed, and its most significant findings were published in 1976. In "Variations in the Earth's Orbit: Pacemaker of the Ice Ages," James D. Hayes, John Imbrie, and Nicholas J. Shakleton concluded that the evidence from deep-sea cores strongly supported Milankovitch's theory that major Pleistocene climate changes fell under orbital control and that variations in orbital geometry directly caused the succession of quaternary ice ages. They confirmed pronounced climate cycles of 100,000 years, 43,000 years, and a twin cycle of 24,000 and 19,500 years. The first two cycles matched Milankovitch's predictions very closely, although he had called for a third single-precession cycle of 22,000 years rather than the twin periods observed. A Belgian astronomer, André Berger, soon refined the precession calculations and confirmed the two periods derived by Hayes, Imbrie, and Shakleton. Ice ages now fit with astronomical theory in a very impressive way.

Attempts to understand fully Milankovitch's climate-forcing theory have not yet found a satisfactory explanation of the prominence of the 100,000-year cycle that fits so well with orbital eccentricity. If tilt and precession have more to do with the distribution of insolation than eccentricity does, why does the eccentricity cycle dominate? One of several speculative theories is that an underlying resonance—a preferred 100,000-year oscillation—occurs in the ocean, atmosphere, and ice sheets. The cycle of orbital eccentricity may force this resonance and thus create a more prominent cycle than tilt and precession.

Among the models that might explain some form of natural resonance is one based on the loading of bedrock. Under the increasing weight of a growing ice sheet, the bedrock must sink, though with a delay, perhaps, of several thousand years. In the meantime, the ice sheet thickens, and its surface reaches a greater height with a greater prospect of snow accumulation. This serves to accelerate the growth of underlying ice. When bedrock has sunk appreciably, the entire ice sheet sinks with it, and melting accelerates. The delayed reaction of bedrock, therefore, encourages later ice-sheet growth and still later melting. Once melting has progressed significantly, the bedrock is elastically restored to its higher level, carrying ice upward, and the cycle of rise and fall is then repeated.

Did the ice age that ended about 10,000 years ago signal the finish of the whole glacial epoch, or is there more cold climate to come? Projected into the future, can the astronomical theory foretell an approaching ice age or a global warming? Both axial tilt and eccentricity are now decreasing—a trend that favors the return of an ice age. Precession, however, is working in the opposite direction, bringing us toward a shorter sun-earth distance in summer that favors a retreat of glaciation. Imbrie's calculation of these combined effects predicts that the present cooling trend will continue and that the next ice age will arrive in 7000 or 8000 years, reaching its maximum about 23,000 years from now. On the basis of the astronomical theory alone, we are on a course of steadily cooling climate.

But perhaps this is too simple. Factors external to those calculated by orbital theory, as well as regulating processes within the climate system, must combine to influence overall variations. The scientific analysis of climate cycles needs further clarification by better dating techniques and better statistical methods.

Extended glaciations are believed to be related to continental drift. Two hundred million years ago, the earth's land masses were joined in a single continent, Pangaea. What are now Brazil, Argentina, Antarctica, and Australia were closer to the south pole and fully glaciated. For the next 200 million years, Pangaea moved northward, the earth warmed, and the ice cover melted. About 55 million years ago, Pangaea began to break up into the present configuration of continents. The separation into northern and southern hemisphere land masses initiated a long cooling trend.

The Very Remote Past

Polar ice caps as extensive as those of the last half-million years have existed only three times in the last billion years. The first of these glaciations occurred about 700 million years ago in Precambrian times; the second appeared about 300 million years ago in the Permo-Carboniferous period; and the present ice age (late Cenozoic, Pleistocene) began about 10 million years ago. When all of the astronomical factors should have been running inexorably through their cycles, why are there no signs of ice ages in the 250 million years before the Pleistocene?

It is suspected that the older, extended glaciations were somehow related to continental drift. When continental land mass is near the poles, ice accumulates in high latitudes. During the Permo-Carbonifer-

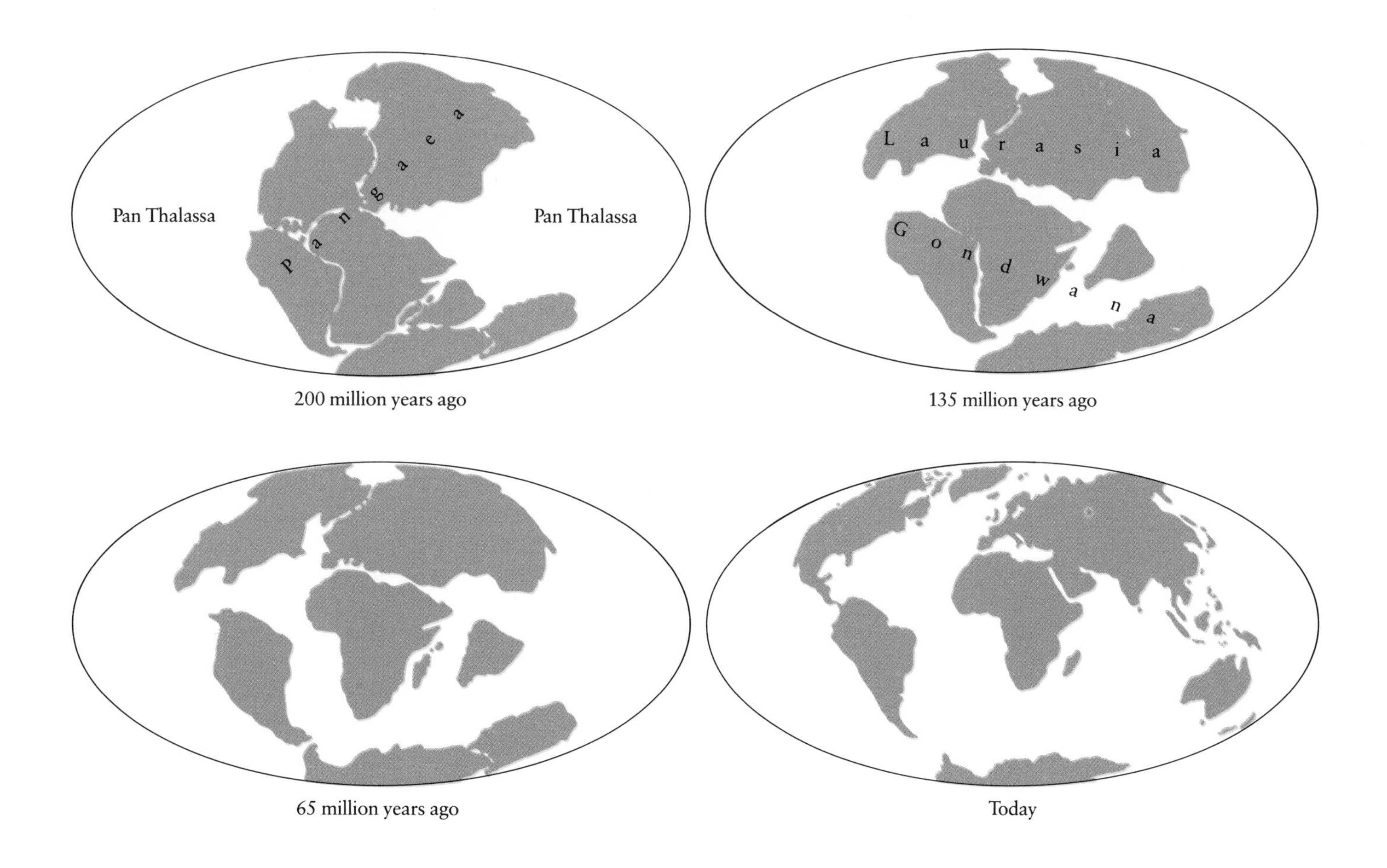

200 million years ago

135 million years ago

65 million years ago

Today

ous period, the earth's land masses were joined in a single great continent, known as Pangea, which reached to the south pole. What are now Brazil, Argentina, Antarctica, and Australia were then located in the high latitudes of the southern hemisphere and were fully glaciated.

As Pangea moved northward during the next 200 million years, the earth warmed and the ice cover disappeared. About 55 million years ago, Pangea began to break up into the present continents. As Antarctica moved to the south pole and North America and Eurasia drifted toward the north pole, a long cooling trend set in. The Arctic ice sheets developed over the northern hemisphere adjacent to the Atlantic Ocean about 3 million years ago. Since then, ice cover has fluctuated in synchrony with the astronomical variations.

Dust, Smog, and the Climate Machine

Any comprehensive model of climatic variation must take account of radiation absorption, scattering, and emission within the atmosphere by dust and aerosols, as well as the variability of the incoming solar energy. At times of volcanic activity, atmospheric particles can exert more immediate influence than any variation of the solar flux itself. Benjamin Franklin was one of the first to appreciate the impact of volcanic dust on climate. During his sojourn in Paris, he attributed the exceptionally cold winter of 1783–1784 to the eruption of Icelandic volcanoes the previous May and June. He sensed a veiling of the sun when he could not set fire to a piece of paper by focusing the sun's light with his magnifying glass, and he correctly attributed the dimming to the volcanoes.

We have ample evidence that volcanic activity affects climate. For example, in 1815, an immense eruption of Mt. Tambora in Indonesia spewed vast amounts of fine dust into the atmosphere. Some 150 cubic kilometers of material blew off the mountain, and the sky was filled with a cloud of ash so dense that pitch darkness prevailed for two days over a distance of 600 kilometers. The aerosol load delivered to the stratosphere amounted to about 200 million tons of sulfuric acid particles. Profound changes in worldwide climate accompanied the dust of Tambora as it circulated about the earth. In middle Europe, the summer of 1816 was the coldest on record in two centuries. New Englanders called it "the year without a summer." Three to six inches of snow fell in June, and crop-killing frosts appeared throughout July and August.

Frederic Edwin Church's 1862 rendering of Ecuador's Cotopaxi volcano. In this view, Cotopaxi spouts straight into the sky. The billowing plume spreads ash in the path of the rising sun, creating a rosy sky cut by a yellow glow that reflects in pink-orange hues from the slopes of the rocks. Church took full advantage of artistic license—a lake appears in the foreground instead of a barren plain—but the painting captures the symbolic force of the volcano.

The earth shudders and shakes more than a million times a year, and historical evidence retains the imprints of some incredibly catastrophic events. Yellowstone National Park, about 600,000 years ago, was the site of perhaps the greatest volcanic eruption in North American history. It blew out so much material that we can still find debris as far east as the Mississippi and beyond. It was about 15,000 times more powerful than the recent eruption of Mt. St. Helens. Even now, there could be enough hot rock and magma stored in the chambers below Yellowstone to bury all of Wyoming some 35 stories deep. A Yellowstone catastrophe could put enough dust and ash into the air to devastate weather and agriculture throughout North America.

The Krakatoa outburst in Indonesia in 1883 is well documented and serves as a classic reference for comparison of volcanic explosions. It threw some 20 cubic kilometers of rock and dust to a maximum height of 80 kilometers and the explosion was heard as far as 3500 kilometers away in Australia. Brilliant sunsets prevailed worldwide. The reduction of direct solar radiation measured at ground level was 20 to 30 percent, although most of the missing heat and light must have been scattered sideways and still reached earth by indirect paths, perhaps only 1 or 2 percent being returned to space. The dust veil persisted for several years, and global temperature dropped about 0.3°C.

The eruptions of Mt. St. Helens in the state of Washington (1980) and El Chichon in Mexico (1982) have alerted us once again to the power of volcanism and its ability to deliver enormous amounts of

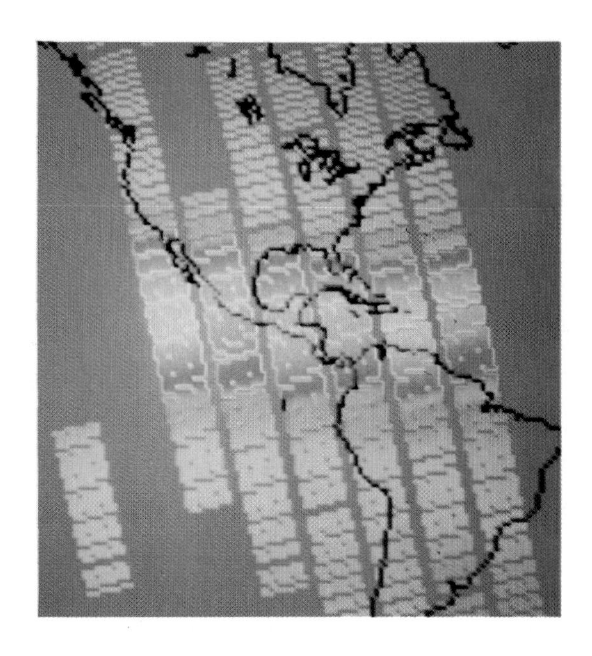

Following the eruption on April 4, 1982, El Chichon's volcanic aerosol cloud drifted through the stratosphere over Central America, as seen in this image produced from Solar Mesosphere Explorer satellite data. The cloud appears here as a bright horizontal band over Latin America. The computer-generated color in the image corresponds to the amount of infrared emission from the cloud, changing from blue to green to red to yellow with increasing radiance.

dust to the high atmosphere. The El Chichon eruption hurled 500 million tons of debris into the stratosphere, the largest volcanic dust cloud in the northern hemisphere since Katmai erupted in Alaska seven decades ago. Mt. St. Helens ejected about six or seven times more matter than El Chichon, but the Mexican eruption had greater atmospheric impact. Mt. St. Helens produced a lateral blast that laid waste 600 square kilometers on land. El Chichon blew straight up, delivering nearly all its ash and gas to the stratosphere. In the first year, its dust lowered direct sunlight by 25 to 30 percent over a wide band of latitude; secondary scattering, however, reduced the total loss of sunlight to 5 percent or less. Global temperature may have been lowered by about one-third to one-half degree Celsius for the next three years.

Much progress is also being made in retrospective analyses of long cores taken from the Greenland ice sheet. Scientists have drilled to depths of nearly 7000 feet in the ice pack and found well-stratified dust chronologies of volcanic activity. Eventually, they hope to read back 100,000 years. Willie Dansgaard, a Danish scientist, has used the ^{16}O-to-^{18}O isotope ratio to date these cores with such precision that trapped dust can be pinpointed in time to within one or two years. Going back to 7000 or 8000 B.C., he has found evidence for seven major eruptions. Beyond 10,000 years ago, the annual layers are less clearly distinguished, but their age can still be set within about 50 years.

According to European accounts, the sun looked reddish and very pale in the years 1601 and 1602. Ice formed from snowfall during those years shows two sharp peaks in volcanic dust content. Virgil and Pliny the Elder relate accounts of the sun going dim after the death of Julius Caesar in 44 B.C. The Greenland ice shows three years of acid fallout about 50 B.C., indicating the occurrence of an extraordinarily great eruption.

The snows that fell in Greenland during the final third of the last ice age were heavily laden with dust. When the ice age came to an end, dust very quickly disappeared from the ice record. Thus the dust may have been either volcanic or, perhaps, a consequence of the growth of ice sheets that removed enough water to lay bare and dry long stretches of wind-blown continental shelves.

Cosmic Encounters with Interstellar Dust

We tend to think of ice ages as the result of processes on a comparatively local scale, within the orbital limits of the earth. But astronomers have proposed a theory of ice ages set on a stage of galactic

A red sunset caused by the high-altitude injection of dust and aerosols from El Chicon. The longer-term effects are produced by sulfur dioxide, which oxidizes and reacts with water to form sulfuric acid aerosols.

dimensions and cosmic time. Our Milky Way galaxy is a huge whirlpool of 100 billion stars. The sun is situated about two-thirds of the distance from the galaxy's center to the trailing edges of its spiral arms. According to one astronomical hypothesis, ice ages occur when the sun is traversing a spiral arm where the concentration of dust and gas increases the possibility of collisions with dense clouds.

All galaxies rotate. The sun, with its planets, orbits the center of the Milky Way at a speed of about 150 miles per second, taking about 200 million years for one orbit. The similarity between astronomical time scales and ice-age patterns is rather suggestive. Major ice ages seem to have occurred about every 250 million years and to have lasted a few million years—a period known as an ice epoch. As the sun circles the galactic nucleus, it drifts slowly through the spiral arms. Because gas and dust pile up on the inside edges of the arms, the sun traverses regions of increasing density up to the edge of an arm and then emerges into a relatively dust-free lane. When the sun collides with a dense cloud, the solar wind may be shut down and the flow of matter reversed. Gas and dust from the cloud can then impact the surface of the sun and raise the photospheric temperature. A change of as much as 1 percent may be possible—enough to have a significant effect on climate.

The number, size, and density of clouds in the path of the sun are very difficult to estimate. Evidence suggests that some clouds contain about 100,000 to 10 million particles per cubic centimeter, but such clouds are so compact that any chance of a collision with the sun is extremely small. More probably, the sun encounters clouds of 100 to 1000 particles per cubic centimeter, which, although diffuse, would still be capable of damping the solar wind appreciably. Such a cloud might envelop the sun for as long as 50,000 years.

A dense cloud is not something that is visible to the eye. Even 10 million particles per cubic centimeter is a trivial density compared to the highest laboratory vacuum. Therefore, there is no simple way to sight ahead and anticipate the next cloud encounter and thus predict an oncoming ice age. Recent estimates of the densities and distribution of interstellar clouds indicate that the sun may have already encountered well over a hundred clouds as dense as 100 particles per cubic centimeter and somewhat more than a dozen clouds as dense as 1000 particles per cubic centimeter. Having only recently emerged from the last ice age, the sun is unlikely to have completely cleared the spiral arm's dust lane, and another cloud encounter could come before long.

The dust-cloud theory derives some support from studies of lunar soil samples brought back by the Apollo missions. Conditions that span more than 3 billion years are revealed by samples from different depths of the moon's surface, and any dust encountered by the moon would also strike the sun. Analyses do show periodic increases in micrometeoroid accumulation, separated by a few hundred million years. Such evidence seems consistent with the passage of the sun through the dust lanes of the spiral arms, although it does not absolutely confirm it.

Anthropogenic Influences on Climate

Human beings have had a constantly growing impact on their environment since their first appearance on earth. As agriculture spread and industry developed, populations expanded over the globe. Human influence on the environment accelerated, but at a pace so slow that it was almost imperceptible in any single generation. In this century, we have begun to compete with nature for control of climate change.

The practical concern of most climatologists for the near future has more to do with fire than ice. By burning oil and coal and by cutting down forests, we have increased the burden of carbon dioxide (CO_2) in the atmosphere. At the turn of the century, the CO_2 concentration was about 300 parts per million. At this rate, it will reach 600 parts per million in the next hundred years. Three hundred additional parts per million may seem very little, but because of the greenhouse effect, any noticeable increase can have a profound influence on global temperature. Without the existing trace of CO_2, the world would be 10°C colder and the oceans would freeze. With twice as much CO_2, global temperature would average 2°C to 3°C warmer 100 years from now.

Whether a 2°C rise in global temperature would be harmful or beneficial overall is hard to say. Shifts in climate zones could disrupt agricultural economies, but a longer growing season could be a positive consequence. The breakup and melting of polar ice caps would be accompanied by a pronounced rise in sea level, however, and the flooding of coastal plains.

As world energy consumption grows, even direct heat release to the atmosphere becomes a matter of concern. New York emits six times more heat than it absorbs, and Moscow three times. The heat domes that envelop great cities already have significant impacts on local climate regimes. Even though the quantity of energy released by humanity globally is now only 0.01 percent of solar input, it is growing rapidly. Another 100 years of increasing energy consumption at the present rate would bring global energy use to a level as high as 1 percent of total solar input, sufficient to melt the ice caps and raise sea level by 185 feet. Most of the great cities of the world would vanish beneath the sea, and great agricultural belts would become dismal swamps. Energy conservation, however, has greatly slowed the rate of growth of consumption in recent years, and this scenario is not likely to occur.

The story of the mechanisms that controlled the ice ages in the Pleistocene and grand speculations about cosmic epochs may seem somewhat remote from present human concerns. But there is practical value in understanding the historical resilience of the climate system to natural perturbations. Confronted by possibilities that human influences may seriously alter our climate, it is helpful to understand something about natural trends, feedback mechanisms, and competing processes. More of such knowledge should help us to assess future dangers and present requirements for environmental protection.

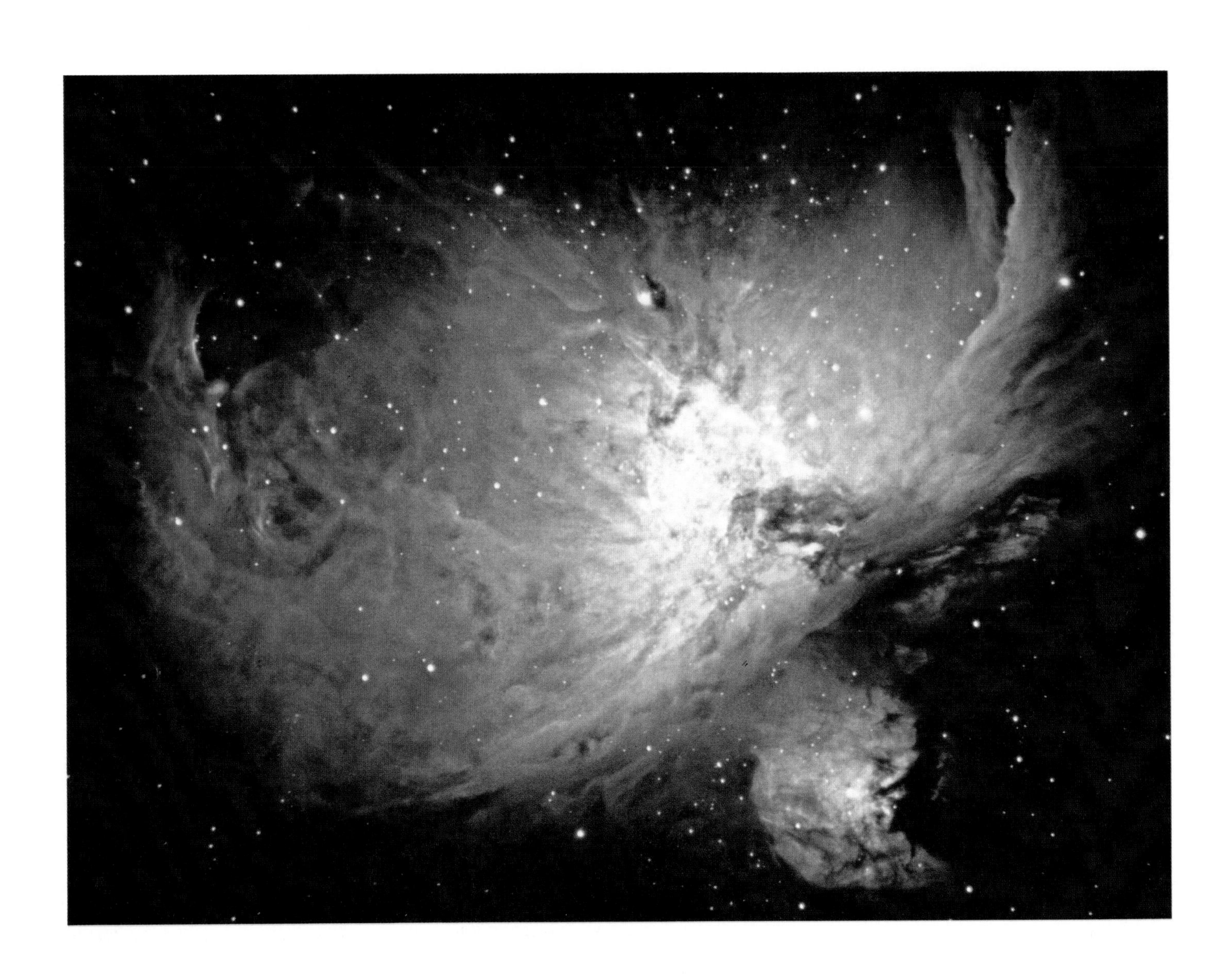

7

Origins and Endings

Gasballs spinning about, crossing each other, passing. Same old dingdong always. Gas, then solid, then world, then cold, then dead shell drifting around, frozen rock like that pineapple rock. The moon.

James Joyce, *Ulysses*

The Great Nebula in Orion is a nursery of newborn stars, 1500 light-years from earth. Some of the very young stars are still embedded in cocoons of dust and gas in which their light is transformed to heat radiation in the infrared.

Galactic Catastrophes

In a universe full of violent upheaval, where stars implode and explode and celestial outbursts of incredible power beset the nuclei of galaxies, the sun has seemed sublimely peaceful. But over aeons of time could it have had close encounters of devastating consequence to terrestrial life? Interactions with exploding stars, or supernovae, may have catastrophic consequences, but we have little evidence of such encounters since the sun was born. The frequency of supernova explosions in the galaxy is on the order of one every few decades. Although the energy release in a supernova is fantastically great (about 10^{50} ergs, or the equivalent of 100 trillion trillion trillion megaton H-bombs), the event must occur within a distance of a few tens of light years to have substantial impact on earth. Perhaps one such concurrence would be expected on the average every hundred million years.

The first consequence for the terrestrial environment would be a serious depletion of stratospheric ozone lasting perhaps as long as ten years. This "flash" effect would be followed by the arrival of an expanding cloud of energetic particles that would continue to destroy ozone for several centuries.

Ozone is our protective shield against harmful short-wavelength ultraviolet that destroys biological material on the surface of the earth and in the upper layers of the oceans. Ozone may also affect climate by influencing atmospheric circulation. Because ozone absorbs solar ultraviolet and thus heats the stratosphere, its loss would mean a cooler stratosphere. As yet we understand the climatic consequences of a cooler stratosphere quite poorly, but we do know that a change in the dynamic coupling between stratospheric circulation and the underlying troposphere might trigger a significant climate change.

Have supernovas occurred nearby in the past? A prominent feature of our galaxy, believed to be a supernova remnant, is the Galactic Spur, a nebulous ring that now emits radio synchrotron radiation and X rays. Its estimated distance is about 80 light-years, and it appears to have exploded about 20,000 years ago. Cro-Magnon man would have seen a brilliant star that shed a thousand times the light of the full moon on earth. Most of its radiation must have been ultraviolet, perhaps ten thousand times the solar ionizing radiation flux. But none of that radiation could penetrate the atmospheric blanket, and its effect on life would have been insignificant.

Two Soviet scientists, Josif S. Shklovsky and Valerian I. Krosovsky, once suggested that the most serious impact of a nearby supernova would be the genetic mutations accompanying a prolonged

The Crab Nebula is the remnant of a supernova explosion that was observed by Oriental astronomers on July 4, 1054. It was so bright it remained visible in daylight for 23 days. The nebula contains a tangled web of gaseous filaments that shine like neon lights. An amorphous glow of synchrotron light extends to six light-years in diameter and is produced by the helical motions of energetic particles in the nebular magnetic field.

increase in cosmic ray intensity. Cosmic rays produce radioactivity in the air, and radioactivity close to ground causes spontaneous mutations. For short-lived species, a very large dose of radiation, a thousand times the present natural level, is necessary to double the mutation rate. Long-lived species, however, would be affected by an increase in exposure of only three to ten times normal.

All mutations are not necessarily destructive. An increase in energetic radiation could enhance the evolution of some species, and may indeed have done so. The two Soviet scientists speculate that this may explain the luxurient vegetation of the Carboniferous period. And still earlier, perhaps, a supernova might have stimulated the origin of life on earth. Scientists, however, tend to know more about the ending of life than its beginnings. It has long been speculated, for example, that the extinction of the dinosaurs was somehow connected to a random cataclysmic event.

A new line of thinking about extinctions began in about 1979, when physicist Luis Alvarez and his son Walter, together with colleagues at the Lawrence Berkeley Laboratory and the University of California, proposed that the demise of the dinosaurs, about 65 million years ago, followed the impact of a large asteroid or comet. Until then, most homely speculations invoked climate change—too much heat or cold, global epidemic, smaller predators with an appetite for dinosaur eggs, and so on.

The California scientists were surprised to discover in clay samples from the top of the Cretaceous layer at many locations around the world an anomalous excess of the element iridium—about 160 times the normal rare abundance in terrestrial rocks. High iridium abundances are found in meteorites, and the Alvarezes guessed that the terrestrial iridium layer might be traced to the debris of a tremendous explosion produced by the crash of a large asteroid. The impact must have been so violent that pulverized asteroidal and terrestrial rock spread a cloud of dust throughout the earth's atmosphere. The sky became dark as night, perhaps for several months. The ground grew so cold that plant life died, and the dinosaurs may have starved or frozen to death.

Over many years, the dust settled down to earth in uniform layers over the land surface of the entire globe. Numerous samplings of the Cretaceous boundary at many points on the earth since 1979 have confirmed a worldwide layering of iridium. Searches for evidence of an iridium connection with other extinctions, however, were unsuc-

An iridium-rich layer found by the Alvarez team at a site of exposed limestone in the Apennine Mountains of Italy. The photograph shows a dark layer of iridium-rich clay sandwiched between underlying white limestone from the late Mesozoic era and overlying grayish limestone from the early Cenozoic. The coin shown for scale is the size of a U.S. quarter. Since this discovery in 1979, similar iridium-rich layers have been found at widely separated sites around the world. Geological dating sets the time of deposit at about 65 million years ago, when the dinosaurs disappeared along with about 65 percent of all other species then alive.

cessful, except for a layer corresponding to an extinction about 37 million years ago.

Recently, evidence has been put forward for at least a dozen great extinctions of living species over the past 250 million years. David Raup and John Sepkoski of the University of Chicago sifted the fossil records of ancient rocks and sediments to time the disappearance of thousands of families of marine organisms. Astonishingly, they discovered a recurrent pattern of species loss every 26 to 28 million years. Such a long cycle is incomprehensible for any earthly biological processes, and the data suggested to Raup and Sepkoski that such destructions followed the bombardment of earth by asteroids or comets that was triggered by some periodic astronomical release.

The Raup and Sepkoski story has brought forth a rash of explanations, two of which are currently receiving favorable attention. Both involve the dislodging of comets from the Oort cloud outside the solar-planetary system. According to Jan Oort, a distinguished Dutch astronomer, comets by the trillions occupy a "parking" orbit perhaps 10 billion miles from the sun, where they fill a vast spherical shell. Occasionally, a passing star may gravitationally stir the comet cloud, shaking many comets loose so that they move inward toward the sun. Two different scientific teams, one led by Richard Muller at Berkeley, the other by Daniel P. Whitmire of the University of Southwestern Louisiana, believe that the sun may have a binary companion star whose highly eccentric orbit penetrates the Oort cloud roughly every 26 million years. Muller's group calls this tentative sun companion Nemesis, alluding to the Greek goddess who meted out retribution to the "excessively rich, proud, and powerful." But intensive searching of old photographic star plates and combing the sky anew in the vicinity

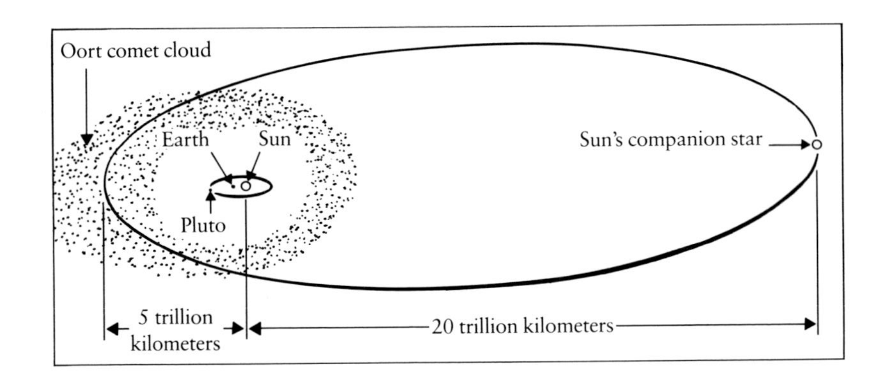

A theory proposed by Dutch astronomer Jan Oort invokes a cloud of a hundred billion comets stored in the form of dirty ices in a sun-centered shell hundreds of times more distant than Pluto. Occasional perturbations nudge material out of the cloud to fall toward the sun. On the way, they develop characteristic comet tails as the ice evaporates. Richard Muller and Daniel Whitmire recently proposed that a companion star in a highly elongated orbit moves through the Oort cloud every 27 million years. Comets are then shaken loose and launched toward the earth.

of the sun have revealed no glimpse of Nemesis as yet, although it may be a small brown dwarf and very difficult to observe.

The second hypothesis, by Michael Rampino and Richard Stothars of the Goddard Institute for Space Studies in New York, attributes abrupt extinctions of species to the undulations of the sun as it tracks in and out of the plane of the Milky Way. As the sun orbits the nucleus of the galaxy every 250 million years, it bobs up and down across the galactic plane every 27 million years like a horse on a very slow merry-go-round. Each crossing of the plane brings the sun in danger of colliding with dense clouds of interstellar matter and stars in the disk of the galaxy. This once again greatly increases the probability of disturbing the Oort cloud.

All these speculations may be misguided, but such controversies stir the scientific community to heated argument about the probabilities of apocalyptic happenings. Many scientists prefer more mundane explanations of extinctions: volcanism as a cause of iridium deposits, or oscillations in global glaciation between icehouse and greenhouse. In these matters, Mark Twain's comment may still be most apt: "There is something fascinating about science. One gets such wholesale returns of conjecture out of such a trifling investment of fact."

The Birth and Death of the Solar System

On the long-term fate of the sun, speculation stands a bit sturdier. The sun is a relative latecomer to the galaxy of stars. Give or take a few billion years, the Milky Way is believed to be about 15 billion years old. Our sun, together with its coterie of planets, asteroids, and other debris, condensed out of a massive, spinning, turbulent cloud of cosmic gas and dust about 4.5 billion years ago. Much of the material that went to make the sun was left over from exploded stars of earlier generations—we are all composed of second-hand atoms.

Interstellar gas is clumped in great clouds unstably balanced between the gravitational forces that drive the clouds toward collapse and the gas pressure that results from the heat of compression. The shrinking process is complicated. Cooler clouds collapse more readily than massive hotter ones. Even in the deep-freeze of interstellar space, gas clouds must contain tens of solar masses of matter to promote collapse. But evidence that the collapse of the nebula from which the sun is descended was initiated by shock waves from the explosion of a nearby supernova does exist.

The Milky Way is a giant spiral galaxy of a hundred billion stars, stretching one hundred thousand light-years across. Our sun is located in one of the spiral arms, two-thirds of the distance from the center to the outskirts of the galaxy.

A supernova is a violent explosion of a star perhaps 10 or more times as massive as the sun. Having burned with a hot blue light for several million years, a star exhausts its nuclear fuel and collapses in a matter of a second. The implosion is so violent that strong shock waves rebound and blow away the outer shell of the star. This implosion-explosion process spawns a whole range of heavier elements and sprays them through interstellar space.

The role of supernova shock waves in the compression of the solar nebula is supported by evidence obtained from the Allende meteorite, a two-ton carbonaceous chondrite that fell to earth in 1969. The fallen rock contained a telltale clue, the presence of the isotope magnesium

The Helix Nebula in the constellation Aquarius (NGC 7293) is an extended planetary nebula that covers an area of the sky half as large as the full moon. At the center of the spherical shell is a white dwarf, the collapsed remnant of the star that blew off its outer layers to form the nebula.

26 (^{26}Mg), which is created by the decay of aluminum 26, an isotope with a half-life of 700,000 years. Gerald Wasserburg and his colleagues at Cal Tech were startled by the large amount of ^{26}Mg in a meteorite that was 4.6 billion years old. Arguing that ^{26}Mg could not have been produced in the early sun, they propose that it must have been left over from a supernova that exploded more than 5 billion years ago. As the shock waves from the explosion ripped through the outer shell of the star, ^{26}Al was formed and fired like a spray of shotgun pellets into the nearby protosun nebula.

With the help of the supernova, gravity forced the cloud to contract; its spin accelerated, and it grew hotter. Gradually, it flattened

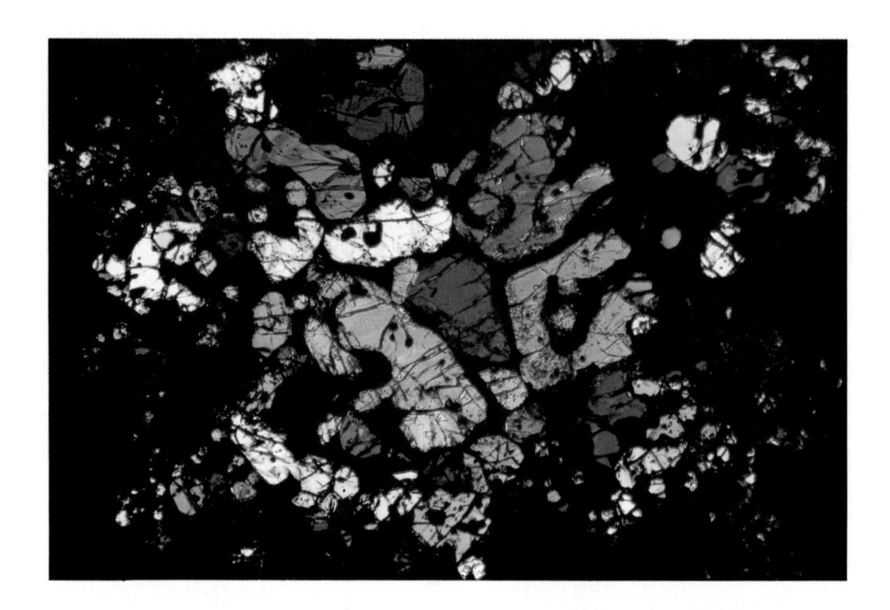

A cross section of a fragment of the Allende meteorite, which fell near Allende, Mexico, in February, 1969. Many small, mineral-rich, crystalline inclusions are embedded in the carbonaceous matrix and were probably formed very early in the condensation of the solar nebula. Their isotopic composition reveals an excess of magnesium 26, which could have derived from the decay of aluminum 26. The aluminum 26 is believed to have formed in the explosion of a nearby supernova and was ejected into the solar nebula at the time that the supernova shock wave accelerated the condensation of the nebula.

out like a fried egg with a bulge near the center. Spinning faster and faster as it collapsed, the protosun had to find ways of shedding its angular momentum. Perhaps magnetized plasma within the cloud interacted with the galactic magnetic field to provide a braking action.

From the beginning of its collapse to the formation of a substantial core may have taken half a million years. The protostar grew steadily hotter and began to glow brightly. When the central temperature reached millions of degrees, helium began to form out of the fusion of lithium and hydrogen—two helium nuclei out of ^{7}Li plus a proton. The outpouring of thermonuclear energy from the core generated so much heat that the pressure of thermal expansion balanced gravity. Contraction slowed to a halt, and the sun was born.

From the ignition of nuclear burning, it took perhaps 50 million years for thermonuclear fusion to approach a steady equilibrium. Spinning rapidly, the sun generated strong magnetic fields, and the solar atmosphere was the scene of violent activity, punctuated by frequent great solar flares and an enormous outpouring of solar wind. Gales of fast particles carried away angular momentum and slowed the rate of rotation. Had this not happened, the early sun would have spun around once every few hours. At half a billion years, the sun was still spinning three times as fast as it is now—one rotation every nine days. As the sun continued to age, it slowed steadily, and its internal generation of magnetic fields decreased substantially. The wind calmed, and its spin settled down to the present 27-day period. Any decrease in spin rate now is almost unmeasurable.

While the sun assumed its final spherical shape, the debris left behind as it collapsed clumped into meteorites and protoplanets. Isotopic evidence in meteorites indicates that their condensation took 50 to 100 million years from the time that thermonuclear burning was ignited in the solar core. Planets required another few hundred million years to solidify.

Since the sun began to shine like a star 4.5 billion years ago, it has been cannibalizing itself at the rate of 11 billion pounds of hydrogen per second. In spite of this ferocious appetite, enough hydrogen remains for another 100 billion years. But catastrophe will overtake the sun long before then. By current evolutionary stellar models, it will survive in approximately its present form for only another 5 billion years.

At the present halfway point in the total life expectancy of the sun, the original hydrogen in its core has been depleted from 75 percent to 35 percent. At the same time, helium has increased from 25 to 65 percent. With these changes, the sun has become a little larger, and its luminosity has increased by about 40 percent. It grows steadily redder and warmer.

The evolutionary picture described thus far is based on astrophysical models that astronomers believe are very close to correct. Atmospheric scientists, however, find it difficult to reconcile these astronomical models with their concepts of climate. According to their views, if the carbon dioxide content and relative humidity of the atmosphere were unchanged over the life span of the earth, ice should have totally covered the earth between 4.5 and 2.3 billion years ago because of the weaker flow of sunlight and heat. But several sources of geological evidence offer convincing proof that life has existed on earth for the past 3.5 billion years and that oceans of water have covered most of the earth for at least 3.8 billion years.

With all we understand of climatology, this discrepancy between solar luminosity and global temperature cannot be easily explained—we cannot model the net response of the climate system to such a change in solar luminosity with any precision. Somehow, some atmospheric adjustment to the lower solar intensity must have prevented global freezing. One possible answer is the evolution of carbon dioxide. With little plant life and vegetation, the conversion of CO_2 into oxygen by photosynthesis would have progressed very slowly, leaving an atmosphere very rich in CO_2. The resulting greenhouse effect would have kept the earth warm like an insulating blanket. But very large changes in CO_2 are required to produce such substantial changes in surface temperature. If, for example, the present CO_2 content were doubled, the surface temperature over most of the earth would increase by only a few degrees Celsius.

When the sun reaches an age of about 6 billion years, it will have grown about three times larger and about 15 percent brighter. Ice and snow will disappear completely from the temperate zones. The borders of the Arctic circle will resemble our Southwest today. Polar ice caps will melt away, and the rising sea level will inundate the coastal zones, submerging all vestiges of our great seaports.

At about 10 billion years, the solar diameter will become about 40 percent larger, and the photosphere will be twice as luminous but cooler and redder. Hydrogen in the core will be almost fully depleted, and the core will contract. Fresh hydrogen from outside the core falling onto the mass of helium ash will trigger fusion, and a new energy-generating shell will be created outside the core. Heating of the outer regions will cause the body of the sun to expand to bloated proportions. In this red subgiant phase, the sun will reach 10 times its present diameter. On the earth, surface temperature will rise above 100°C, and the oceans will boil away. A hot wind will rage furiously out of the sun and scorch the barren surface.

Space exploration studies of Venus and Mars have made us aware of how narrow is the ecological zone that permitted, and continues to permit, a life-supporting atmosphere and a stable environment to evolve on earth. If the earth were only 5 percent closer to the sun, it could have experienced a runaway "greenhouse effect" aeons ago and its surface would resemble that of Venus—hot as an oven. A slightly larger orbit might have led to glaciation and a terrestrial surface that resembled the cold and arid deserts of Mars. While the sun is evolving toward the red-giant phase, the favorable ecological zone gradually will move outward to almost twice the diameter of the earth's orbit, somewhere beyond Mars. With the passage of aeons, humanity will need to depart the earth in space colonies, drawing on solar heat and light at an ever receding distance from the sun. And the ultimate end of life support by solar energy is inevitable as the hydrogen supply is exhausted.

The brilliant red-giant phase will be comparatively short. While hydrogen burns in the shell zone and helium ash accumulates in the core, the core will shrink and its temperature will climb. When it reaches about 100 million degrees, helium will fuse to carbon, and the thermonuclear furnace will spring to life once again. Nuclear burning will accelerate to a great burst, and the red giant sun will expand to

100 times its present diameter, up to the edge of the earth's orbit. From earth, it will appear to fill nearly the entire sky. Its luminosity will increase about 1000 times, and a fierce wind, millions of times stronger than the strongest wind we can imagine, will blast from the sun. Mercury will be engulfed, and the earth's surface, just under the roiling edge of the sun, will fuse to molten lava.

As the last helium fuses to carbon, the sun will enter its death throes. It may blow off its envelope in one great burst or a rapid succession of bursts to form a "planetary nebula," an escaping shell of gas that, viewed from afar, looks like a great smoke ring. About half the original solar mass will be left, a tiny carbon core at a temperature of about 10,000 kelvins, a white dwarf no bigger than earth and of dazzling brilliance. At a density of hundreds of tons per cubic inch, the carbon star will be millions of times as hard as diamond.

Twinkle, twinkle, little star. . . .
Like a diamond in the sky.

Over tens of billions of years, the white dwarf will grow steadily dimmer and redder. Eventually, it will become a black rock, cold as the void of space.

For Further Reading

Chapter 1

G. Abetti, *The Sun,* Macmillan, 1957.

H. Friedman, *The Amazing Universe,* National Geographic Society, 1975.

R. Giovanelli, *Secrets of the Sun,* Cambridge University Press, 1984.

D. H. Menzel, *Our Sun,* Harvard University Press, Cambridge, Mass., 1959.

E. N. Parker, "The Sun," *Scientific American* **233**(3):42–50 (1975).

Chapter 2

J. N. Bahcall and R. Davis, Jr., *Solar Neutrinos: A Scientific Puzzle, Science* **191**:264–267 (1976).

H. Bethe, "Energy Production in Stars," *Physics Today* **21**:36–44 (1968).

A. S. Eddington, *Stars and Atoms,* Yale University Press, New Haven, 1927.

S. Mitton, *Daytime Star,* Scribner, New York, 1983.

P. Morrison, "The Neutrino," *Scientific American* **194**(1):58–68 (1956).

F. Reines and C. L. Cowan, Jr., "Detection of the Free Neutrino," *Physical Review Letters* **90**(492):830–831 (1953).

Chapter 3

J. R. Brooks, G. R. Isaak, and H. B. van der Raay, "Observation of Free Oscillations of the Sun," *Nature* **259**:92–95 (1976).

G. Grec, E. Fossat, and M. Pomerantz, "Solar Oscillations: Full Disk Observations from the Geographic South Pole," *Nature* **288**:541–554 (1980).

G. E. Hale and S. B. Nicholson, *Magnetic Observations of Sunspots,* parts I and II, Carnegie Institution, Washington, 1938.

H. A. Hill, "Seismic Sounding of the Sun," in *The New Solar Physics,* edited by J. A. Eddy, Westview Press, Boulder, Colo., 1978.

B. Lyot, "Étude de la couronne solaire en dehors des eclipses," *Zeitschrift für Astrophysik* **5**(73) 1932.

S. A. Mitchell, *Eclipses of the Sun,* Columbia University Press, New York, 1935.

R. W. Noyes, *The Sun, Our Star,* Harvard University Press, Cambridge, Mass., 1982, pp. 213–230.

J. H. Parkinson, L. V. Morrison, and F. R. Stephenson, "The Constancy of the Solar Diameter Over the Past 250 Years," *Nature* **288**:548–551 (1981).

A. B. Severny, V. A. Kotov, and T. T. Tsap, "Observations of Solar Pulsations" *Nature* **259**:87–89 (1976).

I. I. Shapiro, "Is the Sun Shrinking?" *Science* **208**:51–53 (1980).

C. A. Young, *The Sun,* Appleton, New York, 1895.

Chapter 4

S. Chapman, "Some Phenomena of the Upper Atmosphere," *Proceedings of the Royal Society* **132A**(353):353–374 (1931).

H. Friedman, "Solar Observations Obtained from Vertical Sounding," *Reports on Progress in Physics* **XXV**(163):164–217 (1962).

L. Goldberg, "Ultraviolet and X Rays from the Sun," *Annual Review of Astronomy and Astrophysics* 5(279):279–324 (1967).

J. A. Ratcliffe, ed., *Physics of the Upper Atmosphere,* Academic, New York, 1960.

R. Tousey, "Some Results of Twenty Years of Extreme Ultraviolet Solar Research," *Astrophysical Journal* **149**(2):239–252 (1967).

R. C. Willson, S. Gulkis, M. Janssen, H. S. Hudson, and G. A. Chapman, "Observations of Solar Irradiance Variability," *Science* **211**:700–702 (1981).

Chapter 5

S. I. Akasofu and S. Chapman, *Solar-Terrestrial Physics,* Clarendon Press, Oxford, England, 1972.

S. I. Akasofu and L. J. Lanzerotti, "The Earth's Magnetosphere," *Physics Today* **28**:28–29 (1975).

J. C. Brandt, *Introduction to the Solar Wind,* W. H. Freeman and Co., New York, 1970.

S. Chapman and J. Bartels, *Geomagnetism* (2 vols.), Oxford University Press, New York, 1940.

A. J. Dessler, "Solar Wind and Interplanetary Magnetic Field," *Review of Geophysics* **5**(1):1–41 (1967).

R. H. Eather, *Majestic Lights,* American Geophysical Union, Washington, D.C., 1980.

J. G. Roederer, "Planetary Plasmas and Fields," *EOS Transcripts,* American Geophysical Union, **57**(53) (1976).

C. Stormer, *The Polar Aurora,* Clarendon Press, Oxford, 1955.

L. Svalgaard and J. M. Wilcox, "A View of Solar Magnetic Fields, the Solar Corona, and the Solar Wind in Three Dimensions," *Annual Review of Astronomy and Astrophysics* **16**:429–443 (1978).

J. A. Van Allen, "Radiation Belts Around the Earth," *Scientific American* **200**(3):39–47 (1959).

Chapter 6

W. S. Broecker, D. L. Thurber, J. Goddard, T. Ku, R. K. Matthews, and K. J. Mesolella, "Milankovitch Hypothesis Supported by Precise Dating of Coral Reefs and Deep-Sea Sediments," *Science* **159**:1–4 (1968).

J. A. Eddy, "The Case of the Missing Sunspots" *Scientific American* **236**(5):80–88 (1977).

C. Emiliani, "Paleotemperature Analysis of Caribbean Cores P6304-8 and P6304-9 and a Generalized Temperature Curve for the Past 425,000 Years," *Journal of Geology* **74**:109–126 (1966).

J. D. Hays, J. Imbrie, and N. J. Shackleton, "Variations in the Earth's Orbit: Pacemaker of the Ice Ages," *Science* **194**(4270):1121–1132 (1976).

J. Imbrie and K. P. Imbrie, *Ice Ages: Solving the Mystery,* Enslow Publishers, Short Hills, N.J., 1979.

E. L. Ladurie, *Times of Feast, Times of Famine: A History of Climate Since the Year 2000,* translated by Barbara Bray, Doubleday, New York, 1971.

E. Lurie, *Louis Agassiz: A Life in Science,* University of Chicago Press, 1960.

M. Milankovitch, *Durch ferne Welten und Zeiten,* Koehler and Amalang, Leipzig, 1936.

J. M. Mitchell, Jr., C. W. Stockton, and D. M. Meko, "Evidence of a 22-Year Rhythm of Drought in the Western United States Related to the Hale Solar Cycle Since the 17th Century, in Solar-Terrestrial Influences on Weather and Climate," edited by B. M. McCormac and T. Seliga, Reidel, Boston, 1979.

M. Stuiver and P. D. Quay, "Changes in Atmospheric Carbon-14 Attributed to a Variable Sun," *Science* **207**(4426):11–19 (1980).

W. Sullivan, *Assault on the Unknown,* McGraw-Hill, New York, 1968.

Chapter 7

H. Alfven, *On the Origin of the Solar System,* Oxford University Press, 1954.

L. W. Alvarez, W. Alvarez, F. Asaro, and H. V. Michel, "Extraterrestrial Cause for the Cretaceous-Tertiary Extinction," *Science* **208**(4448):1095–1108 (1980).

A. G. W. Cameron, "The Origin and Evolution of the Solar System," *Scientific American* **233**(3):32–41 (1975).

C. B. Officer and C. L. Drake, "Terminal Cretaceous Environmental Events," *Science* **227**(4691):1161–1166 (1985).

M. R. Rampino and R. B. Stothers, "Terrestrial Mass Extinctions, Cometary Impacts and the Sun's Motion Perpendicular to the Galactic Plane," *Nature* **308**:709–712 (1984).

D. Raup and J. Sepkoski, "Periodicity of Extinctions in the Geologic Past," *Proceedings of the National Academy of Sciences, U.S.A.* **81**:801–805 (1984).

G. Wasserburg, *Short-Lived Nuclei in the Early Solar System. Protostars and Planets,* vol. II, University of Arizona Press, 1985.

Sources of the Illustrations

Cover image
Museum Marmottan, Paris/Art Resource

Chapter 1

Chapter opener
Board of Trustees, Victoria and Albert Museum

page 2
Giraudon/Art Resource

page 3
National Photo & News Service

page 4
From "A Commentary on the Dresden Codex," J. Eric S. Thompson, American Philosophical Society

page 5
Travis Amos

page 6 (top)
C. Falco/Photo Researchers

page 6 (bottom)
Dan McCoy/Rainbow

pages 8 and 9
George V. Kelvin

page 12
Gernsheim Collection, Harry Ransom Humanities Research Center, University of Texas at Austin

page 14 (bottom)
Deutsches Museum

page 15
Deutsches Museum

page 16
Deutsches Museum

page 18
NASA

page 19
NASA

page 20
NASA

Chapter 2

page 22
Yale University Art Gallery, New Haven, Conn. Gift of Katherine S. Dreier to the Collection Société Anonyme

page 24
Ann Ronan Picture Library

page 25
The Granger Collection

page 29
Cornell University

page 33
Raymond Davis, Jr., Brookhaven National Laboratory

page 38
Jon R. Friedman

page 43
National Solar Observatory, Sacramento Peak Association of Universities for Research in Astronomy, Inc.

page 44
Mt. Wilson and Las Campanas Observatories, Carnegie Institution of Washington

page 45
National Solar Observatory, Sacramento Peak Association of Universities for Research in Astronomy, Inc.

page 46
Martin Schwarzchild, Princeton University Observatory

page 47
Robert Leighton, California Institute of Technology

page 51
Martin Pomerantz, Bartol Research Foundation of the Franklin Institute

Chapter 3

page 55
J. Durst

page 56
Archiv/Photo Researchers

page 58
Martin Schwarzchild, Princeton University Observatory

page 59 (top)
Alistair B. Fraser

page 59 (bottom)
Mount Wilson and Las Campanas Observatories, Carnegie Institution of Washington

page 63
Berenice Abbott Collection

page 64
William Livingston, Kitt Peak National Observatory

page 66
NRL

page 68
Royal Observatory, Edinburgh

page 70
Con Edison

page 71
Los Alamos Scientific Laboratory

page 73
Dr. Speer

page 75 (top)
R. Dunn, National Solar Observatory, Sacramento Peak Association of Universities for Research in Astronomy, Inc.

page 75 (bottom)
Big Bear Solar Observatory, California Institute of Technology

page 78
Pic-du-Midi Observatory

page 79
NRL/NASA

page 80 (top)
NRL/NASA

page 80 (bottom)
Big Bear Solar Observatory, California Institute of Technology

page 81
NRL/NASA

page 82
NRL/NASA

Chapter 4

page 86
NRL/NASA

page 88
NASA

page 89
Hugh Hudson, University of California at San Diego

page 90
The Bettmann Archive

page 95 (top)
NRL

page 95 (bottom)
Archives, National Academy of Sciences

page 96
Robert Hutchings Goddard Library, Clark University

page 97
Popperfoto

page 100
R. Tousey, NRL

page 101
R. Tousey, NRL

page 102
R. Tousey, NRL

page 107
NRL

page 108
Emory Kristof © National Geographic Society

page 110
Herbert Friedman, NRL

page 112
NRL

page 120
NRL

page 122
Robert Kreplin, NRL

page 124 (left)
George A. Dulk and Dale E. Gary, National Radio Astronomy Observatory, © by Associated Universities, Inc.

page 124 (right)
Marshall Space Flight Center, NASA

page 125
Dennis di Cicco

page 127
Colonel Richard Gimbel Aeronautics History Collection, U.S. Air Force Academy

page 134
Blair Pittman

page 135
Dr. L. A. Frank, University of Iowa

page 138
Robert Fischer

page 144
NRL

page 145 (left)
NASA

page 145 (right)
NRL

Chapter 5

page 146
National Museum of American Art, Smithsonian Institution, Gift of Eleanor Blodgett

page 148 (left)
Geophysical Institute, University of Alaska, Fairbanks

page 148 (right)
Claude Nicollier

page 149
A. Krieger, American Science and Engineering, Inc.

page 151
A. Krieger, American Science and Engineering, Inc.

page 155 (right)
Neil R. Sheeley, Jr., NRL

page 156
National Center for Atmospheric Research/ High Altitude Observatory/National Science Foundation

page 159
Trustees of the Science Museum, London

page 163
Dr. L. A. Frank, University of Iowa

page 164
Dennis Milon

page 165
Dennis di Cicco

page 166
Alv Egeland/Institute of Physics, University of Oslo

page 169
NASA

page 173
Camera Hawaii

page 177
George V. Kelvin

page 178
Dr. L. A. Frank and J. D. Caven, University of Iowa

page 183
Air Force Geophysics Laboratory and Johns Hopkins University, Applied Physics Laboratory

Chapter 6

page 188
John Brett, *The Glacier of Rosenlaui,* 1856. Tate Gallery, London

page 191
Roger Archibald/Earth Scenes

page 194
Laboratory of Tree-Ring Research, University of Arizona

page 195
David Muench

page 197
ET Archive, Ltd.

page 199
Dr. G. E. Williams

page 201
Original, property of the Library of the University of Neuchâtel. Copy courtesy of John Imbrie

page 202
The Snite Museum of Art, University of Notre Dame, South Bend, Indiana. Gift of Mr. and Mrs. John T. Higgins

page 203
Vaslo Milankovitch

page 210 (top)
Dr. Kozo Takahashi, Woods Hole Oceanographic Institution

page 210 (bottom)
John Clegg/Ardea London, Ltd.

page 211
Deep Sea Drilling Project, Scripps Institution of Oceanography, University of California, San Diego

page 219
The Detroit Institute of Arts, Founders Society Purchase, Robert H. Tannahill Foundation Fund, Gibbs-Williams Fund, Dexter M. Ferry, Jr., Fund, Merrill Fund, Beatrice W. Rogers Fund, and Richard A. Manoogian Fund

page 220
Jet Propulsion Laboratory

page 221
Ronald Laub, Lick Observatory

Chapter 7

page 224
David Malin/Anglo-Australian Observatory

page 226
Palomar Observatory

page 227
Luis Alvarez

page 230
Rob Wood/Stansbury, Ronsaville, Wood, Inc.

page 231
Anglo-Australian Observatory

page 232
Chip Clark

Index

Abbott, Charles Greeley, 78, 87
Accelerator mass spectrometry (AMS), 196
Adhémar, Joseph Alphonse, 203–204
Aerobee rocket, *107, 108,* 112, 144
Aerosols, 196, 218
Agassiz, Louis, *204*
Air, exploration of, 126–127. *See also* Atmosphere
Airglow cycles, 69
Airglow, visible, 144–145
Akasofu, Syun-Ichi, 179
Albedo, 201, 202
  factors contributing to, *202*
Allende meteorite, 230–*232*
Alvarez, Luis, and Walter, 226
Anderson, James, 133
Angstrom, 11, 100
Appleton, Edward, 93
Apollo-16 camera spectrograph, *145*
Apollo-16 mission, 144
Apollo Telescope Mount, *18*
Arc minute, definition of, 50
Arcs, auroral, 163–164
Argon 37, 32
Argus experiment, 172–173
Aristotle, *55, 56,* 67
Asteroids, and species extinction, 226–227, 228
Astronomical units, 156
Astronomy, *5*
  birth of, 2
  early rocket, 86–98, 100–102, 104–105
Atmosphere, 126–143
  colors in (*see* Airglow; Aurora)
  constituents of neutral, *142*
  earth's heat release to, 223
  exosphere of, 140–141, 143
  ionized components of, *8*
  mesosphere of, 137–139
  neutral, 127–130
  polar, 195–196
  pressure of an, 37
  and solar activity, 141
  stratosphere of, 130–133, 136
  temperature variations in, 129
  thermosphere of, 139–140
  troposphere of, 130
Atmospheric pressure, 126–127
Aurora, 162–168
  computer-graphic image of, *163*
  forms of, 163
  green, *165*
  red, *164, 173*
  sunspots and, 67–68
  woodcut of, *68*
Aurora australis, 163, 172
Aurora borealis, in sunlight, *183*. *See also* Northern lights
Auroral activity at poles, 195, 196
Auroral substorm, *178*
Axial precession of earth, 203, 205, 215–216

Bahcall, John, 32
Balloons
  and atmospheric pressure, 126–127
  with payload, *134*
Barnett, Miles, 93
Barometric pressure, 126
Becquerel, Henri, 12–13, 24
Berger, Andre, 215
Beryllium, *28*, 29, 196
Bethe, Hans, *29*
Biermann, Ludwig, 147
Biot, Jean Baptiste, 127
Birkeland, Kristian, 165, *166*
Black-body spectra, *98, 99*
Black drop, 53
Blanchard, Jean Pierre, 127
Bomb tests, 171–174
Boron, formation of, 28, 29
Breit, Gregory, 93, 95
Brewster, David, 57
Brunhes, Bernard, 213
Bumstead, Newman, 108
Bunsen, Robert, 15, 16
Bunsen burner, *16*
Byrd, Richard E., 95

Camera
  electrographic, *145*
  solar, 108
Cameron, Alastair, 33
Carbon cycle, 29, *30*
Carbon dioxide
  in atmosphere, 191–192, 222–223
  and earth's temperature, 233–234
Carbon 14, in tree rings, 193–195
Carrington, Richard C., 117
Carruthers, George, 144
Casa Riconada Tower, 5
Cerenkov radiation, 11
Chaco Canyon, New Mexico, 5
Chadwick, James, 25, 26
Chapman, Sidney, 114, 137, 147, *148*
Chapman layer of electrons, *114*
Charge coupled devices (CCDs), 84
Chlorine, 32, 37
Chlorofluoromethanes, and ozone destruction, 136

Chromosphere, 37, *44*, *75*
  flash spectrum of, 72, *73*
  pattern of, *82*
  prominences of, *80*, *81*
  short-wavelength signals of, 123
  temperature inversion in, 72–76
  temperature rise across, *74*
Chubb, Talbot A., 119, *120*
Climate, 21, 189–223
  changes in, 87–89, 190, 191
  historical data on, 189–190
  human influences on, 222–223
  ice ages and, 197–207, 209, 212–213, 215–218
  sedimentary record and, 207, 209, 212–213
  sources of variability in, 200
  tree rings and, 193–195
  volcanic eruptions and, 218–220
Clouds
  altocumulus, *59*
  nacreous, 131
  noctilucent, 137–*138*
  Oort, 228, 229
Colors, visible, 10, *11*, 12
Comet(s)
  Bennett, *148*
  species extinction and, 228
  sunspots and, 69
  tails, 147
Compass
  and ionospheric currents, *92*
  magnetic, 158, 160
Comte, August, 7
Conrad, Pete, 18
Continental drift, *217*, 218
Convection cells, in altocumulus clouds, *59*
Convection zone, of sun, 45, 46
Copernicus, *5*
Coral-reef terracing, 209
Core drilling, deep-sea, *211*
Corona, *41*, *149*
  blending of chromosphere into, 76
  filaments of, 79
  high temperature of, 73
  holes in (*see* Coronal holes)
  Lyman-alpha glow of, *144*
  magnetic cage of, 148
  prominences of, 79
  properties of, 81, 82
  radio outbursts from, 124
  radio probing of, 76–77
  and solar winds, 148
  temperature of, *75*, 83
  transient bubble in, *155*
  X-ray sources in, 109–112
Coronal holes, 149–150, *151*
Coronal light, 82
Coronal streamers, 147
Coronal transients, 154, *155*, 156
Coronograph, *78*
  artificial eclipses and, 77–79, 81–83
Coronograph mission, orbiting, 155–156
Cosmic rays
  and carbon 14, 194
  and beryllium 10, 196
  global survey of, 169
  origin of, 182
  and radioactivity, 226
Cosmic-ray shower, *181*
Cotopaxi volcano, *219*
Cowan, Clyde L., 31
Crab Nebula, 76–77, *226*
Cristofilos, Boris, 172
Critchfield, Charles, 29
Croll, James, 206–207
Crooke, Sir William, 166
Cryosphere, 200, 201
Current flows and magnetic fields, 150
Curtains of light, 163, 164

Dansgaard, Willie, 220
Davis, Jr., Raymond, 31, 32, 33
de Bort, Léon Philippe Teisserens, 127
Dendrochronology, 193
Density, atmospheric, 128, 140
Deuteron, 26, 27
Diameter of sun, 36, 50–53
Diffusive equilibrium of atmosphere, 128
Dinosaurs, extinction of, 226–227
Dirac, P. A. M., 26
Doppler effect, 44
Douglass, Andrew Ellicott, 193, 198
Drag on satellites, 139–140, 141, 143
Draperies of colored light, 163, 164
Dresden Codex, 3, *4*
Dunham, David, 52
Dust
  climate and volcanic, 218–220
  interstellar, 220–222

Earth
  axial precession of, *203*, 205, 215–216
  distance from sun to, 36
  magnetic shield of, 158–174
  natural resonance of, 216
  orbit of (*see* Earth's orbit)
  sunspots and, 66–67
Earth's orbit
  and climate, 203–204
  and ice ages, 205–207, 215–216
Eccles, W. H., 91
Echo satellite, *139*, 140
Eclipses, 1, 2, 69–79
  chromosphere and, 72–79
  coronographic artificial, 77–79, 81–83
  observing, 69, *70*, *71*–72, 109–111
Eddington, Sir Arthur, *25*, 26, 35
Eddy, John A., 50, 51, 197
Edison, Thomas, 70, 122
Edlen, Bengt, 83
Egyptian temples, and sun, 2, *3*
Einstein's theory on matter and energy, 24
El Chichon, eruption of, 219–*220*, *221*
Electromagnetic radiation, of sun, 87–143
Electromagnetic spectrum, 7, 10–*11*, *15*
Electrons in the ionosphere, *114*–115
Elsasser, Walter M., 160
Emiliani, Cesare, 212–213
Energy
  law of conservation of, 23
  theory on matter and, 24–25
  world consumption of, 223
  See also Solar energy
Energy transfer, from sun's core, 41
Energy transport, modes of, 39
European Space Agency (ESA), 185
Evans, John, 109
Exosphere, 140–141, 143
Explorer satellites, 141, 148, 169, 173
Explorer spacecraft missions, 18–19
Explosive flaring, 112, 117–126
Ezer, Dilhan, 33

Fabricious, Johannes, 55
Faraday, Michael, 62
Ferraro, V., 147
Flamsteed, John, 198
Flares, *119*
  explosive, 112, 117–126
  radio signals from, 124–125
  solar, 141, 154
  solar proton, 180–185
  space shuttles and solar, 141, 143
  and VLA radio images, 126

See also Solar flares
Foraminifera, 209, *210*, 215
Forest, Lee de, 91
Fossils, in deep-sea sediments, *210*
Franklin, Benjamin, 218
Fraunhofer, Johann von, *14*–15
Fraunhofer lines, 14, *15*, 16–17, 44
and corona, 83
Free radicals, 122
Frequencies for communications, *115*
Friedman, Herbert, *108*, *120*
Fusion
birth of sun and, 232
proton-proton, 26–27, 29
in sun's core, 37, 39, 41

Galactic spur, 225
Galileo, 5, *55*, *56*, *57*, 84
Gallium, and neutrino detection, 34
Gamma rays, 11, 13
Gas, solar, 18–20, 41–42
Gaseous plasma, 81
Gassendi, Pierre, 162
Gauss, units of, 62, 63
Gay-Lussac, Joseph Louis, 127
Geocoronoa, 144–145
Geomagnetic storms, terrestrial impacts of, 183–185
Gilbert, William, 159–160
Glacial sediments and sun-climate connection, 199
Glaciers, 21, 198
advance and retreat of, 200–202
earth's orbit and, 205–207
See also Ice Ages
Global temperature, 190–192, 202
rise in, 223
and solar luminosity, 233–234
Glomar Challenger, *211*
G-mode pulsation, 48
Goddard, Robert H., 96
Goddard Space Flight Center, 52
Granulation, solar, *45*, 46
Gravitation, Newton's law of, 36
Gravity wave, 48
Great Nebula in Orion, *224*
Greenhouse effect, 223
Ground wave, 93

Hale, George Ellery, 44, 64
Halley, Edmund, 52, 68
Harmonics from power grids, amplification of, 184
Harriot, Thomas, *55*
Hayes, James D., 215
Heaviside, Oliver, 91
Heaviside layer, electrification of, 94–95
Heliopause, 157
Helioseismology, 37
Heliosphere, model of, *157*
Helium
in exosphere, 140–141
fusion of hydrogen into, 26–27, 29
surging spray of, *79*
Helix nebula, *231*
Herschel, Sir Willaim, 12, 23, 36, 57, 192
Hertz, Heinrich, 13
Hey, J. Stanley, 122
Hill, Henry, 48
Hiorter, Olav Peter, 160–161
Hodgson, R., 117
Huggins, William, 77
Hulburt, Edward O., 93, 94, *95*
Humboldt, Baron von, 204
Hydrogen, ultraviolet resonance line of, 104, 114, 119
Hydrogen bomb explosions, 172
Hydrogen ions, negative, 41–42

Ice Age, Little, *197*, 198, 201
Ice ages
astronomical theory of, 203–204
climate and, 197–207, 209, 212–213, 215–218
continental drift and, 217–218
earth's orbit and, 215–216
interstellar dust and, 220–222
IGY. *See* International Geophysical Year
Imbrie, John, 215
Infrared rays, 12
Infrasound in solar atmosphere, 73
Interferometer, Very Large Array, 125–126
International Geophysical Year (IGY), 119, 169
International Solar Terrestrial Physics (ISTP) program, 185, *186*
International Sun-Earth Explorer (ISEE) program, 185
Interstellar dust, ice ages and, 220–222
Ionized component of atmosphere, 8
Ionizing radiation, sun's, 96–98, 101–*106*, 107–112, 113–116, 117–126
Ionograms, 115
Ionosondes, 115
measurements with, *116*
Ionosphere, 81
currents in, *92*
early probing of, 90–96
electrons in, *114*–115
increasing density of, 109
periodicity in, 109
See also Sudden ionospheric disturbances
Ionospheric disturbances and X-ray flares, 118–122
Ionospheric layering, 113–116
Iridium
in clay, *227*
in rocks, 227–228
Isaak, G. R., 49
Isotope ratios, in sediment layers, 209, *212*
Isotopes
carbon 14, 193–195
iridium, 227–228
measuring natural, 196
Italian Space Agency, 137

Jeans, Sir James, 37
Jeffries, John, 127
Johns Hopkins University Applied Physics Laboratory, 107

Karnak, temples of, 2
Kellogg, Paul J., 171, 172
Kelvin, William Thomson, *24*, 50, 166
Kennelly, Arthur E., 91
Kennelly-Heaviside layer, 93
Kepler, Johannes, 203
Kerwin, Joe, 18
Kirchhoff, Gustav, 15, 16
Krakatoa, volcanic eruption in, 219
Krosovsky, Valerian I., *225*, *226*
Kullenberg, Bjore, 209
Kupperian, James E., Jr., *120*

Landing Ship Dock, 110, 120
Law of conservation of energy, 23
Leighton, Robert, 46
Leverrier, Urbain, 205–206
Light
bands of, 163, 164
coronal, 82
of night sky, 144–145
northern, 162–168
properties of, 7, 10

Lockyer, Sir J. Norman, 2, 193
Lodestone, 158, 159
Lowell, Percival, 193
Ludlum, David, 198
Luminosity of sun, 36, 233
Lyman-alpha glow, *144*
Lyman-alpha line, 104, 114, 119
Lyot, Bernard, 77

McMath telescope, 6
Magnetic disturbances and sunspots, 161
Magnetic elements; 66
Magnetic energy and sunspots, 118
Magnetic fields
  charged particles in, *167, 168, 169*
  and current flows, 150, 153
  Faraday map of, *63*
  models of, 214
  polarity of earth's, 213, 215
  strength of, 62–66
  sunspots and, 62–66
Magnetic field strength, *119*
Magnetic sector pattern, solar, *152*
Magnetic shield, earth's, 158–174
Magnetic storms, 160, *180*
  after solar flare, 143
  terrestrial impact of, 183–185
Magnetite, 158, *159*, 213
Magnetohydrodynamic waves in solar wind, 174–175
Magnetometers, space probe, *175*
Magnetosphere, 174–180
  description of, 20
  of planets, *187*
  schematic view of, 176, *177*
  and solar wind, 176, 178–179
  spacecraft exploration of, 185
  See also Magnetic field, earth's
Magnetospheric storms, 179–180
Magnetotail, computer-modeled stages in, *179*
Marconi, Guilielmo, *90*–91
Mariner 2, 148
Mars, space explorations of, 234
Mass
  conversion into energy of, 24–25
  of the earth, 36
Maunder Minimum, 198
Maunder pattern of sunspots, 62
Maxwell, James Clerk, 13, 62
Mayan civilization, 3, *4*
Mayer, Julius Robert, 23, 24
Menzel, Donald, 77–78
Mercury, and solar diameter, *52*–*53*
Meridan transit telescope, 53
Mesosphere, 137–139
Meteorite, Allende, 230–231, *232*
MHD. *See* Magnetohydrodynamic waves
Milankovitch cycles, 207, *208*
Milankovitch, Milutin, *202*, 203, 207, 215, 216
Milky Way, 1, 229, *230*
Mirror points, 167, 170
Monkey payload, 133, *134*
Monsoon, *191*
Montgolfier, Jacques and Étienne, 126, *127*
Montonauri, Matuyama, 213
Moses, 1, 2
Mt. St. Helens, eruptions of, 219–220
Mt. Tambora, eruptions of, 218
Muir, John, 162
Muller, Richard, 228
Muon neutrino, 34

NASA, 52, 87, 137
National Center for Atmospheric Research, 88
Naval Research Laboratory, 94, 95, 101, 144
Nebula
  compression of solar, 230
  Crab, 76–77, *226*
  Great Orion, *224*
  Helix, *231*
Neutrino detector, 32, *33*
Neutrinos, 26, 31–24
  detection of, 31–32
  kinds of, 34
  production of, 29, 31
Newcomb, Simon, 72
Newton, Sir Isaac, 12, 13
Ney, Edward P., 171, 172
Nicolet, Marcel, 140
Night sky, light of, 144
Nike-Asp rocket, *110*
Nitrate dating, 195–196
Nitric oxide, 114, 136
Noctilucent clouds, *138*
Nordhausen, 97
Norman, Robert, 159–160
Northern lights, 162–168
NRL. *See* Naval Research Laboratory
Nuclear burst, *173*
Nuclear fusion, in sun's core, 37, 39, 41. *See also* Fusion
Nuclear reactions in sun, 26–27, 29
Nuclear transmutations, discovery of, 25
Nuclear weapon tests, 171–174
Nut, sky goddess, 2

Observatory
  McMath-Hulburt, 6
  Pic du Midi, 77, *78*
Occultation, 76
Oort, Jan, 228
Oort cloud, *228*, 229
Opacity of solar gas, 41–42
Orbit. *See* Earth's orbit
Orbital eccentricity and resonance, 216
Oscillations, solar, 46–50
Oxygen ratios and climate history, 209, 212
Ozone, 8, 100, 101
  absorption of radiation by, 131
  and atmospheric temperature, 131
  levels of, *136*
  measurement of, 133
  protective shield of, 122
  and supernova explosions, 225
Ozone column density, distribution of, *135*

Pangea, 218
Parker, Eugene N., 148
Parkinson, John H., 51
Parsons, William S., 96
Particle accelerators, 64
Particles
  elementary, 168
  in magnetic field, *167, 168, 169*
  See also Solar particles
Pascal, Blaise, 126
Pauli, Wolfgang, 31, 34
Payload, monkey, *134*
Penumbra, of sunspot, 57, *58*
Permo-Carboniferous period, 217
Photosphere
  definition of, 41
  heat transfer and, 73
  lower levels of, 44
  spray of (*see* Chromosphere)
Pic du Midi Observatory, 77, *78*
Pickering, William H., 169
Pilgrim, Ludwig, 207
Pioneer 10, tracking of solar wind by, 156–158
Plasma, gaseous, 81, 82
Plasma expulsion, transient forms of, 155

Plasmasphere, 175
P mode pulsation, 47
Polar atmosphere and solar particles, 195–196
Polar-cap absorption events, flares and, 182–183
Poles, solar, 150, 151, 153, 195–196
Pomerantz, Martin, *51*
Potassium-argon dating technique, 215
Poulet, C. S., 36
Precession, earth's axial, *203*, *205*
Precession calculations, 215–216
Pressure, solar, 37, *42*
Pressure wave, in sun, 47–48
Prism, spectroscopic use of, 13, 14
Prominences, chromospheric, *80*, *81*
Proton flares, solar, 180–185
Proton-proton cycle, 26–*27*, *28*, 29
Protons, belts of, 169–170
Pulsation, forms of sun's, 47–50
Pyramids, Egyptian, 2, *3*

Quantum energy, 10

Radar, development of, 96
Radiation, 10
  Cerenkov, 11
  cosmic rays and, 226
  discovery of invisible, 12
  earth's absorption of, 190
  electromagnetic, 87–143
  global balance of, *192*
  particle, 143, 147–185
  synchotron, 123
  thermal, 127
  ultraviolet, 98, 100, 104, 111, 144
  and uranium, 13
  See also Solar radiation
Radiation belts
  doughnut-shaped, 169–171
  Van Allen, 20, 158, 168, 171, 175
Radioactive isotopes, and solar activity, 193–195. *See also* Isotopes
Radio communication
  blackout of, 172
  early, 90–91, 93
  frequencies for, 115
  study of, 94–96
Radioheliograph, *124*
Radiolaria, 209, *210*, 215
Radiometer, active-cavity, 87, 89
Radio noise, solar, 122, *123*–126
Radio probing of corona, 76–77
Radio signals, short-wave, *93*, 118
Radiotelescope, Very Large Array, *125*
Radio waves, *94*, 121
Rampino, Michael, 229
Raup, David, 228
Red-giant phase of sun, 234–235
Reflection frequencies, 116
Reines, Frederick, 31, 34
Ritter, J. W., 12
Rocket(s)
  Aerobee, *107*, *108*, 112, 144
  Nike-Asp, *110*
  second-generation, 106–109
  V-2, 97, *100*, *101*, 102
Rocket astronomy, early, 96–98, 100–102, 104–105
Rockoon, 120
Rocks and sunspot record, 199
Roentgen, Wilhelm, 12
Russell, Henry Norris, 5
Rutherford, Lord Ernest, 25

San Diego–Hi, 120, *124*
Satellite(s)
  astronomical, *120*
  Echo, *139*, 140
  Explorer, 141, 148, 169, 173
  and magnetosphere, 185
  and measuring solar constant, 87
  orbital period of Echo balloon, 139
  and radiation belts, 169–170
  thermospheric drag and, 139–140, 141, 143
Scheiner, S. J., Christopher, *55*, 84
Schneider, Stephen H., 88
Schwabe, Heinrich, 59, 61, 192
Sea-bottom sediments, *211*
  and temperature record, 207, 209, 212–213
Secchi, Father Angelo, 2
Sepkoski, John, 228
Severny, G. B., 48
Shakleton, Nicholas J., 215
Shapiro, Irwin, 53
Ship, deep-sea drilling from, *211*
Shklovsky, Josif S., 225, 226
Shortwave radio signals, *93*, 118
Singer, S. F., 168, 170
Skip distance, 94–96
Skylab, 17, *18*, *124*, 149, 150
Sofia, Sabatino, 52
Solar activity
  and carbon 14, 193–195
  map of, *111*
Solar camera, 108
Solar constant, 87
Solar corona. *See* Corona
Solar eclipses. *See* Eclipses
Solar energy, 23–24
  and carbon cycle, 29, 30
  and proton-proton cycle, 26–27, 29
  source of, 23–34
Solar flares, 141, 154
  proton, 180–185
  and X-ray emission, 119, 121
  See also Explosive flaring; Flares
Solar flux, variation of, 89
Solar gas, 18–20, 41–42
Solar granulation, *45*, 46
Solar interplanetary currents, *154*
Solar ionizing radiation. *See* Ionizing radiation, sun's
Solar magnetic sector pattern, *152*
Solar maximum, frequencies used in, *115*
Solar Maximum Mission (SMM), 17, 87, *88*
Solar neutrinos. *See* Neutrinos
Solar neutrino unit (SNU), 33
Solar Optical Telescope (SOT), 17, 84
Solar particle radiation, 147–185
Solar particles and polar atmosphere, 195–196
Solar poles, 150, 151, 153
Solar pressure, 37, *42*
Solar proton flares, 37, 39
  altitudes and attenuation of, *105*
  and climate, 87–89
  electromagnetic spectrum from (*see* Electromagnetic spectrum)
  intensity of, 36–37
Solar radio noise, 122, *123*–126
Solar spectrum, 98, *99*, *100*, 101–103
Solar system, birth and death of, 229–235
Solar telescope, 6, 17, 84
  film magazine from, *19*
  at south pole, *51*
Solar wind, 18–20, 147–151, 153–158
  corona and, 148
  flow pattern of, 151, 153–154
  magnetosphere and, 176, 178–179
  pressure of, 174
  stagnation boundary of, 157
  superheated, 18–20

Solar X rays, 83, 84, 98, 100
Solwind, 153
Sound waves
  in sun, 47, 48
  and temperature, 73–74, *131*
Soviet space probe, 148
Spacecraft, plans for future, *186*
Space exploration, and sun, 7, 17–19. *See also* Satellites
Space observatories and solar observation, 84
Space shuttle, 17, 137
Space station, first American. *See* Skylab
Species
  extinction of, 226–229
  temperature-sensitive, 209, *210*, 215
Spectral lines, 14, *15*, 16–17
Spectrograph
  design of rocket mounted, 102, *103*
  first rocket carrying, 100, 101
  in spinning rockets, 107, 108
Spectroheliograms, *47, 66*
Spectrohelioscope, 44
Spectrometer, ozone measuring, 133
Spectroscope, components of slit, *14*
Spectroscopy
  and magnetic fields, 64–65
  optical, 13–17
Spectrum. *See* Black-body spectrum; Electromagnetic spectrum
Spectrum lines over sunspot, *64*
Spicules, on sun, *75–76*
Sporer Minimum, 197
Spray, surging helium, *79*
Starfish, 173–174
Stars
  explosion of, 229
  shooting, 138–139
  See also Supernovas
Stebbin, Robin, 48
Stonehenge, 3, *21*, 22
Stormer, Carl, 166–167, 168
Stothers, Richard, 229
Stratoscope I, *46*
Stratosphere, 130–133, 136, 191
Stuiver, Minze, 199
Sudden ionospheric disturbances (SIDs), 118–122
Sullivan, Walter, 133
Sun
  in alpha red line, *43*
  architecture of, 35–37, *38*, 39–43
  birth and death of, 229–235
  and climate, 21, 189–197
  convection wraps in, *65*
  diameter and mass of, 36, *50–53*
  distance to earth from, 36, 203
  edge of, 41–42
  Egyptian pyramids and, 2, *3*
  galactic orbiting of, 221–222
  hot plasma in, *40*
  ionizing radiation of, 96–98, 101–112, 117–126
  images of, *86*
  magnetic field over, 66
  measuring and weighing, 35–36
  oscillations in, 46–48, *49*, 50
  red-giant phase of, 234–235
  shells of, 37, 39, 41–43
  shrinking, *50–53*
  sound on edge of, 47, 48
  space exploration and, 17–19
  spicules on, *75–76*
  spin rate of, 233
  spots on (*see* Sunspots)
  temperature of, 25–26, 32–34, *42*
Sunspot cycle, *60*, 153, 192, 193
Sunspots, 55, *56*, *57–58*, *59*, 60–69
  auroras and, 67–68
  bipolar groups of, 150
  characteristics of, *57–59*
  cold weather and, 197–198
  and comets, 69
  cycle of (*see* Sunspot cycle)
  early discoveries of, *55–57*
  effect on earth's activities by, 66–67
  and magnetic disturbances, *161*
  and magnetic fields, 62–66
  umbra and penumbra of, 57, *58*
Sun worship, 1–2
Superflares, 117–118, 182
Supergranules, 46, *47*
Supernova, 225, 230–232
Synchrotron radiation, 123

Tau neutrino, 34
Taylor, E. Hoyt, 93, 94, 95
Telescope, *5*
  invention of, *55*
  Meridan transit, *53*
  Newcomb's, *72*
  solar, 6, 17, *19, 51,* 84
  Very Large Array radio, *125*
  X ray, 112
Tempel, E. W. L., *155*
Temperature
  atmospheric, 137, 139–140
  exospheric, 141
  hemispheric, 190–192
  global (*see* Global temperature)
  solar, 25–26, 32–34, *42*
  in stratosphere, 130–131
  sun's core, 37
Temperature variations
  in atmosphere, *129*
  and ocean sediments, 207, 209, 212–213
Terrella experiments, 166
Theophrastus, *55*
Thermal radiation and climate, 127, *192*
Thermonuclear reactions
  carbon cycle and, 29, 30
  and neutrinos, 31
  See also Fusion
Thermosphere, 139–140
Thompson, Sir William, *24*, 50, 166
Torricelli, Evangelista, 126
Total-ozone-mapping spectrometer (TOMS), 133
Transients, coronal, *155, 156*
Tree rings, *208*
  and climate, 193–*195*
  frost damage zones in, *194*
  and sunspots, 198
Tropopause, 130
Troposphere, 130
Turbopause, 128–129, 138
Turbulent mixing, 128–129, 130
Tuve, Merle, 93, *95*, 96

Ulrich, Robert K., 47
Ultraviolet radiation, 12, 104
  in night sky, 144
  Schumann region of, 104
  solar, 98, 100
  source of, 111
Ultraviolet shield, earth's, 122
Umbra, of sunspot, 57, *58*
Uranium, radioactivity of, 13
Urey, Harold, 212

Van Allen, James, 107, 120
Van Allen radiation belts, 20 158, 168, 171, *175*
Varves, patterns in, *199*
Velocity of light, 24–*25*

Venus, *4*
  distance to, 35, 36
  space exploration of, 234
Very Large Array (VLA) interferometer, *125*–126
Viking rocket, 106
Volcanic eruptions and climate change, 218–220
Volcano
  Cotopaxi, *219*
  El Chichon, 219–*220, 221*
Von Braun, Wernher, 97, 169
Von Helmholtz, Hermann, 24, 50
Von Triewald, Samuel, 165
V-2 rocket, *97, 100, 101, 102*
  failure with, 105

Waldmeier, M., 149
Wallace, Henry, 78
Wasserburg, Gerald, 231
Wavelengths of light, 7, 10
Weapons tests, nuclear, 171–174
Weather
  comets and, 69
  forecasting, 200
  and sunspots, 67
  See also Climate
Weizsächer, Carl Friedrich von, 29
Whipple, Fred, 147
Whitmire, Daniel P., 228
Williams, George, 199
Williams, Samuel, 71
Willson, Richard C., 87
Wilson, Alexander, 57
Winds. *See* Solar wind
World War II, rockets and, 97

X ray(s)
  in corona, 109–112
  discovery of, 12
  example of, *12, 113*
  flares, 118–122
  flux, solar, *104*
  intensity and flares, 121
  photograph of sun, *112*
  solar, 83, 84, 98, 100
  sunspots and, 109, 111

Yellowstone National Park, volcanic eruption in, 219
Young, C. A., 23, 70, 72, 83
Young, John, 144

Zone of silence, 131

Other Books in the Scientific American Library Series

POWERS OF TEN
by Philip and Phylis Morrison and the Office of Charles and Ray Eames

HUMAN DIVERSITY
by Richard Lewontin

THE DISCOVERY OF SUBATOMIC PARTICLES
by Steven Weinberg

THE SCIENCE OF MUSICAL SOUND
by John R. Pierce

FOSSILS AND THE HISTORY OF LIFE
by George Gaylord Simpson

THE SOLAR SYSTEM
by Roman Smoluchowski

ON SIZE AND LIFE
by Thomas A. McMahon and John Tyler Bonner

PERCEPTION
by Irvin Rock

CONSTRUCTING THE UNIVERSE
by David Layzer

THE SECOND LAW
by P. W. Atkins

THE LIVING CELL, VOLUMES I AND II
by Christian De Duve

MATHEMATICS AND OPTIMAL FORM
by Stefan Hildebrandt and Anthony Tromba

FIRE
by John W. Lyons